The Fiber-Optic Gyroscope

The Artech House Optoelectronics Library

Brian Culshaw, Alan Rogers, and Henry Taylor, *Series Editors*

Acousto-Optic Signal Processing: Fundamentals and Applications, Pankaj Das

Amorphous and Microcrystalline Semiconductor Devices, Optoelectronic Devices, Jerzy Kanicki, ed.

Electro-Optical Systems Performance Modeling, Gary Waldman and John Wootton

Fiber-Optic Gyroscopes, Hervé Lefèvre

Field Theory of Acousto-Optic Signal Processing Devices, Craig Scott

Highly Coherent Semiconductor Lasers, Motoichi Ohtsu

Introduction to Electro-Optical Imaging and Tracking Systems, Khalil Seyrafi and S. A. Hovanessian

Introduction to Glass Integrated Optics, S. Iraj Najafi

Optical Control of Microwave Devices, Rainee N. Simons

Optical Fiber Sensors, Volume I: Principles and Components, John Dakin and Brian Culshaw, eds.

Optical Fiber Sensors, Volume II: Systems and Applicatons, Brian Culshaw and John Dakin, eds.

Optical Network Theory, Yitzhak Weissman

Principles of Modern Optical Systems, Volume I, I. Andonovic and D. Uttamchandani, eds.

Principles of Modern Optical Systems, Volume II, I. Andonovic and D. Uttamchandani, eds.

Reliability and Degradation of LEDs and Semiconductor Lasers, Mitsuo Fukuda

Single-Mode Optical Fiber Measurements: Characterization and Sensing, Giovanni Cancellieri

The Fiber-Optic Gyroscope

Hervé Lefèvre
Photonetics

Artech House
Boston • London

Library of Congress Cataloging-in-Publication Data

Lefèvre, Hervé.
Fiber-optic gyroscopes / Herve Lefevre.
p. cm.
Includes bibliographical references and index.
ISBN 0-89006-537-3
1. Optical gyroscopes. I. Title
TL589.2.06L44 1993 92-28194
681'.753—dc20 CIP

British Library Cataloguing in Publication Data

Lefevre, Herve
Fiber-optic Gyroscopes
I. Title
621.36

ISBN 0-89006-537-3

685 Canton Street
Norwood, MA 02062

International Standard Book Number: 0-89006-537-3
Library of Congress Catalog Card Number: 92-28194

10 9 8 7 6 5 4 3 2 1

To Sophie, Charlotte, and Elliot

"C'est un gyroscope, c'est la clé de l'Absolu"

Michel Tournier
Le Roi des Aulnes

Contents

Foreword

I take particular pleasure as series editor in seeing this very important book on the fiber-optic gyroscope completed and published. The fiber-optic gyroscope has been a particular fascination of mine for well over a decade; I was fortunate to work, albeit briefly, with Hervé at Stanford in the early 1980s—an especially exciting time when the physics of the gyroscope was still being discovered. Hervé's book encapsulates the excitement and the stimulus that, to afficionados, is unique to the gyro. Its particular charm and challenge stems from the need to understand physical optics, guided-wave optics, electronics, signal processing and mechanical engineering all at once.

To the fiber-optic sensor community (or at least to gyroscope fanatics), the gyroscope is regarded as the pinnacle of achievement. There is no one more qualified than Hervé to tell its story. What follows is indeed Hervé's story. The text encapsulates both the excitement of the physics and the challenge of the application while presenting the author's individual insights and solutions. Even engineers with different solutions will acknowledge the quality of the approaches described herein. I found the development of this text technically stimulating and an interesting read. It reflects one man's approach and one man's enthusiasms. I am sure that you will also share the excitement as you delve into this landmark in the history of what promises to be the linchpin technology for twenty-first century navigational systems.

Professor Brian Culshaw
Glasgow, Scotland
June 1992

Preface

Fifteen years of research and development have established the potential of the fiber-optic gyroscope, which is now considered a privileged technology for future applications of inertial guidance and control. Its "solid-state" configuration brings crucial advantages over previous approaches using spinning wheels or gas ring lasers.

Interest in the fiber-optic gyroscope is growing rapidly in many companies around the world. Development, production, and system engineers are now getting involved, in addition to the scientific and technological communities that have conducted the research. Therefore, this is a good time to present a detailed description of the analysis that has been carried out to achieve a practical device. Despite the relative simplicity of the final scheme, the fiber-optic gyro is a sophisticated instrument with many subtle error sources which must be understood and controlled. The subject requires a multidisciplinary approach involving physics, guided optics, opto-electronic technology, signal processing theory, and electronic design. The variety of topics involved is a good example of a thorough system analysis, and the study of the fiber gyro would be a very formative theoretical and experimental program for graduate students in fiber optics and opto-electronics.

To help the reader, I have included detailed appendixes that provide information on optics, single-mode fiber optics, and integrated optics necessary for understanding the fiber gyro, and the vocabulary for communicating with opto-electronic component designers. For the newcomer to the field, this material will help him or her avoid having to go through general text books to find specific basics. However, based on my own experience preparing these basics, the appendixes could also be a useful review for those already involved in the subject area. I have also tried (except in Appendix 4) to avoid as much cumbersome mathematical calculations and formulae as possible. The many figures act as visual aids to simplify the explanations and help the reader grasp the important ideas dictating the design rules.

Because this is a single-author book, the analysis may be slightly influenced by my personal views. However, I have chosen to share with the reader the results of my fifteen years of experience, and I have clearly indicated my preferences instead of giving a strictly impartial description, which would have resulted in a dreariness this subject does not deserve.

This book is based on research experience that has been shared with scientists who, like me, have been fascinated by the technical serendipity of the device. I owe a very special tribute to H. J. Arditty of Photonetics for our continuous fruitful collaboration. My postdoctoral scholarship at Stanford University was an essential experience, marked by the profound influence of Professor H. J. Shaw. I would also like to acknowledge the crucial contributions of M. Papuchon and G. Pircher of Thomson-CSF, R. A. Bergh, now with Fibernetics, and Ph. Martin and Ph. Graindorge of Photonetics. It is important to recall that this research has been carried out with very open exchanges within the international scientific community, which may be one reason for its success. Finally, the efficiency of C. Hervé was essential in the preparation of the manuscript.

Hervé C. Lefèvre
Paris, France
March 1992

Chapter 1
Introduction

The laws of mechanics show that an observer kept locked up inside a "black box" in uniform linear translation has no way to know his or her movement. On the other hand, it is possible to detect a linear acceleration or a rotation. Precise measurements may be performed with mechanical accelerometers and gyroscopes. This is the basis of inertial guidance and navigation. Knowing the initial orientation and position of the vehicle, the (mathematic) integration of the acceleration and rotation rate measurements yields the attitude and the trajectory of the vehicle. Such inertial techniques are completely autonomous and need no external reference: they do not suffer from any shadow effect or jamming. For fifty years, they have been a key technology in aeronautics, naval, and space systems for civilian and military applications.

In 1913, Sagnac [1] demonstrated that it is also possible to detect rotation with respect to inertial space with an optical system that has no moving part. He used a ring interferometer and showed that rotation induces a phase difference between the two counterpropagating paths. The original setup, however, was very far from a practical rotation rate sensor, because of its very limited sensitivity. In 1925, Michelson and Gale [2] were able to measure earth rotation with a gigantic ring interferometer of almost 2 km in perimeter to increase the sensitivity, but the Sagnac effect remained a rarely observed physics curiosity for many decades, because it was not possible to get usable performance from a reasonably compact device.

This possibility of getting a gyroscope without moving parts to replace the spinning wheel mechanical gyro continued to be very attractive, and in 1962 Rosenthal proposed to enhance the sensitivity with a ring laser cavity [3] where the counterpropagating waves recirculate many times along the closed resonant path instead of once in the original Sagnac interferometer. This was first demonstrated by Macek and Davis [4] in 1963, and nowadays ring laser gyro technology has reached its full maturity and is used in many applications of inertial navigation [5,6].

However, because of the huge technological effort devoted to the development of low-loss optical fiber and a solid-state semiconductor light source and detector for telecommunication applications during the 1970s, it became possible to use a multiturn optical fiber coil instead of a ring laser to enhance the Sagnac effect by multiple recirculation. Proposed early in 1967 by Pircher and Hepner [7] and demonstrated experimentally by Vali and Shorthill in 1976 [8], the fiber-optic gyroscope has since attracted a lot of scientific and technical interest because it provides unique advantages due to its solid-state configuration.

Numerous (770!) publications have been devoted to the subject [9], and the most significant contributions have been compiled in a single volume [10], which is very convenient when working in this field. The proceedings of the three conferences specifically dedicated to the subject [11–13] are also a good indication of the progress of the technology over fifteen years of research and development. A critical step has now been passed with several companies starting industrial production [13]. At this stage, it seems useful to present a thorough analysis of the results of the R&D phase, and to emphasize the concepts that have emerged as the preferred solutions.

Chapter 2 describes the general principle of the fiber gyro, which is based on the Sagnac effect. Chapter 3 analyzes the application of the fundamental concept of reciprocity to a ring interferometer and describes the various possible configurations. All-guided approaches using all-fiber components or integrated optics are emphasized. The first noise sources encountered, backreflection and backscattering, are analyzed in Chapter 4, which shows the importance of the low coherence brought by a broadband source. Chapter 5 is dedicated to the important analysis of the reduction of nonreciprocities due to birefringence with the use of polarization-preserving fiber and broadband sources. It also describes optical coherence domain polarimetry (OCDP), which has been found to be an essential tool for understanding and solving these problems. Chapter 6 presents the parasitic effects a fiber gyro may face with temperature transience and vibrations. Chapter 7 analyzes the Faraday and Kerr effects, which may yield true nonreciprocities similar to the Sagnac effect. Chapter 8 is concerned with the crucial problem of scale factor accuracy, which is preferably solved with closed-loop processing techniques using phase-ramp modulation. The delicate problem of wavelength control is also emphasized. Chapter 9 recapitulates the optimal operating conditions and describes the preferred technological choices. Chapter 10 comments briefly on the alternative approaches that have been proposed. Chapter 11 presents the principle of the competing resonant approach, where a passive ring cavity is used instead of a two-wave interferometer. Finally, Chapter 12 concludes with the present state of the art and future trends.

The appendixes detail the basics of optics, single-mode fiber optics, and integrated optics that are important to know to understand the fiber gyro. A fourth appendix is more specialized and describes the electromagnetism approach explaining the relativistic Sagnac effect.

REFERENCES

[1] Sagnac, G., "L'éther lumineux démontré par l'effet du vent relatif d'éther dans un interféromètre en rotation uniforme," Comptes rendus de l'Académie des Sciences, Vol. 95, 1913, pp. 708–710. Sagnac, G., "Sur la preuve de la réalité de l'éther lumineux par l'expérience de l'interférographe tournant," Comptes rendus de l'Académie des Sciences, Vol. 95, 1913, pp. 1410–1413.

[2] Michelson, A. A., and H. G. Gale, Journal of Astrophysics, Vol. 61, 1925, pp. 401.

[3] Rosenthal, A. H., "Regenerative Circulatory Multiple Beam Interferometry for the Study of Light Propagation Effects," J.O.S.A., Vol. 52, 1962, pp. 1143–1148.

[4] Macek, W. M., and D.T.M. Davis, "Rotation-Rate Sensing With Travelling-Wave Ring Lasers," Applied Physics Letters, Vol. 2, 1963, pp. 67–68.

[5] Ezekiel, S., and G. E. Knausenberger, eds., "Laser Inertial Rotation Sensors," SPIE Proceedings, Vol. 157, 1978.

[6] Chow, W. W., J. Gea-Banacloche, L. M. Pedrotti, V. E. Sanders, W. Schleich, and M. O. Scully, "The Ring Laser Gyro," Review of Modern Physics, Vol. 57, 1985, pp. 61.

[7] Pircher, G., and G. Hepner, "Perfectionnements aux dispositifs du type gyromètre interférométrique à laser," French patent 1.563.720, 1967.

[8] Vali, V., and R. W. Shorthill, "Fiber Ring Interferometer," Applied Optics, Vol. 15, 1976, pp. 1099–1100 (SPIE, MS8, pp. 135–136).

[9] Smith, R. B., "Fiber-Optic Gyroscopes 1991: A Bibliography Of Published Literature," SPIE Proceedings, Vol. 1585, 1991, pp. 464–503.

[10] Smith, R. B., ed., "Selected Papers on Fiber-Optic Gyroscopes," SPIE Milestone Series, Vol. MS8, 1989. Note: For references listed in this book appearing in this Milestone Volume, we have inserted (SPIE, MS8, pp. xx–yy).

[11] Ezekiel, S., and H. J. Arditty, eds., "Fiber-Optic Rotation Sensors and Related Technologies," Proceedings of the First International Conference, Springer Series in Optical Sciences, Vol. 32, 1981.

[12] Udd, E., ed., "Fiber Optic Gyros: 10th Anniversary Conference," SPIE Proceedings, Vol. 719, 1986.

[13] Ezekiel, S., and E. Udd, eds., "Fiber Optic Gyros: 15th Anniversary Conference," SPIE Proceedings, Vol. 1585, 1991.

Chapter 2
Principle of the Fiber-Optic Gyroscope

2.1 SAGNAC EFFECT

2.1.1 Sagnac Interferometer

The fiber-optic gyroscope is based on the Sagnac effect, which produces a phase difference $\Delta\phi_R$ proportional to the rotation rate Ω in a ring interferometer [1]. Sagnac's original setup was composed of a collimated source and a beam-splitting plate to separate the input beam into two waves which propagate in opposite directions along a closed path defined by mirrors (Figure 2.1). A pattern of straight interference fringes was obtained with a slight misaligment of one mirror, and a lateral shift of the fringe pattern was observed as the whole system was rotated. This fringe shift corresponds to an additional phase difference $\Delta\phi_R$ between the two counterpropagating waves, depending on the area A enclosed by the path.

This can be explained by considering a regular polygonal path $M_0M_1 \ldots M_{N-1}M_0$. At rest, both opposite paths are equal, but, in rotation around the center, the co-rotating path is increased to $M_0M_1' \ldots M_{N-1}'M_N'$ and the counter-rotating path is decreased to $M_0M_1'' \ldots M_{N-1}''M_N''$ (Figure 2.2). As a matter of fact, for an observer in the inertial rest frame of reference, the points M_i move on a circle of radius R, and light propagates along polygon sides $M_i'M_{i+1}'$ or $M_i''M_{i+1}''$ instead of M_iM_{i+1}. In particular, the first side of the co-rotating polygonal path becomes M_0M_1' (Figure 2.3). Using 2θ to represent the angle M_0OM_1, $\delta\theta$ the angle M_1OM_1', L_M the length M_0M_1, and δL_M the path length increase $M_0M_1' - M_0M_1$, there are:

$$\begin{aligned} \delta L_M &= M_1M_1' \cos\theta \\ M_1M_1' &= R\,\delta\theta \end{aligned} \tag{2.1}$$

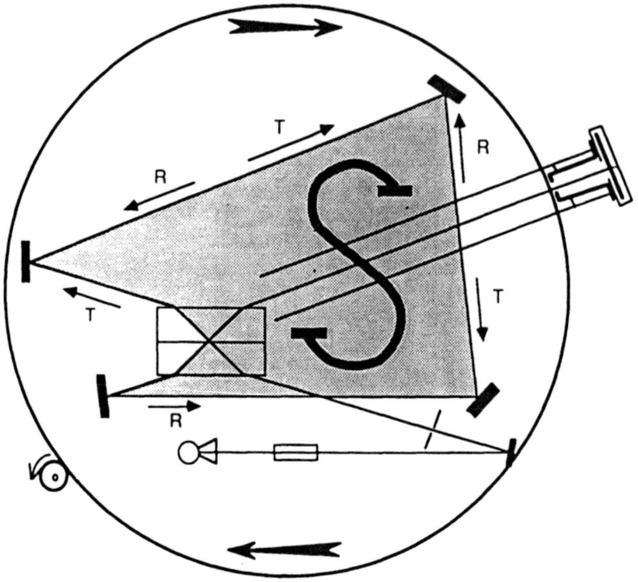

Figure 2.1 Original Sagnac's setup [1] of a ring interferometer to measure sensitivity to rotation rate. (S stands for surface, which means "area" in French.)

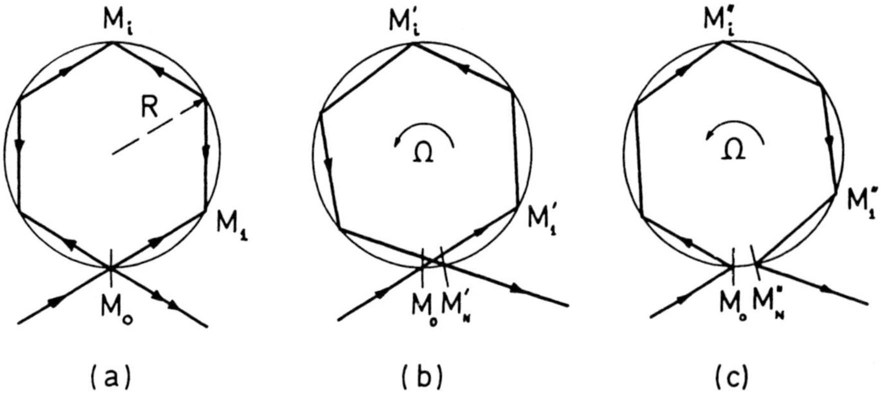

Figure 2.2 Path change in a ring interferometer with a regular polygonal path: (a) at rest; (b) co-rotating path; (c) counter-rotating path.

This angle $\delta\theta$ is, to first order, the angle of rotation during the propagation between M_0 and M_1:

$$\delta\theta = \frac{L_M}{c} \cdot \Omega \tag{2.2}$$

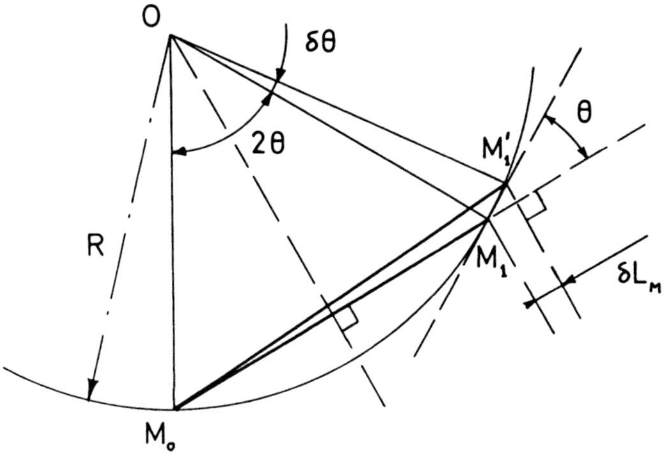

Figure 2.3 Geometrical analysis of the Sagnac effect along one side of a polygonal path.

and since $L_M = 2R \sin \theta$, and the area of the triangle M_0OM_1 is $A_t = (R \sin \theta)(R \cos \theta)$, this yields:

$$\delta L_M = \frac{2A_t\Omega}{c} \tag{2.3}$$

The phenomenon is observed in the rest frame, where light propagates always at velocity c; therefore, the path increase δL_M corresponds to an increase δt^+ of the propagation time:

$$\delta t^+ = \frac{\delta L}{c} = \frac{2A_t\Omega}{c^2} \tag{2.4}$$

There is this same increase for each side of the polygon and the opposite variation $\delta t^- = -\delta t^+$ in the counter-rotating direction. The difference Δt_v of propagation time between the two opposite closed paths in a vacuum is then:

$$\Delta t_v = 2\frac{2\Sigma A_t\Omega}{c^2} = \frac{4A\Omega}{c^2} \tag{2.5}$$

where ΣA_t is the sum of all the triangular areas (i.e., the whole enclosed area A). Measured in an interferometer, this time difference yields the phase difference:

$$\Delta\phi_R = \omega \cdot \Delta t_v = \frac{4\omega A}{2}\,\Omega \tag{2.6}$$

where ω is the angular frequency of the wave. It can be shown that this result is very general and can be extended to any axis of rotation and to any closed path, even if they are not contained in a plane, using the scalar product $\mathbf{A} \cdot \mathbf{\Omega}$:

$$\Delta\phi_R = \frac{4\omega}{c^2}\,\mathbf{A} \cdot \mathbf{\Omega} \tag{2.7}$$

where $\mathbf{\Omega}$ is the rotation rate vector and $\mathbf{A}$ is the equivalent area vector of the closed path defined in terms of the line integral:

$$\mathbf{A} = \frac{1}{2}\oint \mathbf{r} \times d\mathbf{r} \tag{2.8}$$

where $\mathbf{r}$ is the radial coordinate vector. The Sagnac effect appears as the flux of the rotation vector $\mathbf{\Omega}$ through the enclosed area.

To get a better understanding of the Sagnac effect, it is possible to consider the simple case of an "ideal" circular path [2,3], which would be the limit of a polygonal path with an infinite number of sides. Light entering the system is divided into two counterpropagating waves which return in phase after having traveled along the same path in opposite directions (Figure 2.4(a)). Now, when the interferometer is rotating, an observer at rest in the inertial frame of reference sees the light entering the interferometer at point M (Figure 2.4(b)) and traveling with the same vacuum velocity c, in opposite directions; however, during the transit time t_v through the loop, the beam splitter has moved to M′, and our observer sees that the co-rotating wave has had to propagate over a longer path than the counter-rotating one. This path difference $2\Delta l_v$ can be measured by interferometric means.

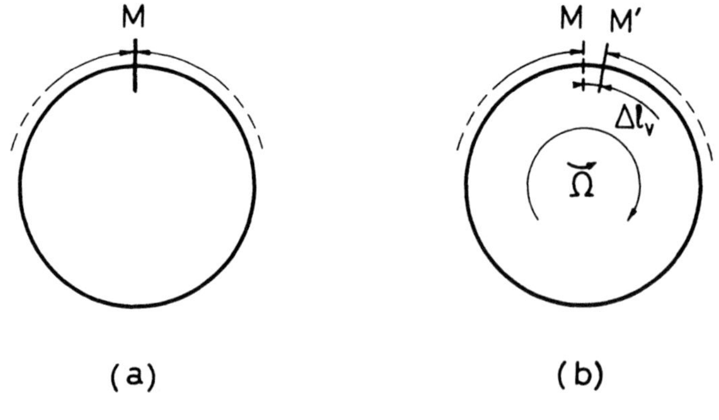

Figure 2.4 Sagnac effect in a vacuum considering an "ideal" circular path: (a) system at rest; (b) system rotating.

This explanation is straightforward, but we must not forget the fundamental point: this is observed in the inertial frame but still observed in the rotating frame, because both events (returns of the co-rotating and counter-rotating waves onto the beam splitter) take place at the same point, and the principle of causality can be applied: if two events take place at the same point in space, their difference of time of occurrence is conserved (to first order in v/c) in any frame of reference. As a matter of fact, it is interesting to compare the Sagnac effect with the well-known problem of relativistic kinematics, which explains that simultaneity of events is not an absolute notion.

Let us consider a system composed of a source S placed at equal distance from two mirrors M_1 and M_2 (Figure 2.5(a)). Light is emitted by the source in opposite directions, and, after reflection, both waves return to the source at the same time. Now, if the system moves laterally (Figure 2.5(b)), an observer in the "laboratory" frame will observe the light hitting first the mirror, M_1, moving toward the incoming wave, and then the other mirror, M_2. The delay between both events is essentially the same as the Sagnac delay, replacing the circumference of the circular path by the distance between the source and the mirrors, and the tangential speed due to rotation by the translation speed. However, in the case of translation, both events take place at two different points, and the principle of causality cannot be applied. An observer in the co-moving system frame has to wait for both returns of the light to the source that will, to the observer, occur at the same time. Then this observer can only deduce that, in his or her moving frame of reference, light hit both mirrors at the same time. Note that the source is also moving for the observer in the "laboratory" frame, and he or she sees that the light returns from both ways at the same time. This is consistent with what was previously said, because the returns on the source are two events taking place at the same point, and if they are simultaneous, this is observed in any frame of reference.

Note: The Sagnac effect can alternatively be interpreted as a double doppler effect on the beam splitter. Instead of the temporal approach we detailed above, this can be analyzed spatially, considering the system "frozen" at a given instant (Figure 2.6). The observer in the laboratory frame measures one wave transmitted twice and keeping the same wavelength, while the opposite wave is reflected twice on

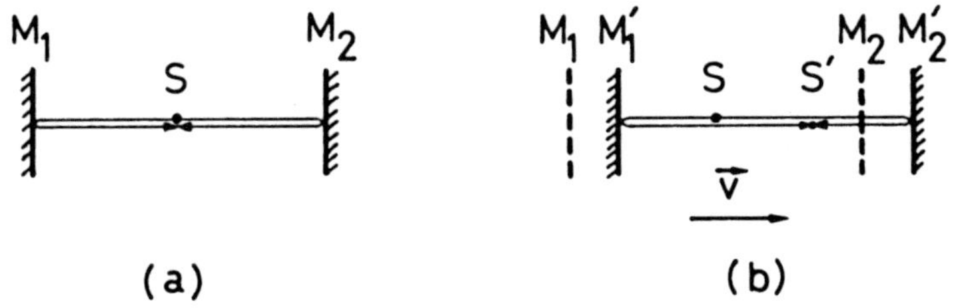

Figure 2.5 Problem of synchronization in relativistic kinematics: (a) system at rest; (b) system in uniform linear translation.

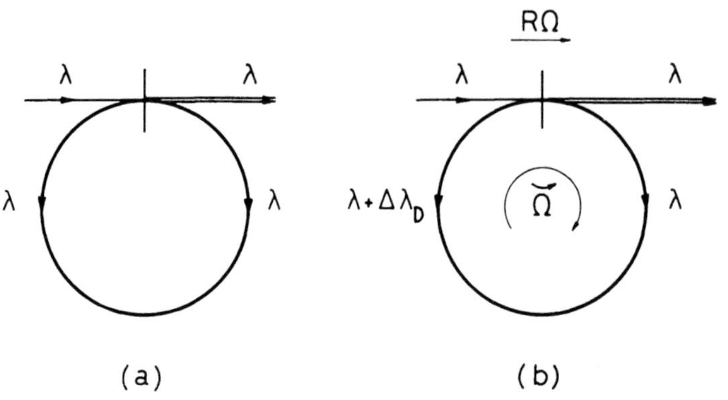

Figure 2.6 Sagnac effect interpreted as a double doppler effect: (a) system at rest; (b) system rotating.

the moving splitter, resulting in a doppler shift of wavelength along the circular path and a second opposite shift at the output, resulting in a return to the original wavelength. The temporal approach considers two waves that propagate at the same velocity along different paths; this spatial approach considers two waves along the same path but with different wavelengths. These two explanations are equivalent, but care must be taken not to use both at the same time!

2.1.2 Case of a Medium

Now if light propagates in a medium, as in the case of the fiber-optic gyroscope, it can be demonstrated that the Sagnac phase difference remains unchanged [2,3,4]. Considering again a circular path for simplicity, at rest both waves propagate at the velocity $v = c/n$, where n is the index of the medium, and return on the beam splitter after the same time $t_m = 2\pi R/v = 2\pi nR/c = n \cdot t_v$. When the interferometer is rotating, the beam splitter moves along a length $\Delta l_m = R\Omega t_m$ during the propagation time t_m (Figure 2.7). This length is n times longer than Δl_v, but, in this case, the light velocity is no longer the same in both directions. Indeed, this experiment is observed in the inertial frame of reference and a Fizeau drag occurs because of the motion of the medium. It depends on the relative directions of light propagation and medium motion. In the inertial "laboratory" frame, the velocities of the co-rotating and counter-rotating waves are, respectively:

$$
\begin{aligned}
v_{cr} &= \frac{c}{n} + \alpha_F R\Omega \\
v_{ccr} &= \frac{c}{n} - \alpha_F R\Omega
\end{aligned}
\tag{2.9}
$$

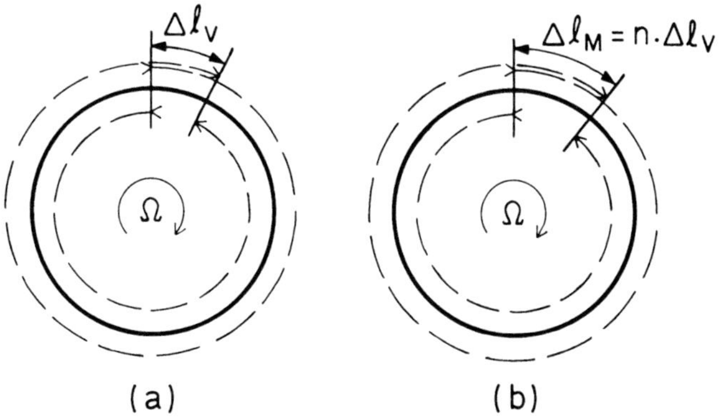

Figure 2.7 Sagnac effect in a medium: (a) case of a vacuum; (b) case of a corotating medium.

where α_F is the Fizeau drag coefficient and $R\Omega$ is the tangential speed of the medium. The difference of propagation time becomes:

$$\Delta t_m = \Delta t_v n^2 (1 - \alpha_F) \tag{2.10}$$

where Δt_v would be the value in a vacuum. Since $\alpha_F = 1 - n^{-2}$, the Fizeau drag compensates for the effect of the index n, and there is:

$$\Delta t_m = \Delta t_v$$

This perfect compensation is not as fortuitous as it may look in this analysis. A rigorous approach [2,5,6] (detailed in Appendix 4) has to consider the laws of electromagnetism in a rotating frame and solve the propagation equation in this frame. Such an analysis shows without any ambiguity that the Sagnac effect is a pure temporal delay that does not depend on the medium, nor on waveguidance if an optical fiber is used.

Note: It is necessary to be careful because the value of the Fizeau drag coefficient is often given as

$$\alpha_F = 1 - \frac{1}{n^2} - \frac{\omega}{n}\frac{dn}{d\omega} \tag{2.11}$$

As a matter of fact, in the original Fizeau experiment the light velocity v is first measured with the medium at rest, and

$$v = \frac{c}{n(\omega_0)} \tag{2.12}$$

where ω_0 is the light frequency in the laboratory frame. When the medium is moving in the laboratory frame at a speed v_m, the light velocity becomes

$$v_F = \frac{c}{n(\omega_0)} + \left[1 - \frac{1}{n^2} - \frac{\omega_0}{n}\frac{dn}{d\omega}\right] v_m \tag{2.13}$$

There is a dispersion term $\left[\frac{\omega_0}{n}\frac{dn}{d\omega}\right]$ because the frequency "seen" by the medium in its proper frame of reference is not ω_0 anymore, but a frequency ω_p, which is shifted because of a doppler effect:

$$\frac{\omega_p - \omega_0}{\omega_0} = \frac{v_m}{v} = \frac{n v_m}{c} \tag{2.14}$$

However, the pure Fizeau effect is actually

$$v_F = \frac{c}{n(\omega_p)} + \left[1 - \frac{1}{n^2}\right] v_m \tag{2.15}$$

where ω_p is the light frequency "seen" by the medium. In the case of the Sagnac interferometer, the light frequency has the same ω_p value for both counterpropagating waves in the rotating frame, where the medium is at rest and one has to use [3]

$$\begin{aligned} v_{cr} &= \frac{c}{n(\omega_p)} + \left[1 - \frac{1}{n^2}\right] R\Omega \\ v_{ccr} &= \frac{c}{n(\omega_p)} - \left[1 - \frac{1}{n^2}\right] R\Omega \end{aligned} \tag{2.16}$$

2.2 ACTIVE AND PASSIVE RING RESONATORS

2.2.1 Ring Laser Gyroscope (RLG)

The Sagnac effect is very small—the original experiment required a high rotation rate to demonstrate the phenomenon. Assuming an enclosed area as large as 1 m^2, a rate as high as 2π rad/s, and a wavelength of 1 μm, the phase difference is only 0.5 rad, that is, about one-tenth of a fringe. The effect has to be greatly enhanced to make a practical rotation sensor with good sensitivity and compactness.

In the early sixties, it was proposed to use a ring laser cavity to increase the Sagnac effect, because the light recirculates many times around the cavity [8]. This technology is now fully mature [9,10], and the ring laser gyroscope (or RLG or laser gyro) has brought major improvements of performance and reliability to inertial navigation systems.

In an ordinary laser, the emission wavelength is an integral submultiple of the double of the Fabry-Perot cavity length (see Appendix 1). It is possible to make a ring cavity working on the same principle of optical resonance (Figure 2.8). The cavity has mirrors with a quasi-total reflectivity and one output mirror with a small transmissivity. The two counterpropagating beams are emitted through the output mirror. At rest, the emitted frequencies (or wavelengths) are equal, since the cavity length is the same in both directions. When it is rotated, there is a small difference of cavity lengths because of the Sagnac effect, which yields a frequency difference between both output beams:

$$\Delta f_R = 4\frac{A}{\lambda \mathfrak{P}} \cdot \Omega \tag{2.17}$$

where A is the area enclosed by the ring cavity, $\mathfrak{P}$ is the perimeter, and λ the wavelength at rest. This frequency difference is measured by combining the two output beams to get interferences. Since the beams have different frequencies, their phase difference varies as

$$\Delta\phi = 2\pi \Delta f_R t \tag{2.18}$$

and the interference intensity I is modulated at the beat frequency Δf_R:

$$I = I_1 \, [1 + \cos(2\pi \, \Delta f_R t)] \tag{2.19}$$

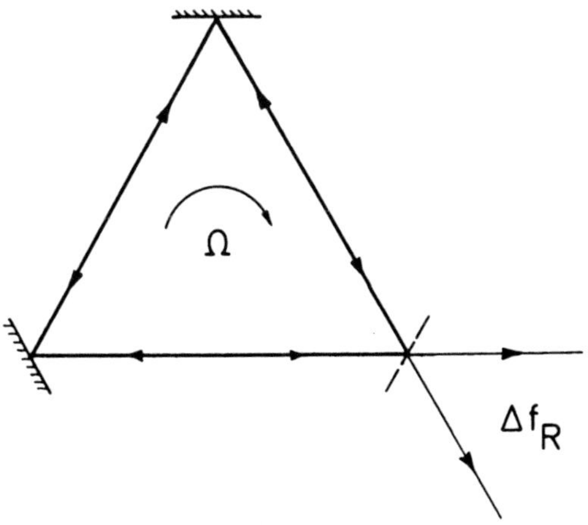

Figure 2.8 Ring laser cavity.

The counting of the beats gives the angle of rotation, since Δf_R is proportional to the rotation rate Ω. The angle value corresponding to one modulation period is called the angular increment θ_{inc}, with

$$\theta_{inc} \cdot \Delta f_R = \Omega \qquad (2.20)$$
$$\theta_{inc} = \frac{\lambda \mathfrak{P}}{4A}$$

Most high-performance laser gyros have a triangular or square cavity with a perimeter of about 30 cm. They operate at a wavelength of 633 nm with an He-Ne amplifying medium. We have

$$\theta_{inc} \approx 10^{-5} \text{ rad} \approx 2 \text{ arcsec}$$

A rotation of 1 deg/h (i.e., 1 arcsec/s) gives a beat frequency of 0.5 Hz.

The effect can be understood very simply by considering an "ideal" circular cavity. Both counterpropagating beams create a standing wave with a space of $\lambda/2$ between nodes (Figure 2.9). When the gyro is rotating, the standing wave remains at rest in the inertial space, but the detector rotates and gives one count each time it is passing a length of $\lambda/2$. Therefore, the angular increment θ_{inc} is simply

$$\theta_{inc} = \frac{\lambda}{2R} \qquad (2.21)$$

where R is the radius of the cavity (which is consistent with the general formula $\theta_{inc} = \lambda\mathfrak{P}/4A$, since in this case $\mathfrak{P} = 2\pi R$ and $A = \pi R^2$).

The main problem of the laser gyro is the phenomenon of mode-locking between the counterpropagating beams. As a matter of fact, these are oscillators with a very high resonance frequency (in the range of 5×10^{14} Hz) and a very small frequency difference. If there is some weak coupling between both oscillators, they get locked together and oscillate at the same frequency, creating a dead zone at low rotation rate. The main source of coupling is the backscattering of the mirrors. An intense technological effort was required to improve the quality of the reflective coating. However, even with very low scattering mirrors, there is still a dead zone (typically several tens of degrees per hour) much wider than the potential sensitivity of the device, and this is solved with a mechanical dither to vibrate the gyro at a rate outside of the dead zone. Present dithered laser gyros have excellent performances (bias stability better than 10^{-2} deg/h and scale factor accuracy better than 1 ppm over a dynamic range of ± 400 deg/s); but it is a very complex technology, and with the progress of low-loss optical fibers during the

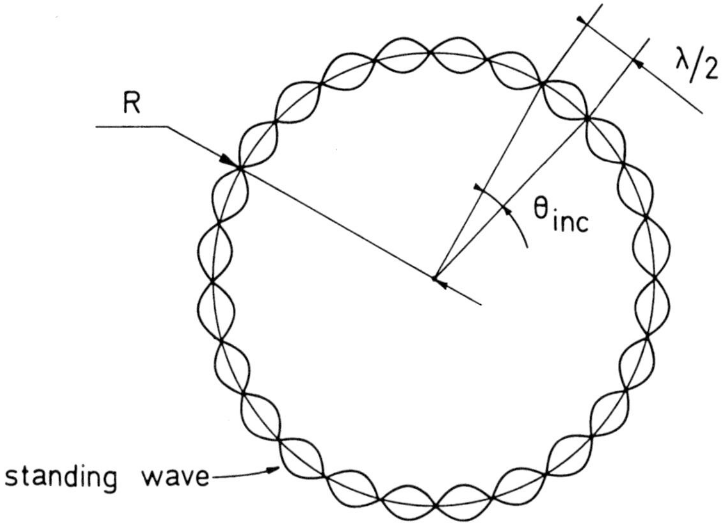

Figure 2.9 Simple case of an "ideal" circular cavity.

seventies, it appeared that fiber-optic gyroscopes could offer an interesting alternative.

2.2.2 Resonant Fiber-Optic Gyroscope (R-FOG)

To avoid the problem of locking, the use of a passive ring cavity instead of an active system was proposed [11]. An external source is fed in both directions into the cavity. Such a cavity is very similar to a Fabry-Perot interferometer (Figure 2.10), with resonance frequencies or wavelengths that are transmitted when the cavity length is equal to a multiple number of wavelengths. When the cavity is rotated, the Sagnac effect yields a difference Δf_R between the resonance frequencies of the opposite directions. This difference has the same value as the one in the laser gyro:

$$\Delta f_R = \frac{4A}{\lambda \mathfrak{P}} \Omega \tag{2.22}$$

The width of the resonance is given by the finesse and the free spectral range, just as for Fabry-Perot cavities (see Appendix 1). The rotation sensitivity is amplified by the finesse, since it corresponds approximately to the number of recirculations in the ring cavity.

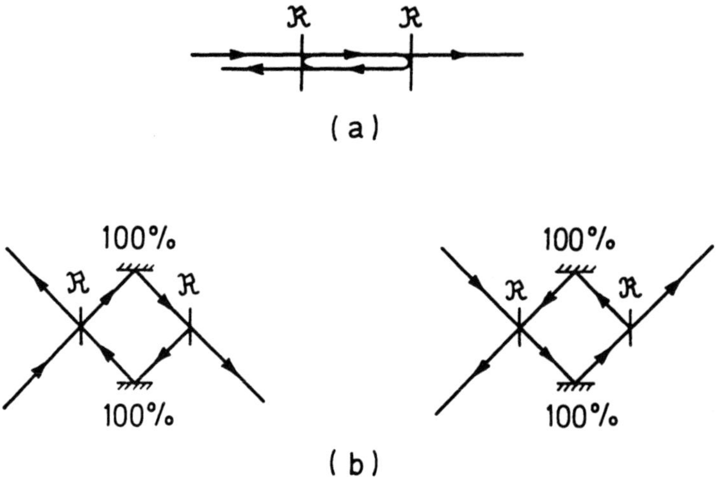

Figure 2.10 Similarity between a Fabry-Perot cavity and a ring cavity: (a) Fabry-Perot cavity; (b) counterpropagating resonant paths in a ring cavity.

This passive approach in a bulk form does not, however, bring significant advantages over the active device. On the other hand, the use of a single-mode fiber permits increasing the sensitivity further with a multiturn coil [12]. The frequency difference is still

$$\Delta f_R = \frac{4A}{n\lambda \mathfrak{P}} \Omega$$
$$\Delta f_R = \frac{D}{n\lambda} \cdot \Omega \tag{2.23}$$

where D is the diameter of the coil, and n is the index of refraction. But increasing the number of turns (N) increases the length of the cavity and then reduces its free spectral range, which decreases the width of the resonance for a given finesse. Compared to the original Sagnac interferometer, the potential improvement is the product of the finesse $\mathfrak{F}$ by the number N of turns.

An analysis of this resonant fiber-optic gyroscope, often abbreviated R-FOG, will be provided in Chapter 11. This gyroscope faces a very difficult problem: to fully exploit the system, a very narrow source spectrum is required, and the related large coherence length induces various sources of noise which degrade the performance.

2.3 PASSIVE FIBER RING INTERFEROMETER

2.3.1 Principle of the Interferometric Fiber-Optic Gyroscope (I-FOG)

Since the Sagnac effect is proportional to the flux of the rotation rate vector Ω, it can be enhanced with a multiturn path, just as the flux of a **B** field is enhanced in a multiturn inductance coil. With a low-loss single-mode fiber, the enhancement can be made so large that it does not require the use of a resonant cavity, and a two-wave ring interferometer with a multiturn single-mode fiber coil may provide adequate sensitivity [13,14] (Figure 2.11). The Sagnac phase difference becomes

$$\Delta\phi_R = \frac{2\pi LD}{\lambda c} \cdot \Omega \quad (2.24)$$

where λ is the wavelength in a vacuum, D is the coil diameter, $L = N\pi D$ is the fiber length, and N is the number of turns. A constant rate yields a constant phase difference. The response is (co)sinusoidal, just as for any two-wave interferometer (see Appendix 1), with intensity I given by

$$I = I_1[1 + \cos\Delta\phi_R] \quad (2.25)$$

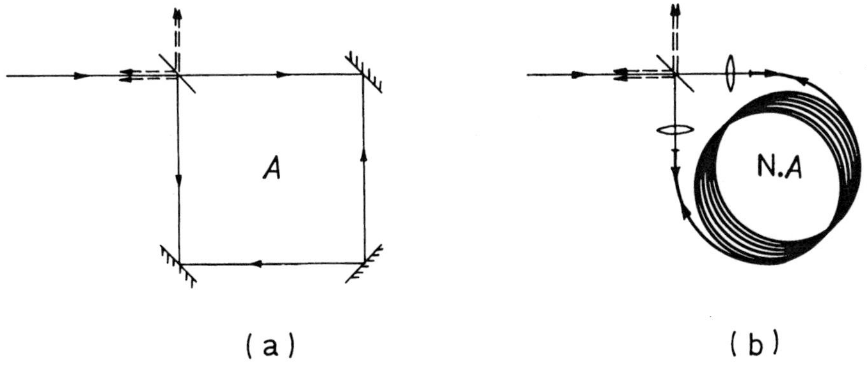

Figure 2.11 Two-wave ring interferometer: (a) bulk form with an enclosed area A; (b) enhanced sensitivity with a multiturn fiber coil that has an enclosed area $N \cdot A$.

There is an unambiguous range of phase measurement of $\pm\pi$ rad around zero which corresponds to an unambiguous operating range of $\pm\ \Omega_\pi$ for the rotation rate (Figure 2.12):

$$\Omega_\pi = \frac{\lambda c}{2L \cdot D} \tag{2.26}$$

Let us give some orders of magnitude. A high-sensitivity fiber gyro would have a coil length of 1 km and a coil diameter of 10 cm. With a wavelength of 850 nm, this yields

$$\Omega_\pi = 1.275 \text{ rad/s} = 73 \text{ deg/s}$$

A phase difference of a micro-radian being a good order of magnitude of sensitivity, this corresponds to a rate of

$$\Omega_\mu = \frac{\Omega_\pi}{\pi\ 10^6} = 0.084 \text{ deg/h}$$

For applications requiring a larger operating range, a shorter and smaller coil can be used. For example, a high operating range fiber gyro would have a coil length of 100m and a coil diameter of 3 cm, which yields

$$\Omega_\pi = 2400 \text{ deg/s and } \Omega_\mu = 2.8 \text{ deg/h}$$

This geometrical flexibility is an important advantage of the FOG technology, since, as we shall see, the same basic components and the same assembly techniques

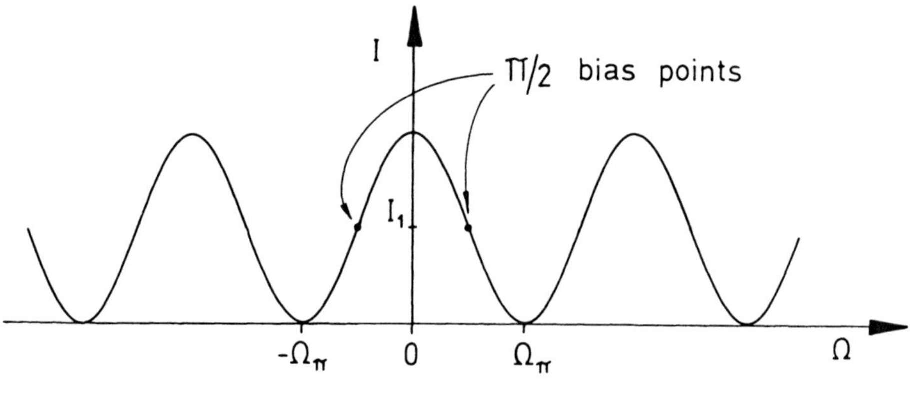

Figure 2.12 Response of an interferometric fiber-optic gyroscope.

may be used for various devices without a complete redesign. By simply scaling up or down the effective area of the fiber coil, the actual operating range may be shifted while conserving a dynamic range of more than 130 dB (corresponding to a phase difference of π rad to less than 1 μrad).

In practice, the I-FOG works over a few fringes about zero path difference, and thus does not require the use of a very narrow spectrum source, contrary to the case of the resonant fiber gyroscope. As we shall see, this is a fundamental advantage, since many parasitic effects are greatly reduced with the low temporal coherence obtained with a broad-spectrum source. Furthermore, a fiber ring interferometer behaves like a vacuum interferometer despite propagation in a medium. The Sagnac phase difference $\Delta\phi_R$ may be expressed with an equivalent geometrical path length difference ΔL_R without any dispersion effect that could be encountered with a broad spectrum:

$$\Delta\phi_R = 2\pi\frac{\Delta L_R}{\lambda} \qquad\qquad \Delta L_R = \frac{LD\Omega}{c} \tag{2.27}$$

The interference intensity is

$$I = I_1\left[1 + \gamma_c(\Delta L_R)\cos\left(2\pi\frac{\Delta L_R}{\lambda}\right)\right] \tag{2.28}$$

where $\gamma_c\,(\Delta L_R)$ is the coherence function of the source (this problem of temporal coherence in interferometers is detailed in Appendix 1).

2.3.2 Theoretical Sensitivity

As in any passive optical system, the theoretical sensitivity of the I-FOG is limited by the photon shot noise. Assuming a phase bias of $\pi/2$ rad to operate the interferometer at the inflection point of the cosine response for maximum sensitivity, the detected power is

$$P(\Delta\phi_R) = P_o[1 + \cos(\Delta\phi_R - \pi/2)] \tag{2.29}$$

However, an optical beam may be regarded as a stream of photons, which behaves statistically like any ensemble of uncorrelated discrete particles. Any flow $\dot{N} = dN/dt$ yields a random counting with a standard deviation $\sigma_{\dot{N}}$ of $\dot{N}$ following

$$\sigma_{\dot{N}}^2 = 2\dot{N}\Delta f_{bw} \tag{2.30}$$

where Δf_{bw} is the counting bandwidth (i.e., the inverse of the duration of the counting). For photons with an energy $h \cdot f = hc/\lambda$ (where $h = 6.63 \times 10^{-34}$ J · s is the Planck constant), the optical power P and its standard deviation σ_P follow

$$\left[\frac{\sigma_P}{hc/\lambda}\right]^2 = 2\frac{P}{hc/\lambda}\Delta f_{bw}$$
$$\sigma_P/\sqrt{\Delta f_{bw}} = \sqrt{2 \cdot (hc/\lambda) \cdot P} \qquad (2.31)$$

For $\lambda = 850$ nm,

$$\sigma_P/\sqrt{\Delta f_{bw}} = 0.7 \times 10^{-9} \sqrt{P} \quad (P \text{ in watts}) \qquad (2.32)$$

A detector converts the flow of photons $\dot{N}_p$ into a primary current of electrons $\dot{N}_e$, which is also shot-noise limited. Its quantum efficiency $\eta = \dot{N}_e/\dot{N}_p$ must be as close as possible to 1 to limit the degradation of the theoretical photon noise. A perfect detector would have a responsivity of 0.68 A/W at 850 nm, and practical semiconductor PIN diodes have a responsivity of 0.55 A/W, which yields a slight increase by a factor $\sqrt{0.68/0.55} \approx 1.1$ of the actual detected photon noise:

$$\sigma_{PD}/\sqrt{\Delta f_{bw}} = 0.77 \times 10^{-9} \sqrt{P} \quad (P \text{ in watts}) \qquad (2.33)$$

On bias, the signal slope is unity with respect to the phase difference, and the noise equivalent phase difference $\sigma_{\Delta\phi}/\sqrt{\Delta f}$ is

$$(\sigma_{\Delta\phi} \text{ in radians}) \ \sigma_{\Delta\phi}/\sqrt{\Delta f_{bw}} = \sigma_{PD}/P\sqrt{\Delta f_{bw}} = \frac{0.77 \times 10^{-9}}{\sqrt{P}} \ (P \text{ in watts}) \qquad (2.34)$$

Present technology gives a typical returning bias power of 1 to 10 μW; therefore:

	Anticipated Bias Power	Equivalent Shot Noise	Noise Equivalent Phase Difference
$\lambda = 850$ nm	1 μW	10^{-12} W/$\sqrt{\text{Hz}}$	1 μrad/$\sqrt{\text{Hz}}$
	10 μW	3×10^{-12} W/$\sqrt{\text{Hz}}$	0.3 μrad/$\sqrt{\text{Hz}}$

Comparing this result with the Ω_μ values, which range, as we have seen, between 0.1 and 3 deg/h, it can be seen that the fiber-gyro technology has a very good theoretical sensitivity, which has motivated an important R&D effort around the world over the last fifteen years.

Note that the wavelength has, in practice, a little influence on the theoretical sensitivity (if, of course, it is within a transparency window of the fiber). For the same coil dimension and the same returning power, the Sagnac phase difference is inversely proportional to the wavelength and the signal-to-noise ratio is proportional to the square root of the wavelength, since, as the wavelength increases, the photon energy decreases, increasing their number for a given power. Any of the usual transparency windows (i.e., 850, 1060, 1300, and 1550 nm) of silica fibers have been used with little difference in performance. The advantage of the very low attenuation of the 1300- or 1550-nm windows is not as significant as for telecommunications, since the fiber length usually remains below 1 km. The criteria of choice are mainly economic and based on cost, availability, and standardization of the components, unless the gyro has to withstand radiation, which then makes the long wavelength range necessary to avoid any increase of the fiber attenuation.

Note 1: For a given fiber attenuation per unit length α (in decibels per km), it is possible to define an optimal fiber length L_{op}. As a matter of fact, for a given diameter, the Sagnac phase difference increases proportionaly to the length L, but the power decreases as $10^{-\alpha L/10}$, which accordingly reduces the signal-to-noise ratio of phase detection to $(10^{-\alpha L/10})^{1/2} = 10^{-\alpha L/20}$. The optimal length is defined by

$$\frac{df}{dL}(L_{op}) = 0 \tag{2.35}$$

where $f(L)$ is a function defined by

$$f(L) = L \cdot 10^{-\alpha L/20} \tag{2.36}$$

which yields

$$L_{op} = \frac{8.7}{\alpha}$$

where L_{op} is in km and α is in dB/km; and typically with silica fibers:

	850 nm	1060 nm	1300 nm	1550 nm
α	2 dB/km	1 dB/km	0.4 dB/km	0.25 dB/km
L_{op}	4 km	8 km	20 km	35 km

This optimal length is much longer than what is used in practice, because this would significantly reduce the unambiguous dynamic range $\pm\Omega_\pi$, and the coil volume would become too large to fit within reasonable overall dimensions. With coil lengths ranging between 100m to 1 km, the I-FOG has adequate performance for most applications, but, as we can see, it still remains possible to improve sensitivity if there are relaxed size and dynamic range constraints for some specific applications.

Note 2: Operation around the $\pi/2$ radian bias point yields the highest sensitivity, and the theoretical photon noise has been calculated for this value. However, the optimal performance is obtained for the best signal-to-noise ratio, which is not a priori this $\pi/2$ bias.

Let us consider a perfectly contrasted interferometer. The response is a raised cosine. The sensitivity at a given phase bias ϕ_b is proportional to the slope (i.e., the derivative $\sin\phi_b$ of the raised cosine $(1 + \cos\phi_b)$), but the theoretical photon noise is proportional to the square root of the bias power (i.e., $\sqrt{(1 + \cos\phi_b)/2} = \cos(\phi_b/2)$). Thus, the theoretical signal-to-noise ratio, which is proportional to the ratio between the sensitivity and the noise, follows [15] (Figure 2.13)

$$\frac{\sin\phi_b}{\cos(\phi_b/2)} = 2\sin(\phi_b/2) \tag{2.37}$$

Therefore, the theoretical signal-to-noise ratio is optimal for $\phi_b = \pi$ rad (i.e., on a black fringe) and not for $\phi_b = \pi/2$ rad, where it is $\sqrt{2}$ times lower.

In practice, it is not desirable to operate very close to the black fringe, because of the thermal noise of the detector, but this shows that the bias point may be chosen between $\pi/2$ and typically $3\pi/4$ without degradation of the signal-to-noise ratio.

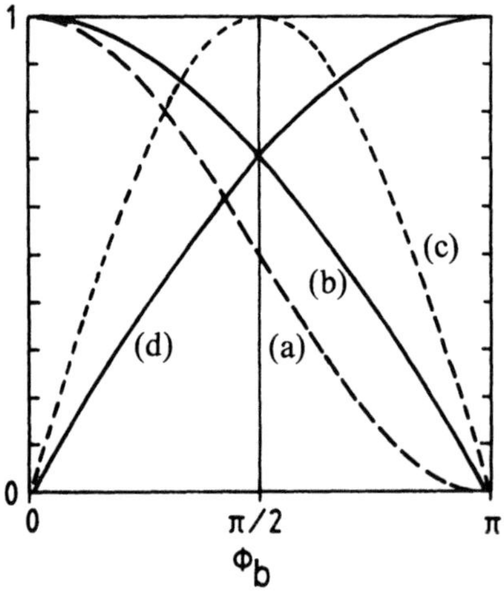

Figure 2.13 Optimal signal-to-noise ratio as a function of the phase bias ϕ_b with (a) the actual optical power; (b) the photon noise; (c) the sensitivity; (d) the signal-to-noise ratio. (The vertical coordinates are normalized.)

2.3.3 Noise, Drift, and Scale Factor

At rest, the output signal of a fiber gyro is a random function that is the sum of a white noise (with the theoretical limit of the photon shot noise) and a slowly varying function to take into account the long-term drift of the mean value. The white noise is expressed in terms of the standard deviation of equivalent rotation rate per square root of bandwidth of detection (i.e., degrees per hour per square root of hertz or (deg/h)/$\sqrt{\text{Hz}}$). Equivalent noise power spectral density may be used instead by taking the square of the standard deviation (i.e., (deg/h)^2/Hz). The third possible definition is the so-called random walk performance in deg/$\sqrt{\text{h}}$, which has the same dimension as the standard deviation, but 1(deg/h)/$\sqrt{\text{Hz}}$ is equal to 1/60 deg/$\sqrt{\text{h}}$. Random walk yields more "impressive" numbers, but it is only a particular unit to evaluate the white noise.

In particular, random walk must not be confused with the drift, which evaluates the peak-to-peak boundaries of the long-term variations of the mean value of the output signal (Figure 2.14). Drift is usually expressed in ±deg/h. In fiber gyro, the noise limit is the detection noise, which depends mainly on the amount of returning optical power; but the drift, which can be theoretically null, corresponds to a residual lack of "reciprocity," which we are going to analyze.

Noise and drift are different requirements which depend on applications. Low noise is important for fast response stabilization and control, but, for navigation, drift is a more fundamental parameter. The rotation rate signal is integrated to get the change in angular orientation, and this process of integration produces an

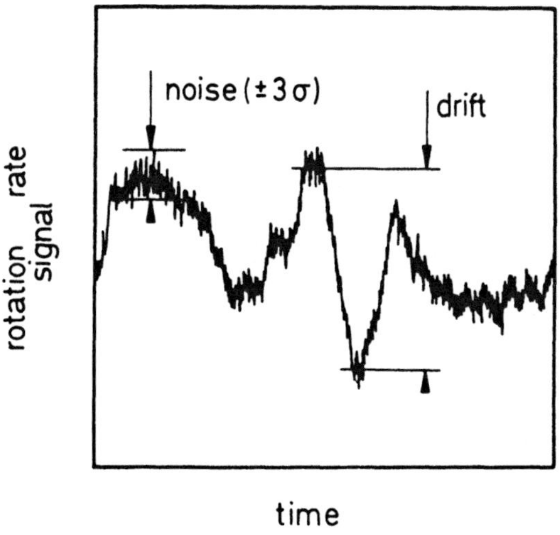

Figure 2.14 Example of bias variation with noise and long-term drift.

averaging of the white noise which renders the effect of the drift predominant in the long term.

Another very important characteristic of a gyro is the scale factor. As a matter of fact, compared to other sensors, a gyroscope needs a much better accuracy over a much wider dynamic range: the important measurement is the integrated rotation angle, and any past error degrades future information. It is important to have low noise and drift to measure very low rate, but it is also important to have an accurate measurement of high rates (i.e., an accurate scale factor). The required performances depend on the kind of trajectory, and they are precisely defined by complex system analysis and modeling, but most applications are usually classified into three grades:

	Bias White Noise or Random Walk	Bias Drift (σ Value)	Scale Factor Accuracy (σ Value)
Inertial grade	<0.001 deg/$\sqrt{h}$	<0.01 deg/h	<5 ppm
Tactical grade	0.5 to 0.05 deg/$\sqrt{h}$	0.1 to 10 deg/h	10 to 1000 ppm
Rate grade	>0.5 deg/$\sqrt{Hz}$	10 to 1000 deg/h	0.1% to 1%

As will be seen, fiber gyro technology is particularly suitable for tactical grade applications. Inertial grade performance is possible, but it will be more difficult to compete with the laser gyro, which has reached a very strong position in this market segment.

2.3.4 Bandwidth

The minimum response time of the interferometric fiber gyroscope is the transit time through the fiber coil (i.e., 1 μs for a length of about 200m). This yields a very high theoretical bandwidth of several hundreds of kilohertz. As will be seen later, signal processing techniques have to be used, which reduces the bandwidth, but in practice frequency ranges of several kilohertz are reached, which is a very significant improvement over previous technologies.

It is important to note that the rate signal is averaged over the transit time. Therefore, signal sampling with this periodicity does not yield any loss of information for the integrated angle of rotation [16], and subsequent averaging does yield the exact averaged rate. Some proposed signal processing techniques use signal gating and sampling with a longer periodicity (see Section 10.2), and it is important to be aware that may yield error if the frequency band of the rate signal is too high, since the variation of the rate during the gating is not taken into account.

As pointed out [17], the interferometric fiber gyro is usually viewed as a rate gyro; that is, the basic measurement is a rotation rate signal. However, considered only over the transit time through the fiber coil, the rate is averaged and the fiber gyro may be viewed in this case as a rate integrating gyro; that is, the basic measurement is an angle of rotation, since, mathematically, an average is equivalent to an integration.

REFERENCES

[1] Sagnac, G., "L'éther lumineux démontré par l'effet du vent relatif d'éther dans un interféromètre en rotation uniforme," Compte-rendus de l'Académie des Sciences, Vol. 95, 1913, pp. 708–710. Sagnac, G., "Sur la preuve de la réalité de l'éther lumineux par l'expérience de l'interférographe tournant," Comptes rendus de l'Académie des Sciences, Vol. 95, 1913, pp. 1410–1413.

[2] Post, E. J., "Sagnac Effect," Review of Modern Physics, Vol. 39, 1967, pp. 475–494.

[3] Arditty, H. J., and H. C. Lefèvre, "Sagnac Effect in Fiber Gyroscopes," Optics Letters, Vol. 6, 1981, pp. 401–403 (SPIE MS 8, pp. 105–107).

[4] Leeb, W. R., G. Schiffner, and E. Scheiterer, "Optical Fiber Gyroscopes: Sagnac or Fizeau Effect," Applied Optics, Vol. 18, 1979, pp. 1293–1295 (SPIE MS 8, pp. 85–87).

[5] Post, E. J., "Interferometric Path-Length Changes Due to Motion," J.O.S.A., Vol. 62, 1972, pp. 234-239 (SPIE MS 8, pp. 79–84).

[6] Arditty, H. J., and H. C. Lefèvre, "Theoretical Basis of Sagnac Effect in Fiber Gyroscopes," Fiber-Optic Rotation Sensors, Springer Series in Optical Sciences, Vol. 32, 1982, pp. 44–51.

[7] Rosenthal, A. H., "Regerative Circulatory Multiple Beam Interferometry for the Study of Light Propagation Effects," J.O.S.A., Vol. 52, 1962, pp. 1143–1148.

[8] Macek, W. M., and D. T. M. Davis, "Rotation Rate Sensing With Travelling-Wave Ring Lasers," Applied Physics Letters, Vol. 2, 1963, pp. 67–68.

[9] Ezekiel, S., and G. E. Knausenberger, eds., "Laser Inertial Rotation Sensors," SPIE Proceedings, Vol. 157, 1978.

[10] Chow, W. W., J. Gea-Banacloche, L. M. Perdrotti, V. E. Sanders, W. Schleich, and M. O. Scully, "The Ring Laser Gyro," Review of Modern Physics, Vol. 57, 1985, p. 61.

[11] Ezekiel, S., and S. R. Balsamo, "Passive Ring Resonator Laser Gyroscope," Applied Physics Letters, Vol. 30, 1977, pp. 478–480 (SPIE MS 8, pp. 457–459).

[12] Meyer, R. E., S. Ezekiel, D. W. Stowe, and V. J. Tekippe, "Passive Fiber-Optic Ring Resonator for Rotation Sensing," Optics Letters, Vol. 8, 1983, pp. 644–646 (SPIE MS 8, pp. 467–469).

[13] Pircher, G., and G. Hepner, "Perfectionnements aux dispositifs du type gyromètre interférométrique à laser," French patent 1.563.720, 1967.

[14] Vali, V., and R. W. Shorthill, "Fiber Ring Interferometer," Applied Optics, Vol. 15, 1976, pp. 1099–1100 (SPIE MS 8, pp. 135–136).

[15] Lefèvre, H. C., S. Vatoux, M. Papuchon, and C. Puech, "Integrated Optics: A Practical Solution for the Fiber-Optic Gyroscope," SPIE Proceedings, Vol. 719, 1986, pp. 101–112 (SPIE MS 8, pp. 562–573).

[16] Lefèvre, H. C., Ph. Graindorge, H. J. Ardity, S. Vatoux, and M. Papuchon, "Double Closed-Loop Hybrid Fiber Gyroscope Using Digital Phase Ramp," Proceedings of OFS 3/'85, OSA/IEEE, San Diego, postdeadline paper 7, 1985 (SPIE MS8, pp. 444–447).

[17] Fidric, B., D. Tazartes, A. Cordova, and J. Mark, "A Rate Integrating Fiber Optic Gyro: From the Theoretical Concept to System Mechanization," SPIE Proceedings, Vol. 1585, 1991, pp. 437–448.

Chapter 3
Reciprocity of a Fiber Ring Interferometer

3.1 PRINCIPLE OF RECIPROCITY

3.1.1 Reciprocity of Wave Propagation

As we have seen, the theoretical noise of a fiber ring interferometer is, in practice, on the order of 1 μrad/$\sqrt{\text{Hz}}$. Through integration, phase differences of 10^{-7} to 10^{-8} rad should be measurable while the absolute phase accumulated by the wave along 100m to 1 km of fiber is 10^9 to 10^{10} rad. The sensitivity limit is 16 to 18 orders of magnitude below the actual propagation path when the temperature dependence is already as high as 10^{-5}/K. Such a performance could look unrealistic, but the fundamental principle of reciprocity of light propagation in a linear medium is the key to the solution of this problem.

As a matter of fact, in a linear medium, the propagation equation of the wave amplitude A is (see Appendix 1).

$$\nabla^2 \cdot A - \frac{n^2}{c^2}\frac{\partial^2 A}{\partial t^2} = 0 \tag{3.1}$$

Looking for harmonic solutions, $A(x,y,z,t) = A_s(x,y,z)e^{i\omega t}$, where A_s depends only on the spatial coordinates and ω is the angular frequency, the propagation equation is reduced to

$$\nabla^2 \cdot A_s + \frac{n^2\omega^2}{c}A_s = 0 \tag{3.2}$$

Therefore, any solution $A(x,y,z,t) = A_s(x,y,z)e^{i\omega t}$ has a perfectly "reciprocal" solution $A'(x,y,z,t) = A_s(x,y,z)e^{-i\omega t}$, since the reduced propagation equation depends on the square $\omega^2 = (-\omega)^2$. Physically, this mathematical change of sign corresponds to a propagation in the opposite direction with exactly the same propagation delay and the same attenuation of the phase front. In free space, difficult alignments are required to excite both reciprocal opposite solutions, and it is never perfect (except when phase-conjuguate mirrors are used! [1]). However, if the system is single-mode (i.e., "single-solution"), alignments are only needed to optimize the throughput power; but once it is coupled, both opposite waves are automatically reciprocal. The single-mode waveguide filters the exact part of both opposite input waves corresponding to the same unique propagating solution.

The Sagnac effect is a very small first-order effect (in $D\Omega/c$) which should be buried in the changes of the zero order, the absolute phase accumulated in the propagation; but single-mode reciprocity provides perfect "common-mode rejection" and will permit the nulling out of this zero-order and its variations almost perfectly, yielding a sensitive measurement of the rotation-induced nonreciprocal phase difference.

3.1.2 Reciprocal Behavior of a Beam Splitter

Reciprocity of counterpropagation is fundamental, but it is also important to understand the behavior of a beam splitter. Considering a simple ring interferometer in bulk optics (Figure 3.1), the two waves returning through the common input-output port are perfectly in phase because they have propagated along the same path, but also because they have both experienced a reflection and a transmission on the beam splitter.

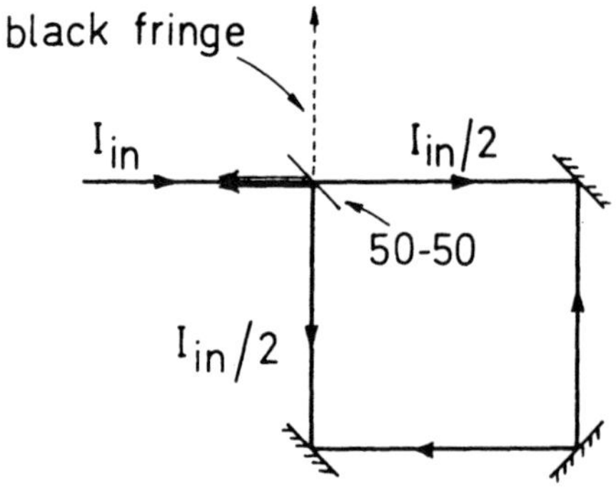

Figure 3.1 Reciprocity of a bulk-optic ring interferometer.

Assuming a 50-50 (or 3-dB) separation, the intensity of these returning waves is $I_{in}/4$, if I_{in} is the input intensity. Since they interfere in phase (i.e., $\Delta\phi = 0$), the intensity I_{int} of the total interfering wave is equal to the input intensity I_{in}:

$$I_{int} = I_{in}/4 + I_{in}/4 + 2\sqrt{I_{in}/4 \cdot I_{in}/4}\cos 0 = I_{in} \tag{3.3}$$

and in the absence of a nonreciprocal effect, a ring interferometer behaves like a mirror. Therefore, because of conservation of energy, the intensity at the second output port (the free port) of the ring interferometer must be zero. As stated early on by Sagnac in his original publication [2], there is always a "black fringe" at the free port, independent of the exact nature of the 3-dB splitter. This implies that the waves have a π rad phase difference at this port. Since their propagation paths are identical, this difference is due to a basic phase shift between the reflected and transmitted waves on the beam splitter.

This basic phase difference may be viewed directly, considering a symmetrical system with two input waves which arrive in phase on an ideal symmetrical splitter with zero thickness (Figure 3.2). Because of symmetry, the power of both interference waves must be equal, which implies that the split waves that interfere are in phase quadrature at both output ports. Since there is no difference of path length, this phase shift must be a basic $\pi/2$ difference between the reflected and the transmitted waves. Since this $\pi/2$ difference is experienced twice at the free port of the ring interferometer, a π difference is yielded at the output.

It is interesting to compare this effect with the Huygens principle, which states that the propagation of a wave may be analyzed with virtual sources that

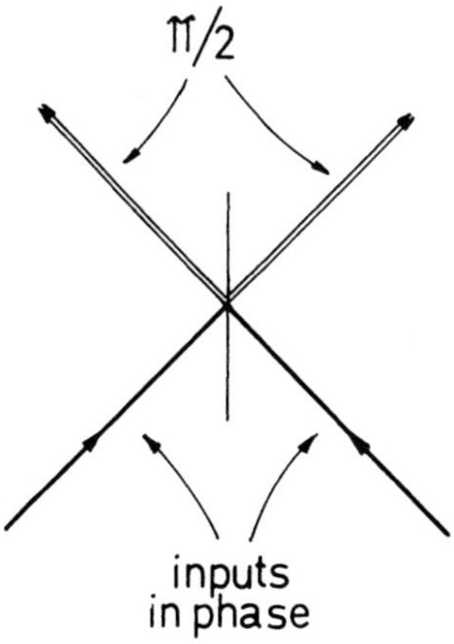

Figure 3.2 Intrinsic phase shift on a beam splitter

re-emit spherical waves from the phase front. It is well-know that in order to restore the exact phase of the pursuant wave front, a $-\pi/2$ phase shift must be added to the virtually re-emitted spherical waves. With an actual reflection from a splitter, "matter" is re-emitting "real" waves without this $-\pi/2$ phase shift; therefore, reflection has a phase quadrature with respect to transmission.

3.2 MINIMUM CONFIGURATION OF A RING FIBER INTERFEROMETER

3.2.1 Reciprocal Configuration

The first experimental demonstration of a fiber gyro [3] showed that a simple fiber ring interferometer is not intrinsically reciprocal. Complementary interference fringes were observed at both ports of the interferometer, depending on the alignments of the fiber ends (Figure 3.3). At the input, a parallel Gaussian laser beam is split and focused on both fiber ends which filter the unique propagation mode, and, at the output, the beams are recombined to interfere. A small change in the alignments does not strongly modify the input power coupling, but it has a significant effect on the matching of both output phase fronts, modifying the fringe pattern and consequently producing a high parasitic variation of the measured phase difference.

All these problems could have been a severe limitation to high performance, but can be solved very simply with a so-called reciprocal configuration. It is sufficient to feed light into the interferometer through a truly single-mode waveguide and look at the returning interference wave, which is filtered through this same

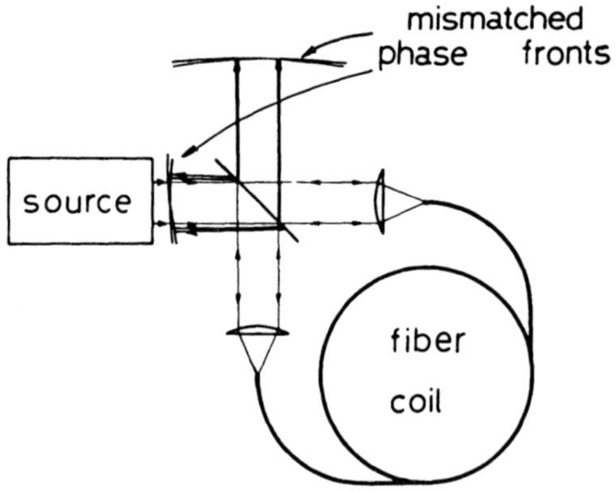

Figure 3.3 Free-space interference pattern in a ring fiber interferometer.

waveguide in the opposite direction (Figure 3.4) [4]. In this case, the alignments are needed solely to optimize the throughput power (and its related signal-to-noise ratio), which requires difficult but nevertheless reasonable mechanical tolerances (see Appendix 2). It is now ensured that both returning waves have propagated along exactly the same path in the opposite direction and that they interfere perfectly in phase when the system is at rest. This simple modification has made both opposite paths identical, zero rotation giving zero phase difference. This common input-output port is called the reciprocal port, and the other free port is called the nonreciprocal port.

A reciprocal operation of the fiber gyro does not require a continuous single-mode propagation, but merely a single-mode filter at the common input-output port. A reasonable amount of power must remain in the primary mode to get some light through the output filter, but any undesirable signals are perfectly eliminated. For example, assuming that 90% of the light remains in the primary mode, there is only a slight decrease by a factor of $\sqrt{0.9}$ in the theoretical signal-to-noise ratio, whereas the rejected 10% may carry a spurious signal equivalent to a phase difference as high as 0.1 rad (i.e., at least six orders of magnitude above the theoretical sensitivity).

Spatially, a short length (about 1 m) of single-mode fiber can be considered as a perfect filter, but there is also a need for polarization filtering [4,5], since a single-spatial-mode fiber is actually a dual-polarization-mode fiber and the fiber birefringence (see Appendix 2) may yield a spurious phase difference. However, polarizer rejection is limited in practice, and we shall explain further how this important problem may be solved with a compromise between polarization filtering, polarization conservation, and statistical depolarization. In addition to the coil splitter, the complete reciprocal configuration needs a second splitter (so-called source splitter) to tap off part of the interference wave returning through the filtered input-output port of the ring interferometer. This gives an intrinsic 6-dB loss, 50%

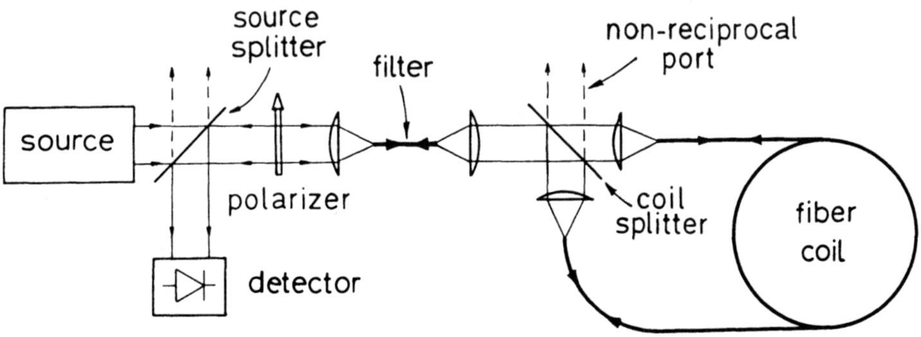

Figure 3.4 Reciprocal configuration of a fiber ring interferometer.

of the useful light being lost at the input and at the output, but it is a very moderate drawback compared to the crucial improvement brought about by reciprocity. Note that the returning power is maximum when both splitters are 50-50 (or 3 dB); but this value does not have to be very precise. For example, with a 60-40 splitter, the transmission of the returning power is 24% instead of 25% with 50-50.

Note: Reciprocity is a much more powerful principle than this simple case of single-mode reciprocity. Considering a multimode filter at the common input-output port, it is possible to show that light entering in any mode M_i and leaving the interferometer through this same mode M_i does not carry any spurious phase difference. This result is simply the generalization of the single-mode case.

However, general linear network theory [6] shows that there is also a reciprocity phenomenon on the crossed terms. Light entering in any mode M_i and leaving through another mode M_j because of parasitic couplings is carrying a spurious phase difference signal $\Delta\phi_{ij}$; but light entering in M_j and leaving through M_i is carrying exactly the opposite phase difference:

$$\Delta\phi_{ji} = -\Delta\phi_{ij}$$

Therefore, if M_i and M_j have equal power at the input and equal attenuation in the system, both detected signals cancel out the spurious term due to $\Delta\phi_{ij}$.

This applies to spatial modes with multimode fibers [7], but also to polarization modes with an unpolarized source [8]. The result of this reciprocity of crossed modes is, however, very difficult to reduce to practice, because it requires a very good equalization of energy between the modes. It is easier to filter a truly single mode with a rejection of 10^{-x} of the spurious terms than to ensure that several modes can be excited and transmitted equally within 10^{-x}.

3.2.2 Reciprocal Biasing Modulation-Demodulation

The reciprocal configuration provides an interference signal of the Sagnac effect with perfect contrast, since the phases as well as the amplitudes of both counter-propagating waves are perfectly equal at rest. The optical power response is then a raised cosine function, $P(\Delta\phi_R) = P_0\,[1 + \cos\Delta\phi_R]$, of the rotation-induced phase difference $\Delta\phi_R$, which is maximum at zero. To get high sensitivity, this signal must be biased about an operating point with a nonzero response slope:

$$P(\Delta\phi_R) = P_0\,[1 + \cos(\Delta\phi_R + \phi_b)] \tag{3.4}$$

where ϕ_b is the phase bias. However, ϕ_b must be as stable as the anticipated sensitivity; that is, significantly better than 1 μrad. For example, the use of the nonreciprocal Faraday effect (explained in Section 7.1), controlled with an electrical current, was proposed [9], but this would require a control of the biasing current with an accuracy better than 1 ppm.

The problem of drift of the phase bias is completely overcome with the use of a reciprocal phase modulator placed at one end of the coil that acts as a delay line (Figure 3.5) [10]. Because of reciprocity, both interfering waves carry exactly the same phase modulation $\phi_m(t)$, but shifted in time. The delay is equal to the difference $\Delta\tau_g$ of group transit time (see Appendix 1) between the long and short paths that connect the modulator and the splitter. This yields a biasing modulation $\Delta\phi_m(t)$ of the phase difference:

$$\Delta\phi_m(t) = \phi_m(t) - \phi_m(t - \Delta\tau_g) \tag{3.5}$$

and the interference signal becomes

$$P(\Delta\phi_R) = P_0[1 + \cos[\Delta\phi_R + \Delta\phi_m(t)]] \tag{3.6}$$

This technique may be implemented with a square-wave modulation $\phi_m = \pm(\phi_b/2)$, where the half-period is equal to $\Delta\tau_g$ (this corresponds to the so-called proper or eigenfrequency f_p of the coil, $f_p = 1/(2\Delta\tau_g)$, and the product, coil length $\times$ proper frequency, is about 100 m · MHz with silica fibers). This yields a biasing modulation $\Delta\phi_m = \pm\phi_b$. At rest, both modulation states give the same signal (Figure 3.6):

$$P(0, -\phi_b) = P(0, \phi_b) = P_0(1 + \cos\phi_b) \tag{3.7}$$

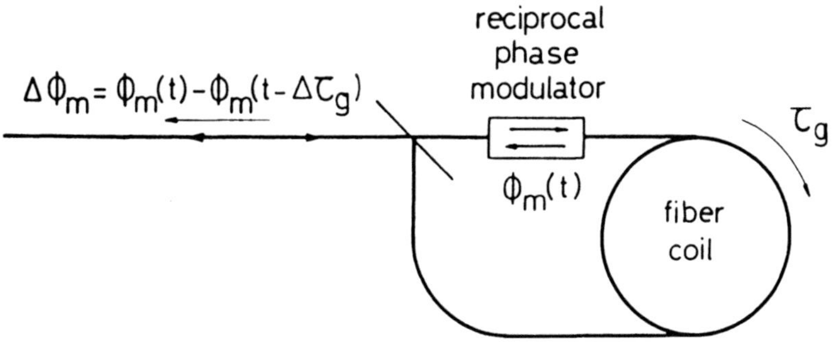

Figure 3.5 Generation of the biasing phase modulation using the delay through the fiber coil.

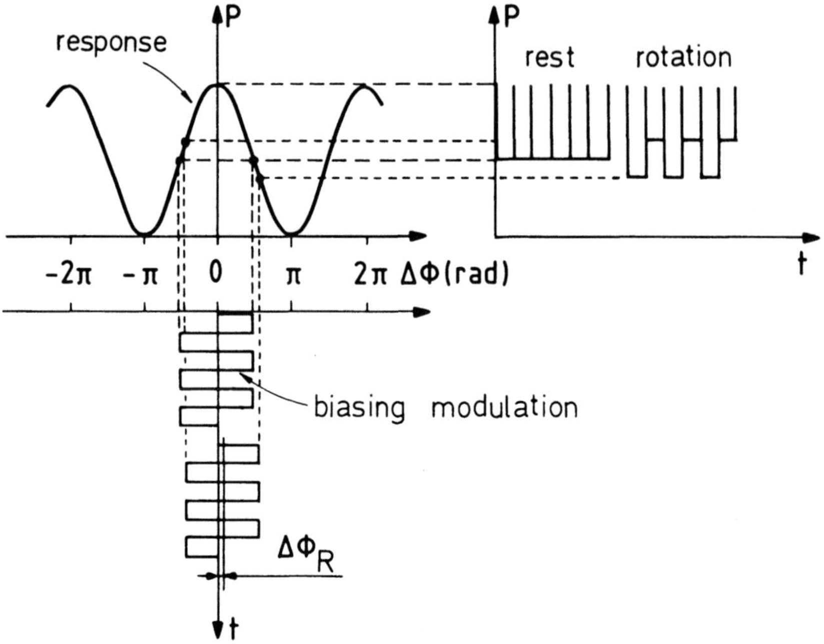

Figure 3.6 Square-wave biasing modulation.

but in rotation,

$$\begin{aligned} P(\Delta\phi_R, \phi_b) &= P_0[1 + \cos(\Delta\phi_R + \phi_b)] \\ P(\Delta\phi_R, -\phi_b) &= P_0[1 + \cos(\Delta\phi_R - \phi_b)] \end{aligned} \tag{3.8}$$

and the difference between both states becomes

$$\Delta P(\Delta\phi_R, \phi_b) = P_0[\cos(\Delta\phi_R - \phi_b) - \cos(\Delta\phi_R + \phi_b)] \tag{3.9}$$

$$\Delta P(\Delta\phi_R, \phi_b) = 2P_0 \sin\phi_b \sin\Delta\phi_R \tag{3.10}$$

This "biased" signal ΔP is measured by demodulating the detector signal with a lock-in amplifier, and the maximum sensitivity is obtained for $\phi_b = \pi/2$, where $\sin\phi_b = 1$.

This modulation-demodulation technique is now widely accepted as the optimal biasing technique, since it yields a sine response (derivative of the unmodulated cosine response) with a very stable bias (Figure 3.7). The power dependence P_0 and the phase bias dependence $\sin\phi_b$ are multiplicative, and therefore have no influence on the bias stability. The use of a reciprocal phase modulator

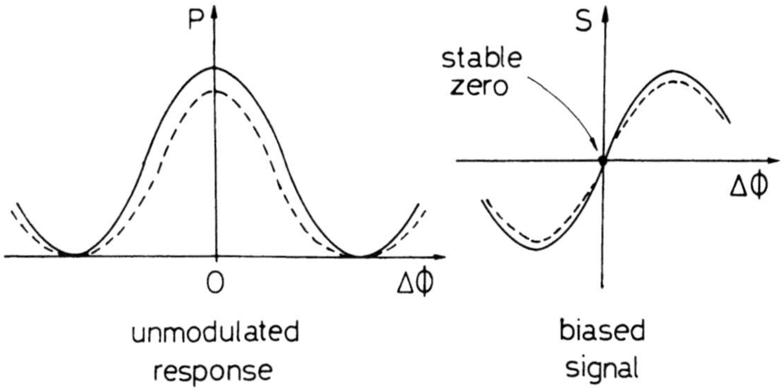

Figure 3.7 Unmodulated response and demodulated biased signal of a fiber ring interferometer.

is fundamental, since it yields a modulation of the phase difference perfectly centered around zero. As a matter of fact, for any square-wave modulation $\phi_m(t) = \begin{cases}\phi_{m1}\\ \phi_{m2}\end{cases}$, the modulation of the phase difference is always $\Delta\phi_m(t) = \pm \phi_b = \pm(\phi_{m1} - \phi_{m2})$.

This reciprocal biasing technique may be alternatively implemented with a (co)sine modulation $\Delta\phi_m(t) = \phi_b \cos(2\pi f_m t)$ [4,10], which does not require a phase modulator with a flat frequency response. The detected signal may be decomposed in harmonic components of the modulation frequency f_m

$$P(\Delta\phi_R) = P_0[1 + \cos(\Delta\phi_R + \phi_b \cos(2\pi f_m t))] \tag{3.11}$$

$$P(\Delta\phi_R) = P_0[1 + \cos(\Delta\phi_R)\cos(\phi_b \cos(2\pi f_m t)) - \sin(\Delta\phi_R)\sin(\phi_b \cos(2\pi f_m t))] \tag{3.12}$$

Using J_n Bessel function expansion, this becomes

$$\begin{aligned} P(\Delta\phi_R) = {} & P_0 + P_0 \cos(\Delta\phi_R)\,[J_0(\phi_b) + 2J_2(\phi_b)\cos(4\pi f_m t) + \ldots\,] \\ & + P_0 \sin(\Delta\phi_R)\,[2J_1(\phi_b)\sin(2\pi f_m t) \\ & + 2J_3(\phi_b)\sin(6\pi f_m t) + \ldots\,] \end{aligned} \tag{3.13}$$

The even harmonics are still proportional to $\cos(\Delta\phi_R)$ as the unbiased response, but the odd harmonics and particularly the fundamental frequency are proportional to $\sin\phi_R$. With a synchronous demodulation this yields a biased signal:

$$P_1(\Delta\phi_R) = 2P_0 J_1(\phi_b)\sin(\Delta\phi_R) \tag{3.14}$$

The maximum sensitivity is now obtained for $\phi_b \approx 1.8$ rad (instead of $\pi/2 \approx 1.5$ with square wave) and $J_1(1.8) = 0.53$. At rest, the detector signal is mainly composed of the second harmonic component, but, in rotation, it appears to be an unbalanced modulation that contains a signal at the fundamental frequency f_m (Figure 3.8). (Note that we call first harmonic the fundamental frequency, and second harmonic the double of fundamental frequency.)

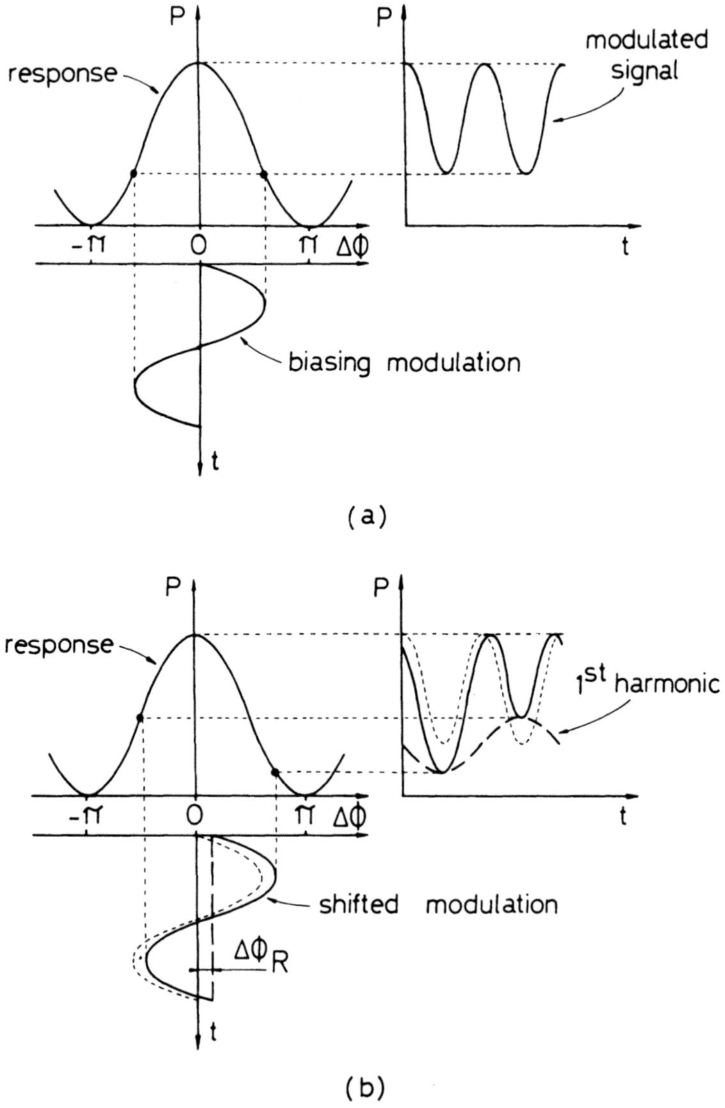

Figure 3.8 Sine wave biasing modulation: (a) at rest; (b) in rotation.

3.2.3 Proper (or Eigen) Frequency

For a sinusoidal phase modulation $\phi_m = \phi_{m0}\sin(2\pi f_m t)$ applied on both counter-propagating waves, the modulation of the phase difference is

$$\Delta\phi_m(t) = \phi_{m0}\left[\sin(2\pi f_m t) - \sin(2\pi f_m(t - \Delta\tau_g))\right] \tag{3.15}$$

and after applying trigonometric identities,

$$\Delta\phi_m(t) = 2\phi_{m0}\sin(\pi f_m \Delta\tau_g)\cos\left[2\pi f_m\left(t - \frac{\Delta\tau_g}{2}\right)\right] \tag{3.16}$$

The ring interferometer behaves like a perfect delay line filter with a sinusoidal transfer function $2\sin(\pi\phi_m\Delta\tau_g)$ that is maximum (in absolute value) at the so-called proper (or eigen) frequency f_p (defined previously as $f_p = 1/(2\Delta\tau_g)$) and its odd harmonics, and that is null at dc and all the even harmonics (Figure 3.9).

From this result, it seems that the choice of the operating frequency is not critical, but spurious nonlinearity or amplitude modulation in the phase modulation may degrade the quality of the bias [11]. As a matter of fact, the use of an additional modulation to detect the maximum of a response curve is very common, but an accuracy of a few tenths of percent of the width is usually sufficient. In a fiber gyro, we are looking for less than 10^{-6}, and potential parasitic effects must be evaluated more carefully.

A basic point deserves to be repeated: since the interference signal is an autocorrelation function, the response is perfectly symmetrical about zero, even with an asymmetrical spectrum (see Appendix 1).

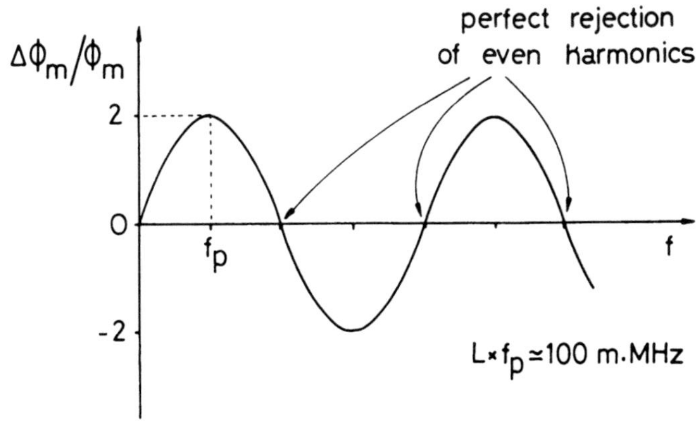

Figure 3.9 Normalized transfer function of the modulation of the phase difference.

Now, a first cause of spurious bias is the nonlinearity of the phase modulator or of the electronic generator. A pure (co)sine modulation centered about zero yields only even harmonics in the modulated interferometer response, and the demodulated response at the fundamental (and also odd harmonic) frequency is perfectly zero. However, if there is a spurious even harmonic content due to a nonlinear response of the modulation chain, additional odd frequency components are generated in the response that are seen as a nonzero bias in the demodulation. This can be easily seen in Figure 3.10, which shows that a second-harmonic content yields an unbalanced modulation, as with an offset, due to rotation. In this worst case, the second harmonic content must be less than 120 dB (electrical) to limit the bias below 10^{-6} rad! This effect also depends on the respective phase of the main fundamental-frequency modulation and on the spurious even harmonic.

A bias offset is not a priori detrimental to the gyro performance if it is stable, but in practice it is susceptible to drift, particularly with changing environmental conditions. As will be seen throughout this analysis, the best way to suppress the drift is, after all, to suppress the offset! In the case of a nonlinear modulator response, a simple solution is to operate the system at the proper frequency [11] (or its odd harmonics), and any spurious even harmonics of the modulation $\phi_m(t)$ will be nulled out in the phase difference $\Delta\phi_m(t)$ because, as we saw, the interferometer behaves as a delay line filter. Note that since this modulation propagates on a very high-frequency carrier, the optical wave, there is no dispersion for these

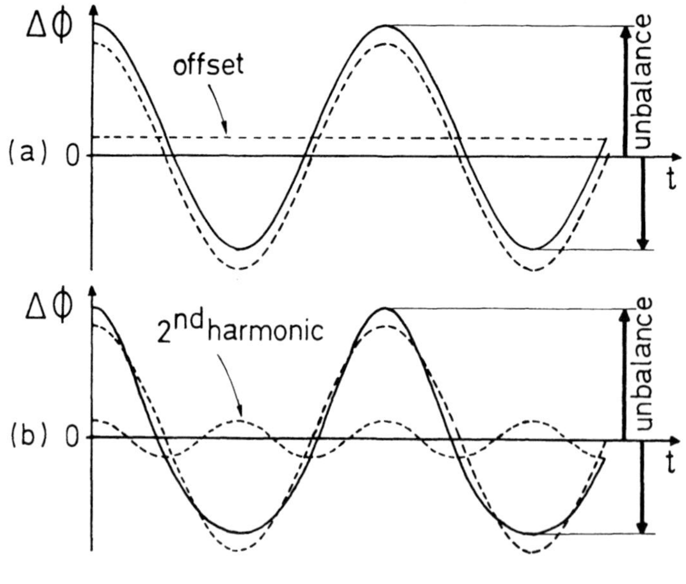

Figure 3.10 Biasing modulation imbalance: (a) with a constant offset; (b) with an additional second-harmonic modulation. (The solid curve in each graph is the sum of the two dashed curves.)

subcarrier harmonic components, which all propagate at the same group velocity, and the notch filtering is perfectly periodic.

With a square-wave modulation, a nonlinear phase modulation response is not harmful, since ϕ_m takes only two values, ϕ_{m1} and ϕ_{m2}, and the modulation of the phase difference $\Delta\phi_m = \pm(\phi_{m1} - \phi_{m2})$ is always balanced around zero. If the modulation frequency f_m is not equal to the proper frequency f_p, the modulation of the phase difference takes four values (Figure 3.11):

$$\Delta\phi_m(t) = \begin{cases} +\phi_b = \phi_{m1} - \phi_{m2} \\ 0 \\ -\phi_b = -(\phi_{m1} - \phi_{m2}) \\ 0 \end{cases} \tag{3.17}$$

but it is still balanced about zero without any demodulated bias offset.

However, a spurious offset may arise with an asymmetry of the duty cycle of the square-wave modulation [12]:

$$\phi_m(t) = \begin{cases} \phi_{m1} \text{ during } (1 - \epsilon)\, T_m/2 \\ \\ \phi_{m2} \text{ during } (1 + \epsilon)\, T_m/2 \end{cases} \tag{3.18}$$

where T_m is the period of the modulation. At the proper frequency (where the period $T_m = 2\Delta\tau_g$), the phase difference $\Delta\phi_m(t)$ takes four values (Figure 3.12):

$$\Delta\phi_m(t) = \begin{cases} \phi_b = \phi_{m1} - \phi_{m2} \text{ during } (1 - \epsilon)\, \Delta\tau_g \\ 0 \text{ during } \epsilon \cdot \Delta\tau_g \\ -\phi_b = -(\phi_{m1} - \phi_{m2}) \text{ during } (1 - \epsilon)\, \Delta\tau_g \\ 0 \text{ during } \epsilon \cdot \Delta\tau_g \end{cases} \tag{3.19}$$

When the gyro is at rest, the output signal is composed of spikes of equal width $\epsilon \cdot \Delta\tau_g$, but when the modulation frequency f_m is not equal to f_p, one spike width is reduced while the other one is enlarged (Figure 3.13), which yields a very strong parasitic signal at the modulation frequency, and thus a spurious offset on the demodulated signal. This is a very sensitive way to measure the proper frequency [12]. However, the duty cycle of a square wave cannot be made perfectly equal to 50-50 (particularly because rise time and fall time are usually not equal in electronics), and it yields a spurious bias which depends on the modulation frequency and goes through zero at the proper frequency. Since this proper frequency depends on temperature (10^{-5}/°C) because of the thermal dependence of the index of refraction, and therefore of the transit time, this spurious bias offset is not stable. To eliminate this defect, an electronic gate has to be used to suppress the spikes that carry the spurious effect of an imperfect 50-50 duty cycle.

This problem makes the use of square-wave modulation more delicate than that of sine-wave modulation, but, as will be seen later, closed-loop digital pro-

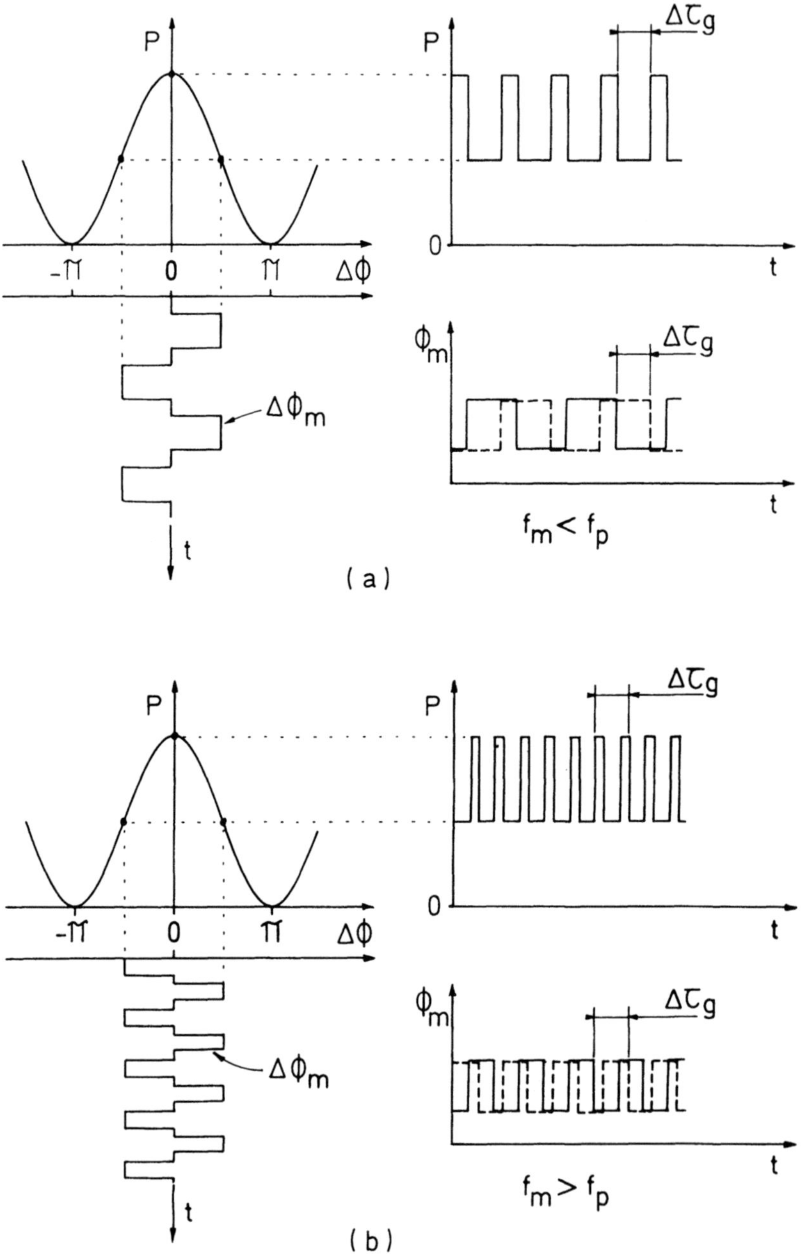

Figure 3.11 Square-wave biasing modulation with (a) $f_m < f_p$; (b) $f_m > f_p$.

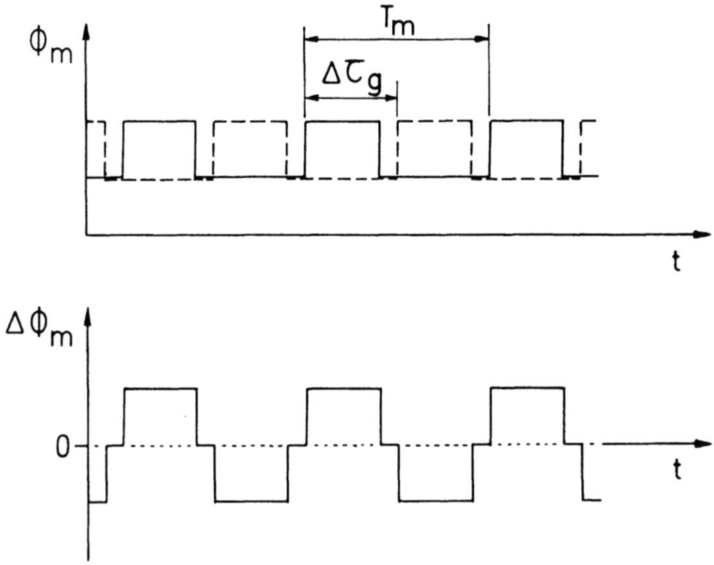

Figure 3.12 Modulated phase difference with an asymmetrical duty cycle of the square wave.

cessing techniques that use square-wave biasing are so advantageous that this additional complexity is a reasonable drawback.

A last parasitic effect is due to residual intensity modulation in the phase modulator and may be detected as a spurious signal at the fundamental frequency [11]. This modulation is also reciprocal and is seen equally by both counterpropagating waves, but the interference signal is derived from the difference of the wave phases, whereas the intensities are added. For intensity modulation, the system also behaves as a delay line filter, but in this case it rejects the proper frequency and its odd harmonics. For intensity modulation, the transfer function is $2 \cos(\pi f_m \Delta\tau_g)$; therefore, operation at the proper frequency (or at an odd harmonic) also eliminates the effect of residual intensity modulation. Note that this is strictly valid only at rest, where the rotation-induced phase difference is zero. While rotating, intensity modulation may yield some parasitic effects even at f_p, which modifies the scale factor [13].

To summarize, the use of a truly single-mode filter at the common input-output port of the interferometer renders a fiber ring interferometer sensitive only to nonreciprocal effects as the Sagnac effect. Furthermore, a modulation-demodulation at the proper or eigenfrequency of the coil (or one of its odd harmonics) provides a biased signal that does not degrade the original perfection of the system. These two simple conditions, combined in the so-called minimum configuration

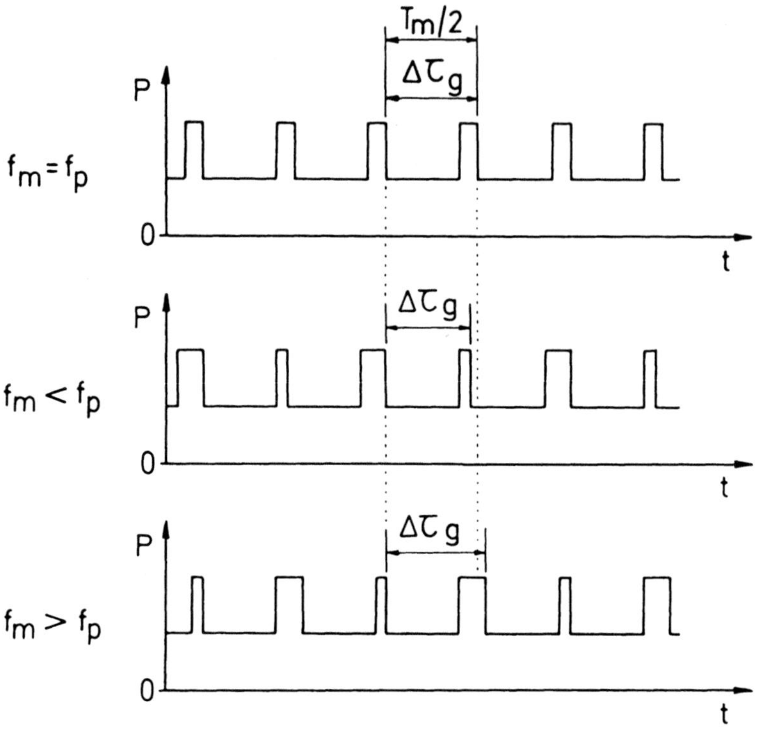

Figure 3.13 Effect of an asymmetrical square-wave modulation.

[14], make high performance possible despite the various defects of the components.

Note: As is discussed in Appendix 1, phase modulation propates at the so-called group velocity v_g as any modulation and not at the phase velocity v, with

$$v = \omega/k = c/n_e \qquad (3.20)$$
$$v_g = d\omega/dk = v\left(1 + \frac{\lambda}{n}\frac{dn_e}{d\lambda}\right)$$

where n_e is the equivalent index of the mode, which depends mostly on the material but also on the guidance characteristics. When a broad spectrum is used, it is possible to consider that the modulation propagates at the group velocity of the mean wavelength and that there is a temporal spreading $\Delta\tau_s$ of the modulation waveform, because $v_g(\lambda)$ is not constant over the spectrum width $\Delta\lambda$ when the second-order derivative $d^2\omega/dk^2$ or $d^2n_e/d\lambda^2$, called the propagation dispersion, is not zero:

$$\Delta\tau_s = \frac{L}{c}\lambda\frac{d^2n}{d\lambda^2}\Delta\lambda \qquad (3.21)$$

This value is often given with $\Delta\tau_s/(\Delta\lambda \cdot L)$ in ps · nm^{-1} · km^{-1}. The silica material dispersion is

λ (in nm)	850	1060	1300	1550
$\frac{\Delta\tau_s}{\Delta\lambda \cdot L}$ (in ps · nm^{-1} · km^{-1})	−100	−30	≈0	+20

This effect is very important in telecommunication applications where modulation frequency reaches the gigahertz range with propagation along kilometers of fiber; but it is negligible in the fiber gyro case.

The modulation period is preferably the double of the group transit time $\Delta\tau_g$ through the coil, and since the temporal spreading $\Delta\tau_s$ is also proportional to the coil length, the ratio $\Delta\tau_s/\Delta\tau_g$ is independent of the coil length. At 850 nm, where dispersion has the highest value, and with a spectrum width as broad as 50 nm, this ratio is only 0.1%. In the case of square-wave modulation, this induces finite rise and fall times, but this does not influence the performance of the fiber gyro.

3.3 RECIPROCITY WITH ALL-GUIDED SCHEMES

3.3.1 Evanescent-Field Coupler (or X-Coupler or Four-Port Coupler)

To avoid the difficulties of the coupling stability of free-space waves into single-mode fibers, it is desirable to use an all-guided scheme which improves ruggedness. This requires the duplication in a guided form of the various functions required in the interferometer. In particular, an evanescent-field coupler may replace the 3-dB splitter. The principle of such couplers, also called X-coupler or four-port coupler, is usually explained with the coupling overlap of the evanescent tail of a waveguide's fundamental mode with a second adjacent waveguide (see Appendix 2). These couplers may be realized in an all-fiber form, but also on an integrated optic substrate.

However, to understand its reciprocity behavior, it is interesting to use an alternative explanation [15]. Two parallel single-mode waveguides may be regarded as a two-mode waveguide. When light is fed into one input port, it excites the fundamental symmetrical mode of the coupler and the second-order antisymmetrical mode (Figure 3.14). The lobes of the modes are in phase in the input lighted waveguide and π (or 180 deg) out of phase in the other waveguide without light. As the two modes propagate with different velocities, their phase difference varies linearly, and the amplitude in each waveguide can be evaluated using a phasor diagram (Figure 3.15). At the half-coupling length there is a 50-50 splitting: both modes are in phase quadrature, and the modulus of the wave amplitude is the same in both waveguides. At the coupling length L_{cp}, light has been completely transferred into the second waveguide. Since the wave amplitudes in both waveguides are the two diagonals of the rhombus describing the vector sum of the mode amplitudes, they are always perpendicular, which means that the coupled wave always has a phase quadrature with respect to the transmitted wave.

The use of an evanescent-field coupler in a ring fiber interferometer (Figure 3.16) seems very advantageous, since there is no fringe pattern in free space and the free port could be exactly complementary to the reciprocal port, with a stable π rad phase difference, getting twice the $\pi/2$ shift of the coupler. This could allow the use of the more complex minimum configuration to be avoided. However, as will be shown later, polarization still has to be filtered at the same reciprocal input-output port, and, furthermore, the phase shift of the coupled wave in the coupler is not perfectly equal to $\pi/2$ in practice. As a matter of fact, there is always a residual differential loss between the symmetrical and antisymmetrical modes of

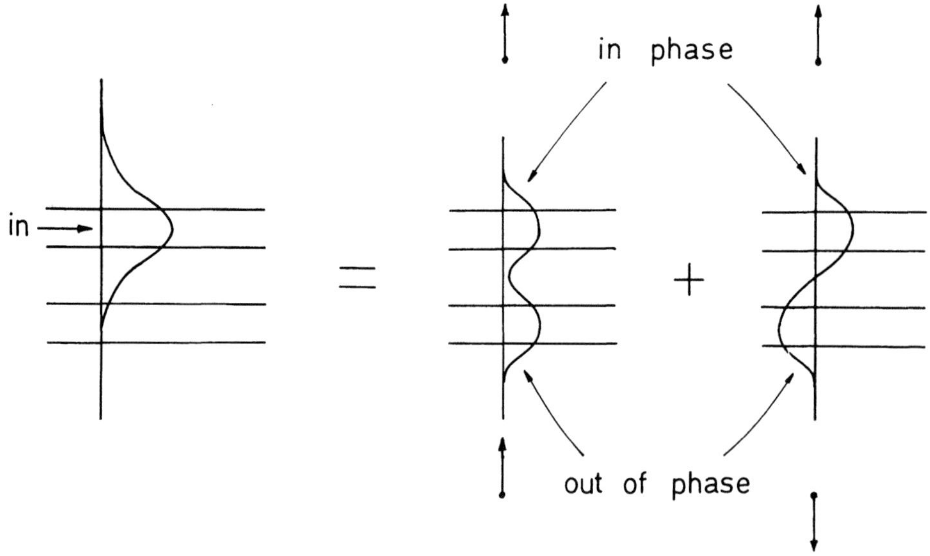

Figure 3.14 Decomposition of the input light with the modes of a two-waveguide coupling structure.

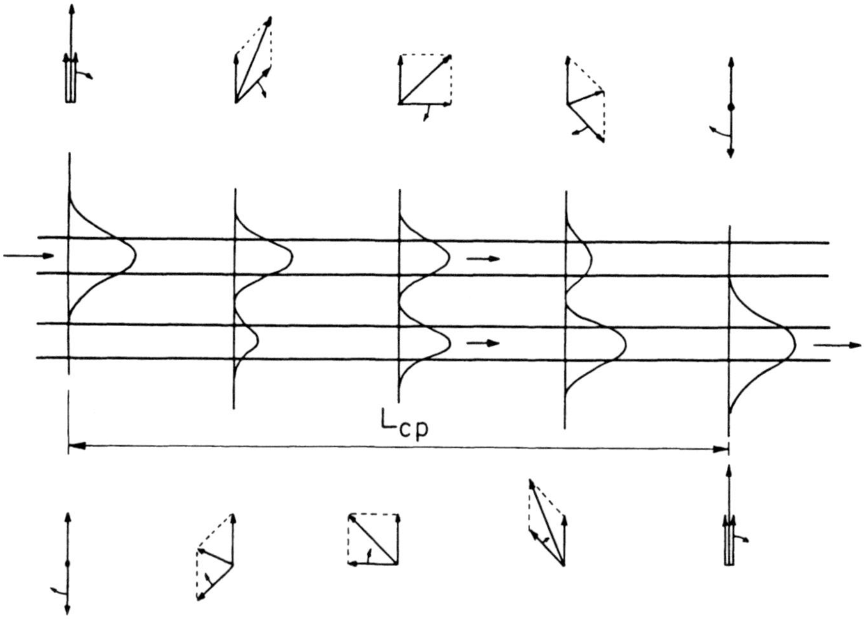

Figure 3.15 Propagation of the modes in a two-waveguide coupling structure with a phasor diagram for light propagating in each waveguide.

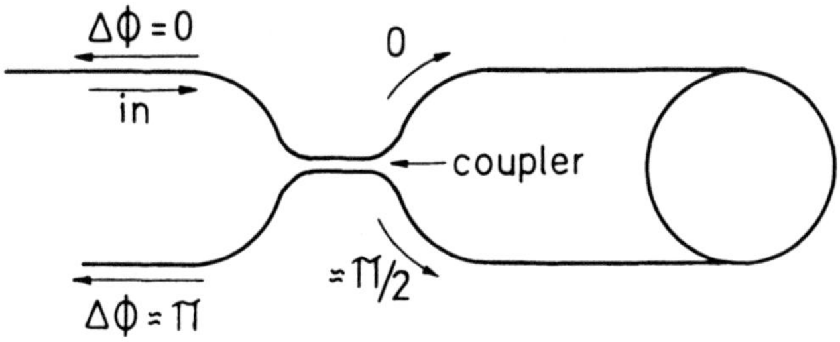

Figure 3.16 All-fiber ring interferometer with a 3-dB coupler.

the coupler regarded as a two-mode waveguide. The ideal rhombus of the phasor diagram is transformed into an ordinary parallelogram with unequal sides, where the diagonals are not perpendicular anymore (Figure 3.17) [15]. This yields a spurious phase difference at the free port of a ring interferometer. With a very low-loss coupler, this effect is small, but it remains significant compared to the very small bias change (less than 1 μrad, which corresponds to less than 10^{-6} in differential loss) which is looked for to get high performance. Therefore, even with

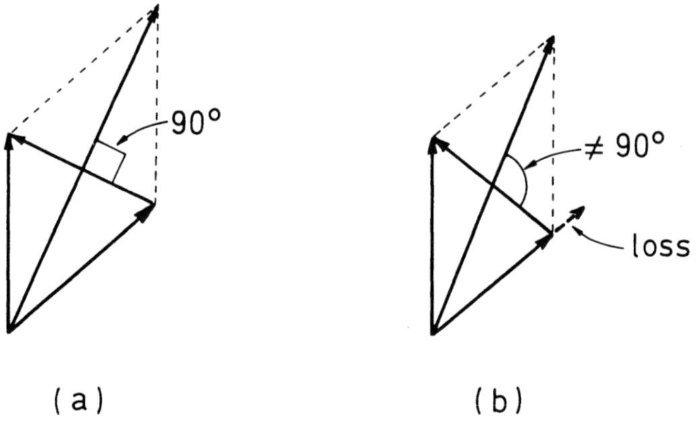

Figure 3.17 Phasor diagram of the transmitted and coupled waves of a coupler: (a) lossless coupler; (b) differential loss.

an evanescent-field four-port coupler, a minimum configuration has to be used to get a low bias drift.

3.3.2 Y-Junction

As we have seen, only three ports are actually useful in an all-guided ring interferometer, so evanescent field couplers can be replaced by Y-junctions. In an all-fiber form, evanescent-field four-port X-couplers are easier to fabricate than three-port Y-couplers; however, with integrated optics (see Appendix 3), which, as will be seen, is a crucial technology in implementing high-performance signal processing schemes, Y-junctions are preferred because of their simplicity and stability.

The Y-junction, also called a branching waveguide, was proposed early on [16] as a very useful integrated optic component. It is composed of a base single-mode waveguide connected to two single-mode branch waveguides (Figure 3.18). This is fabricated very simply with a Y-mask, and symmetry ensures 3-dB splitting, while evanescent-field couplers require careful control of the diffusion process to get the adequate coupling ratio. The principle of direct operation is simple: light that propagates in the single-mode base waveguide is split equally into the two symmetrical single-mode branch waveguides, which diverge with a very small angle (typically 1 deg) to minimize the loss. The behavior of the reverse operation [17,18] is not as straightforward to explain, but as 50% of the base waveguide light is coupled into each branching waveguide, reciprocity arguments show that the same percentage has to be coupled from one branching waveguide into the base waveguide in the opposite direction.

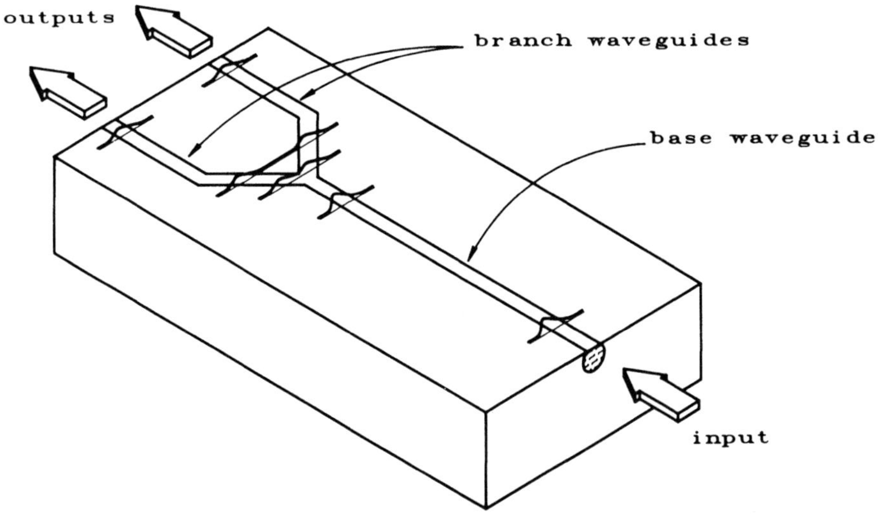

Figure 3.18 Integrated optic Y-junction.

This can be better understood by considering an all-guided Mach-Zehnder interferometer composed of two Y-junctions (Figure 3.19). Light is split at the first junction, and with two equal optical paths the two waves are recombined in phase at the second junction. This is equivalent to a bulk Mach-Zehnder interferometer, in which the two waves interfere in phase at one port and there is no light at the other port because of destructive interference. Now, if there is an additional π rad phase difference between the two paths, they are recombined into a two-lobe second-order antisymmetrical mode which radiates into the substrate because it cannot be guided into the single-mode base waveguide of the output junction (Figure 3.20). This is equivalent to what happens in a bulk Mach-Zehnder, in which the induced phase difference switches the output light at the fourth port of the output splitter. A Y-junction is a four-port device, as is any 3-dB splitter, but in this case there are three guided ports and the fourth port is radiated into the substrate.

The reverse operation can be understood by regarding the pair of branch waveguides as a two-mode waveguide, similar to the case of an evanescent-field coupler. To couple light into one of the branches can be interpreted as the superposition of the fundamental symmetrical mode and the second-order antisymmetrical mode. Both modes are in phase in the lighted waveguide, while they are π rad (or 180 deg) out of phase in the other one, which is not lighted (Figure 3.21). At the junction, the symmetrical mode that carries 50% of the optical power

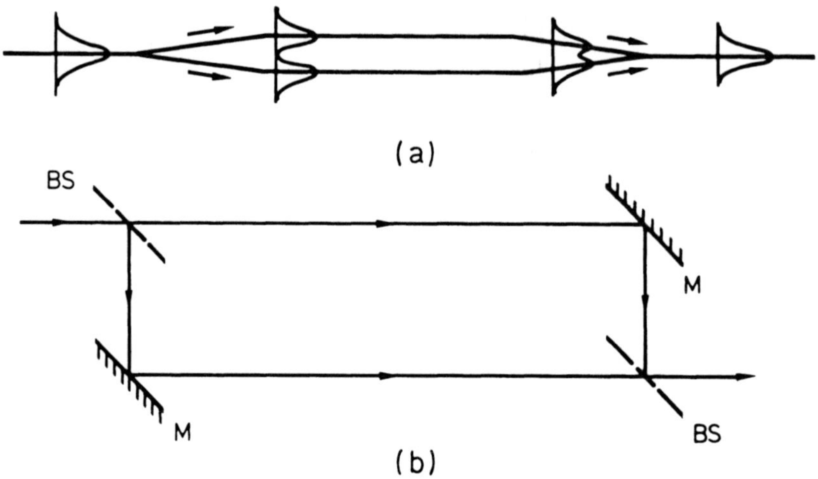

Figure 3.19 Mach-Zehnder interferometer with equal paths: (a) with two Y-junctions; (b) in a bulk form.

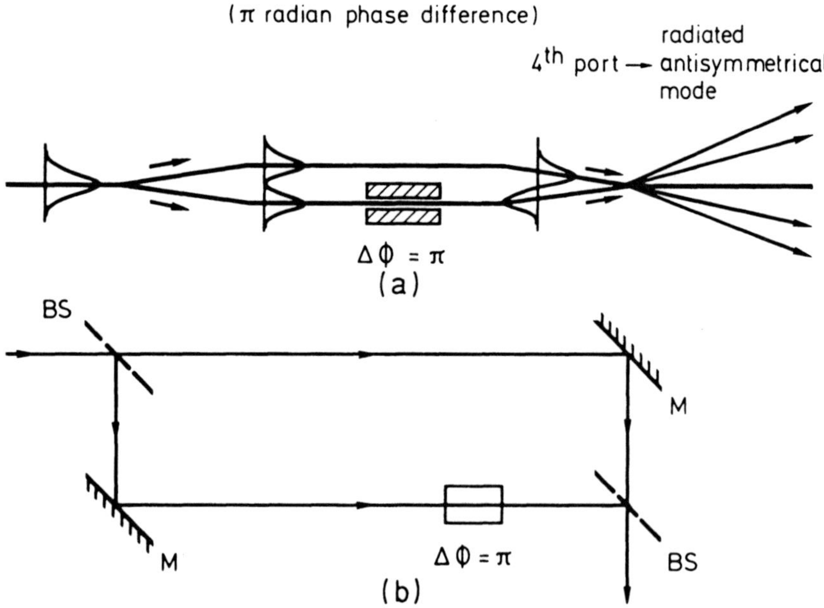

Figure 3.20 Mach-Zehnder interferometer with π rad phase difference: (a) with two Y-junctions; (b) in a bulk form.

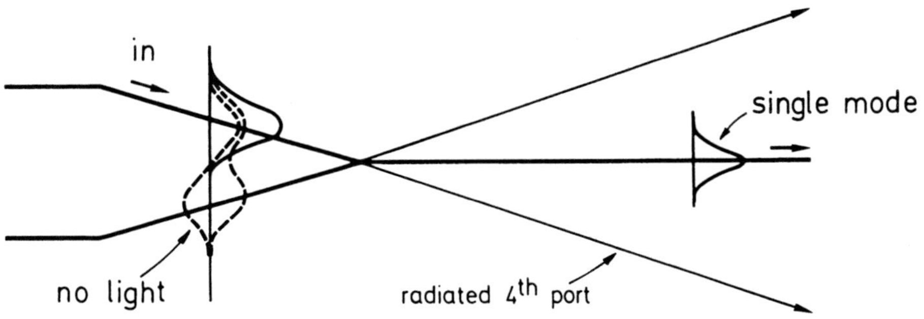

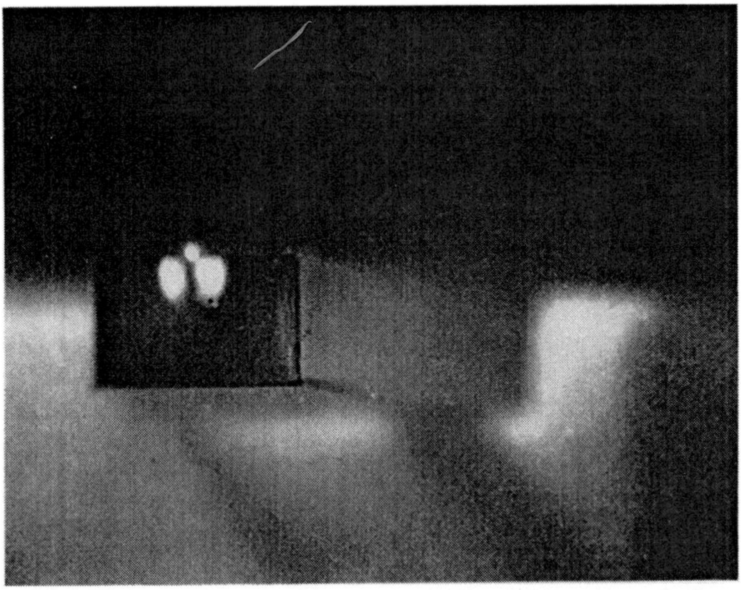

Figure 3.21 Reverse operation of a Y-junction: (a) principle; (b) photograph of the output face of the circuit with the guided single mode and the radiated antisymmetrical wave.

can be coupled to the base single-mode waveguide, while the antisymmetrical mode above the cutoff is radiated into the substrate.

Since, as we have seen, the gyro architecture requires the use of only three ports, the Y-junction is perfectly adequate. It naturally yields a reciprocal configuration because the base waveguide must be used as the common input-output port, since the nonreciprocal free port is lost in the substrate (Figure 3.22). It is the optimal technological solution because it is much more stable and much easier

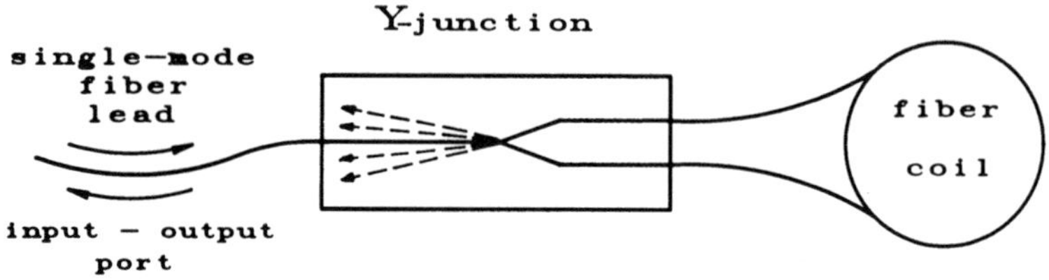

Figure 3.22 Reciprocal configuration with a Y-junction and its single-mode fiber lead.

to fabricate than an evanescent-field coupler in integrated optics. Such couplers are used for fast switching, but the simplicity of Y-junction makes it optimal for permanent splitting.

3.3.3 All-Fiber Approach

The all-fiber approach first appeared as the ideal technological choice for the fiber gyro because of the very low loss of the components, which provides a very good signal-to-noise ratio, due to the high returning power [11,19]. The all-fiber architecture (Figure 3.23) makes use of a first "coil" coupler to split and recombine the interfering wave and of a second "source" coupler to send the signal coming back through the common input-output port onto a detector. The polarization is filtered at this reciprocal port with an all-fiber polarizer, which was originally made with a birefringent crystal facing a fiber polished laterally to extract one polarization

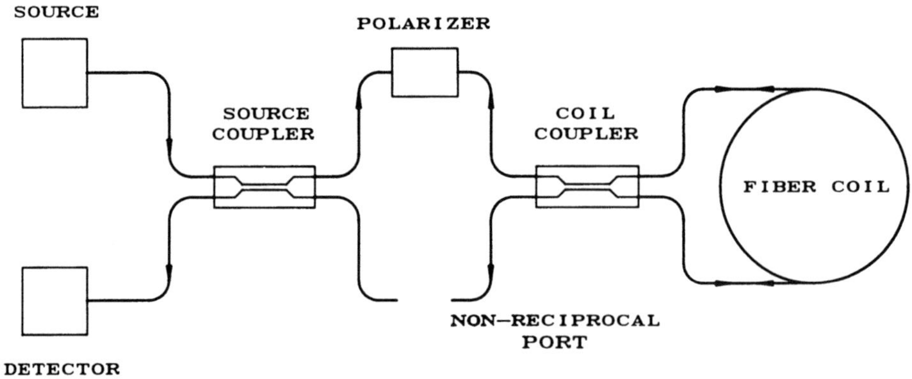

Figure 3.23 All-fiber reciprocal configuration.

by prism out-coupling of the evanescent tail of the mode [20]. Present all-fiber systems now prefer a "coiled" polarizer [21], which works on the differential curvature loss that occurs under special conditions between polarizations in stress-induced high-birefringence fibers. If the early brass-boards were using "ordinary" single-mode fiber with in-line polarization controllers, in particular $\lambda/4$ and $\lambda/2$ loops [22], this constraint is now avoided with the availability of polarization-preserving fibers (see Appendix 2).

The main limitation of this approach is the phase modulator. The only practical technique is to wind a fiber around a piezoelectric tube (or disk) [23] (see Appendix 2), which modifies the fiber length by controlling the tube diameter with a driving voltage. This method is perfectly adequate for the biasing modulation-demodulation, but obtaining an accurate scale factor requires the use of more sophisticated signal processing techniques. Numerous publications have described processing schemes that are compatible with piezoelectric modulators (see Section 10.2), but by far the highest scale factor performances are obtained with phase-ramp closed-loop techniques (see Section 8.2), which require a broad modulation band, while piezoelectric modulators experience narrow mechanical resonances.

An all-fiber approach yields very good bias performances, but the scale factor accuracy is limited in practice to about 1000 ppm.

3.3.4 Hybrid Architectures With Integrated Optics: Optimal "Y-Tap" or "Y-Coupler" Configuration

Integrated optics, particularly on a lithium niobate ($LiNbO_3$) substrate (see Appendix 3), was recognized early on as a very promising technology for the fiber-optic gyroscope [24], because a single multifunction circuit could be used to implement all the functions needed to make the device work, thus yielding a very simple all-guided configuration with a sensing fiber coil connected to an integrated optic circuit. However, the decisive advantage of integrated optics over the all-fiber approach is its phase modulator with a flat response over a large bandwidth, which permits the use of efficient signal-processing techniques that yield high-performance over the whole potential dynamic range of the fiber gyro.

It is possible to use an all-fiber approach and limit the use of integrated optics to a phase modulator on a straight waveguide, but this technology provides the useful advantage of permitting the integration of several other functions onto a single circuit, which improves the compactness and reduces the connections. Nevertheless, the pursuit of maximum integration is perhaps a technological challenge, but it is not a goal in itself, and an optimal hybrid compromise had to be found to take advantage of this possibility of integration without degrading the performance.

As we have already seen, the splitter-combiner of an interferometer can be realized very simply with a Y-junction, and the use of two junctions connected by their base branch [17] has been proposed. However, this so-called double-Y configuration (Figure 3.24) led to disappointing performances because of the limited rejection of the common base waveguide that acts as the spatial filter to ensure reciprocity [25]. As a matter of fact, the input light is split in the first "source" junction, and half the power remains in the base branch to be split again in the second "coil" junction. The rest of the input power forms an antisymmetrical mode that is radiated from the first junction, since it is above cutoff; but it is partially recoupled in the second junction, which acts as a "receiving antenna," since it may again guide this antisymmetric diverging wave (Figure 3.25). This recoupling is

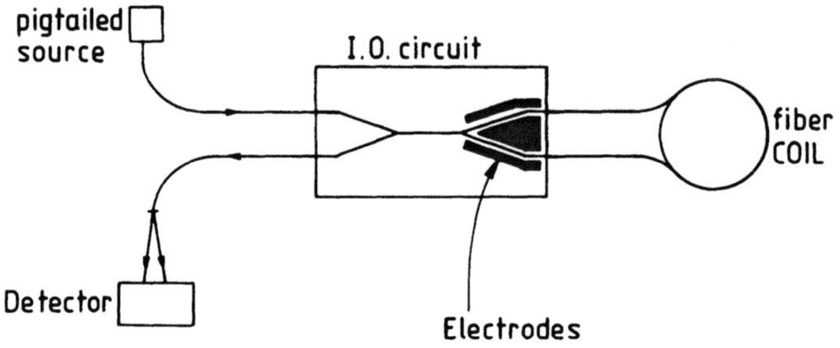

Figure 3.24 "Double-Y" configuration.

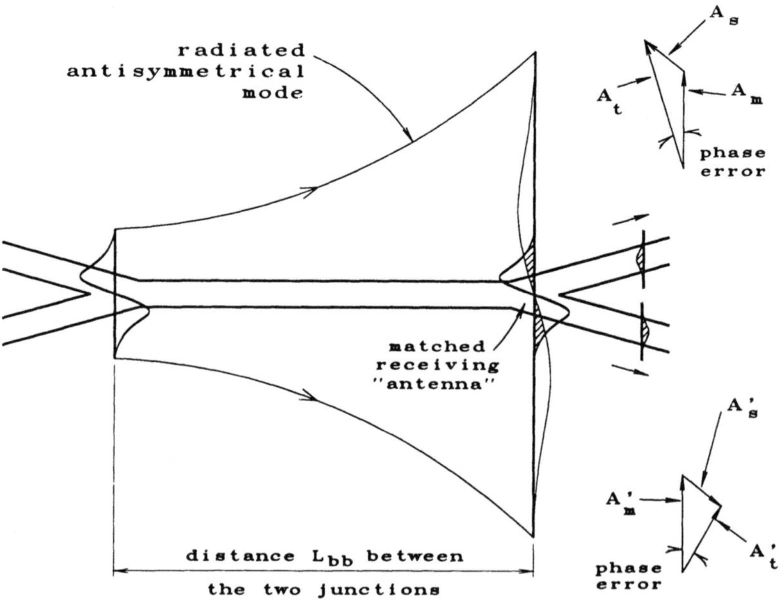

Figure 3.25 Limitation of single-mode rejection in a double-Y-junction.

very small, but the parasitic waves added in the two output branches of the "coil" junction are π (or 180 deg) out of phase because of the antisymmetry of the diverging wave. The total wave amplitude in each branch is the sum of the main amplitude term A_m and the additional parasitic term A_s coming from the spurious recoupling. Considering a phasor diagram (Figure 3.25), it can be seen that the phase of total amplitude A_t or A'_t in each branch is modified and that it also yields a spurious phase difference $\Delta\phi_e$ between both branches while it should be zero with a perfect rejection. The maximum phase error is proportional to the amplitude ratio

$$\Delta\phi_e \leq 2A_s/A_m \tag{3.22}$$

To limit this error below 10^{-7} rad, the rejection should be better than 146 dB! In practice, a common base branch of 10 to 20 mm yields a rejection limited to about 60 to 70 dB, and therefore a phase bias offset as high as several 10^{-4} rad [25]. As we have already seen, an offset is not, strictly speaking, detrimental, but in practice it is susceptible to drift. In this case, the actual phase error $\Delta\phi_e$ depends on the phase difference $\Delta\phi_{ms}$ between the main amplitude A_m, which has been continuously guided, and the spurious term A_s, which has propagated freely between both junctions:

$$\Delta\phi_e = 2(A_s/A_m)\sin\Delta\phi_{ms} \tag{3.23}$$

(It is maximum for $\Delta\phi_{ms} = \pi/2$, where A_s is perpendicular to A_m, and it is zero for $\Delta\phi_{ms} = 0$ or π, where A_s is parallel to A_m.) The phase difference $\Delta\phi_{ms}$ is related to the path difference $\Delta n_{ms} \cdot L_{bb}$, where the difference of relative index Δn_{ms} may be estimated as half the index step of the waveguide (i.e., a few 10^{-3}). With 10 mm of common base-branch length L_{bb}, this path difference is a few tens of micrometers (i.e., a few tens of wavelengths), and this changes with temperature, which induces drift.

This "double-Y" configuration looked at first like a very tempting solution, but it is actually showing that the search for maximal integration may encounter drastic performance limitations. At this stage, one may choose to improve the technology by implementing filtering devices about the central common base waveguide or by using evanescent-field couplers, which avoids the radiating fourth port of Y-junctions; but this increases drastically the complexity of the process and of its control.

The optimal compromise between integration and performance is obtained with the so-called Y-tap or Y-coupler configuration [25,26,27], where the coil splitter is an integrated optic Y-junction and the source splitter is an all-fiber 3-dB tap (Figure 3.26). The spatial filtering required for reciprocity is simply obtained with the fiber lead connected on the base waveguide of the Y-junction, and the 3-dB tap is used to extract 50% of the returning power with low additional loss and send it onto a detector. Phase modulators are fabricated on both branches of the

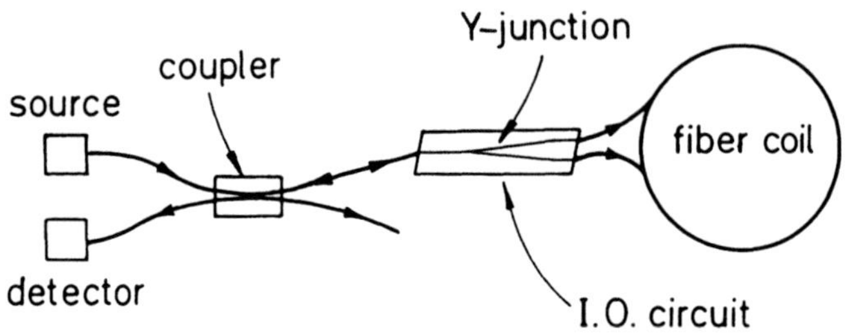

Figure 3.26 Y-tap (or Y-coupler) configuration with a 3-dB coupler as the source splitter.

Y-junction. It is possible to use one modulator for the biasing modulation and the other one for the feedback modulation required with closed-loop processing techniques, but it is preferable to connect both modulators in a push-pull configuration. The voltage is applied between the central electrodes and both external electrodes, which automatically drives the two modulators with opposite polarity (Figure 3.27).

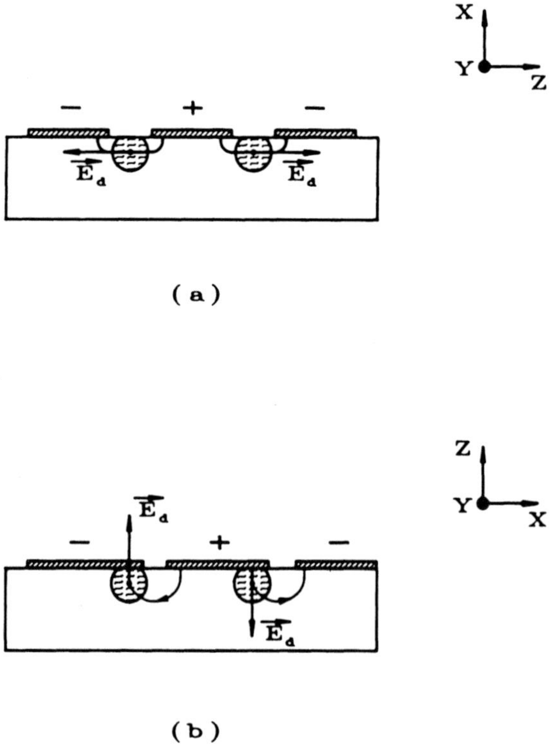

Figure 3.27 Push-pull connection of two integrated optic phase modulators: (a) x-cut and y-propagating $LiNbO_3$ circuit; (b) z-cut and y-propagating $LiNbO_3$ circuit.

This doubles the efficiency, but, above all, this eliminates the second-order nonlinearity of the response of each single modulator, which is very useful to get a good scale-factor linearity with closed-loop schemes.

Note that if a single-mode 3-dB coupler is commonly used for the tap function, it is not strictly required. The light returning through the input fiber lead of the Y-junction has already interfered, and the signal is carried by the intensity modulation. Therefore, the coupler may be replaced with a 3-dB tap, which has a single-mode output port and may be easier to fabricate than a coupler [27].

This Y-tap (or Y-coupler) configuration is nowadays widely accepted [28] as the optimal technological compromise for high-performance fiber gyroscopes which have to use integrated optic phase modulators to get a good scale factor with closed-loop processing schemes.

The preferred integrated optic technology is $LiNbO_3$ (see Appendix 3), which yields very efficient phase modulation. The waveguides are usually fabricated with titanium in-diffusion, and a polarizer can be realized with a metallic overlay that absorbs the TM mode. Proton exchange is another very interesting alternative [29], since it yields single-polarization guidance that provides a very high polarization rejection. This point is very important, as we shall see, for getting good bias stability.

The optimal multifunction gyro circuit (Figure 3.28) has typical dimensions of 1 mm in thickness, a few millimeters in width, and 20 to 35 mm in length. It is composed of a Y-junction with a separation of 200 to 300 μm between the two branches in order to connect the two fiber coil ends. Push-pull phase modulators are fabricated on the branches, and a metallic overlay polarizer is placed on the base waveguide, or polarization rejection is directly obtained with proton exchange. Finally, as will be shown later, the gyro circuit has a parallelogram shape to avoid backreflection at the interfaces between the circuit and the fibers [27].

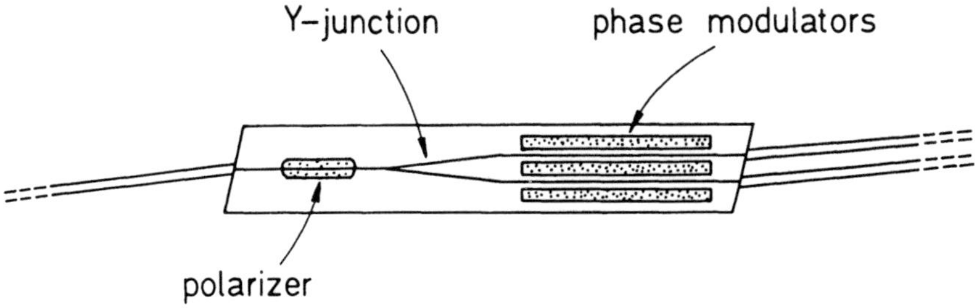

Figure 3.28 The multifunction "gyro circuit."

3.4 PROBLEM OF POLARIZATION RECIPROCITY

3.4.1 Rejection Requirement With Ordinary Single-Mode Fiber

A single-mode fiber has actually two polarization-modes. With an "ordinary" fiber these two modes are almost degenerate, but there is a residual birefringence that modifies the state of polarization as the light propagates. In a ring interferometer, at a given position, both counterpropagating waves have a different state of polarization and therefore do not see exactly the same index of refraction, because of birefringence, which yields a spurious phase difference at the output. Applying the general linear network theory, it is possible to show that if a polarizer is placed at the input and at the output of a fiber, the phase of the waves transmitted in opposite directions are then perfectly equalized [6]. To take into account the birefringence of the splitter-combiner of the ring interferometer, a single polarizer has to be placed at the common reciprocal port [4,5,6]: light is filtered at the input, and the two waves that are coming back through this same polarizer at the output are perfectly in phase. Some polarization control is necessary, since in the worst case, light may come back in the crossed state, yielding signal fading.

However, the rejection of a practical polarizer is not infinite, and there remains a residual phase difference between both counterpropagating waves. It was shown that with ordinary single-mode fibers the bias error is limited by the amplitude rejection ratio ϵ of the polarizer and not by its intensity ratio ϵ^2 [30]: a maximum phase measurement error of 10^{-7} rad would then require a rejection on the order of 140 dB and not 70 dB, as might at first be hoped!

As a matter of fact, the input amplitude A is filtered by the polarizer at the input, and its parallel component A_1 is transmitted while its crossed component A_2 is attenuated (Figure 3.29). The primary component A_1 is split and propagates in opposite directions along the ring interferometer and yields at the output two

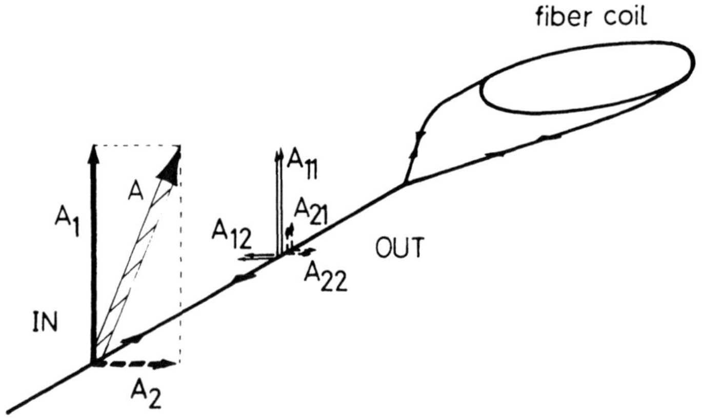

Figure 3.29 Polarization coupling in a ring interferometer (for clarity the "A'_{ij}" amplitudes are not shown).

waves having, respectively, an amplitude component A_{11} and A'_{11} in the transmitted state of the polarizer, and a component A_{12} and A'_{12} in the crossed state:

- The two primary components A_{11} and A'_{11}, which are in the same state at the output as A_1 at the input, interfere in phase (in absence of rotation) and with the same modulus $|A_{11}| = |A'_{11}|$ because of reciprocity.
- The two secondary components A_{12} and A'_{12} that have been coupled in the crossed mode have the same modulus ($|A_{12}| = |A'_{12}|$) but they have a spurious phase difference $\Delta\phi_{12}$ since the waves have not followed exactly the same path in opposite direction.

Similarly, the input crossed component A_2 yields two amplitude components A_{22} and A'_{22} in the same crossed state and two components A_{21} and A'_{21}, which correspond to light that has been coupled back in the transmitted state:

- The two components A_{22} and A'_{22} are also in phase because of reciprocity; but since they are attenuated twice (at the input and at the output), these terms may be ignored.
- The two secondary components A_{21} and A'_{21} also have the same modulus, but they also interfere with a spurious phase difference $\Delta\phi_{21}$, which is equal to $-\Delta\phi_{12}$ because of reciprocity of crossed coupling (see Section 3.2.1).

These two parasitic couples, (A_{12}, A'_{12}) and (A_{21}, A'_{21}), behave very differently. The first couple, (A_{12}, A'_{12}), is cross-polarized with respect to the main couple, (A_{11}, A'_{11}), and their intensities are simply added. Since the main interference signal is $2|A_{11}|^2 \sin\Delta\phi_R$ and the spurious signal is $2\epsilon^2|A_{12}|^2 \sin(\Delta\phi_R + \Delta\phi_{12})$, a phase error $\Delta\phi_e$ is yielded in the total signal:

$$|\Delta\phi_e| < \epsilon^2\rho_{cr} \tag{3.24}$$

where $\rho_{cr} = |A_{12}|^2/|A_{11}|^2$ is the polarization intensity cross-coupling in the ring interferometer. This error term is reduced by the intensity rejection ratio ϵ^2 of the polarizer.

On the other hand, the second couple, (A_{21}, A'_{21}), has the same state of polarization as the main couple, (A_{11}, A'_{11}), and there is interference between ($A_{11} + A_{21}$) and ($A'_{11} + A'_{21}$) instead of between A_{11} and A'_{11}. A phasor diagram (Figure 3.30) shows that it yields a phase error $\Delta\phi_e'$ bound by

$$|\Delta\phi_e| < 2\frac{|A_{21}|}{|A_{11}|} = 2\frac{|A'_{21}|}{|A'_{11}|} \tag{3.25}$$

Note that its exact value depends on $\Delta\phi_{21}$, but also on the phase difference between the primary components, (A_{11}, A'_{11}), and the spurious components, (A_{21}, A'_{21}).

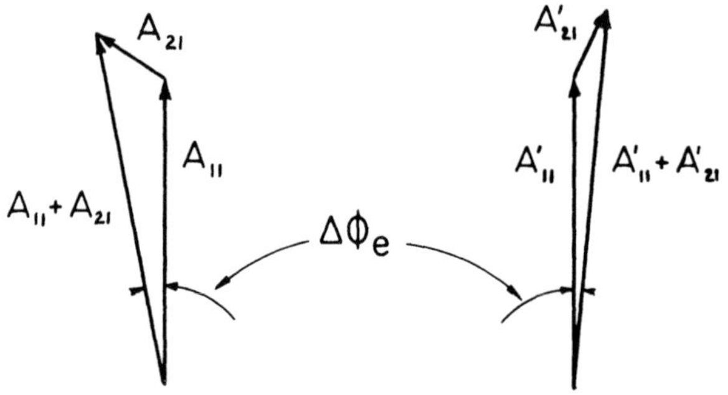

Figure 3.30 Phasor diagram of the nonreciprocity induced by A_{21} and A'_{21}.

As a matter of fact, it is also possible to consider that the main interference wave, $|A_{11} + A'_{11}|^2$, serves as a local oscillator for the coherent detection (see Appendix 1) of the spurious signal, $|A_{21} + A'_{21}|^2$ [31]. Using $\rho_{in} = |A_2|^2/|A_1|^2$ to represent the intensity polarization ratio at the input, we have $|A_{21}|/|A_{11}| = \epsilon\sqrt{\rho_{in}\rho_r}$ and

$$|\Delta\phi_e'| < 2\epsilon\sqrt{\rho_{in}\rho_r} \tag{3.26}$$

This second error term is reduced only by the amplitude rejection ratio ϵ instead of ϵ^2 in the first case, which makes a big difference, since ϵ is much smaller than unity.
The first error term, $\Delta\phi_e$, is often classified an as "intensity-type" error, while the second one, $\Delta\phi_e'$, is classified as an "amplitude-type" error. Both effects may be further decreased with polarization control to optimize the polarization alignment at the input and at the output, which reduces ρ_{in} and ρ_r; however, this improvement is limited, since it is difficult to limit their value below 10^{-2} in the long term.

3.4.2 Use of Polarization-Preserving Fiber

In addition to the problem of lack of polarization rejection, the use of ordinary fiber requires polarization control to avoid signal fading, which increases the complexity and the overall dimensions of the system. The development of high-quality polarization-preserving fibers (see Appendix 2) was an important step toward a compact and practical device. Single-polarization fibers would be even better in theory, but they have to be used with great care to get good rejection and avoid bending loss at the same time, which does not make them very practical, except for making an all-fiber coiled polarizer at the common input-output port.

Nevertheless, the conservation of polarization in polarization-preserving fibers used in coils or with in-line components is typically in the 20-dB range; that is, the intensity ratio ρ_{in} and ρ_r are still on the order of 10^{-2}, which does not relax very much the requirement on polarization rejection. To limit the phase error to 10^{-7} rad, this requires a polarizer rejection of about 100 dB, which is still much too high to be readily achieved in practice. However, polarization-preserving fibers are birefringent in principle: the two states of polarization propagate with a slightly different velocity, and when a broadband source is used, the spurious crossed-state waves are losing their coherence with the main primary signal, which brings a major reduction of the actual phase error signal, even with a medium rejection of the polarizer. This applies mainly to the amplitude-type error, since it is coming from a coherent detection process, but also to the intensity-type term through averaging effects.

As already stated, a broadband source is very advantageous in an interferometric fiber gyroscope, and its influence on birefringence-induced phase error is crucial for good performance. This very important point will be discussed in Chapter 5.

3.4.3 Use of Depolarizer

If a polarization-preserving fiber is obviously the optimal technical choice, its potential for very significant cost reduction with mass production is not fully accepted by the entire community of fiber gyro developers, and some teams still prefer to use an ordinary fiber. The problem of signal fading is solved with the use of Lyot depolarizers [32], which yield a random state of polarization that is evenly distributed over all the possible states. When it is filtered by a polarizer, half the power is always transmitted, independent of the effect of the propagation along a fiber.

A first depolarizer has to be used at the input before the polarizer and a second one in the coil. There is no more signal fading, but 50% of the power is lost through the polarizer at the input and again at the output, which induces an additionnal loss of 6 dB. This degrades the signal-to-noise ratio that depends on the returning power, but there is another important drawback: this does not take full-advantage of the relaxation of the requirement for the polarizer rejection brought by polarization-preserving fiber. The comparison between both approaches will be presented in Chapter 5.

3.4.4 Use of an Unpolarized Source

An unpolarized source has a completely random state of polarization. It is possible to consider it the combination of two incoherent sources with orthogonal states of polarization. With a fiber gyro we can use the same analysis as in Section 3.4.1, but now the two components A_1 and A_2 are incoherent. At the output, there is no more coherent detection of the spurious term, (A_{21}, A'_{21}), coming from A_2 with (A_{11}, A'_{11}) coming from A_1, which was serving as a local oscillator. This (A_{21}, A'_{21}) term still yields its own intensity-type effect, as does the term, (A_{12}, A'_{12});

however, we have seen that the two spurious phase differences are opposite, $\Delta\phi_{12} = -\Delta\phi_{21}$, and since their power is balanced because each parasitic wave passes once through the polarizer (at the input for (A_{21}, A'_{21}) and at the output for (A_{12}, A'_{12})), the two effects complement each other, which cancels out the total error signal.

In theory, this could even avoid the use of a polarizer [8], but this requires that the light remains perfectly unpolarized, which implies that there is no differential loss in the propagation or in components. As stated earlier, in the general multimode case, it is easier in practice to eliminate the crossed polarization with a rejection of 10^{-x} than to ensure that the differential loss is smaller than 10^{-x}, particularly in components.

In any case, the effect of an unpolarized source and that of the polarizer are cumulative, and this applies to any scheme, particularly to the use of polarization-preserving fiber coil. A real source has a certain degree of polarization P defined with the respective intensities of its polarized components. For example, a superluminescent diode (SLD), which is a very popular broadband source for the fiber gyro, emits partially unpolarized light, with the most powerful polarization being parallel to the junction. It can be regarded as the sum of two incoherent sources with an intensity $I_{//}$ for the parallel polarization and $I_{\perp}$ for the perpendicular polarization. In a gyro, it must be regarded as the sum of a perfectly unpolarized source of intensity $2I_{\perp}$ and of a perfectly polarized source of intensity $I_{//} - I$ (Figure 3.31), the degree of polarization being

$$P = \frac{I_{//} - I_{\perp}}{I_{//} + I_{\perp}} \tag{3.27}$$

The perfectly polarized component of intensity $I_{//} - I_{\perp}$ has to be analyzed according to the previous explanation of Section 3.4.1, while the perfectly unpolarized component of intensity $2I_{\perp}$ does not yield any spurious birefringence-induced phase error. This shows that an unpolarized source is advantageous for reducing the phase error caused by a lack of polarization rejection, although it has the drawback of an additional 3 dB of loss because half the power is attenuated through the polarizer at the input.

It also simplifies the system, since the fiber pigtail and the source coupler that send the light to the polarizer of the ring interferometer do not have to preserve

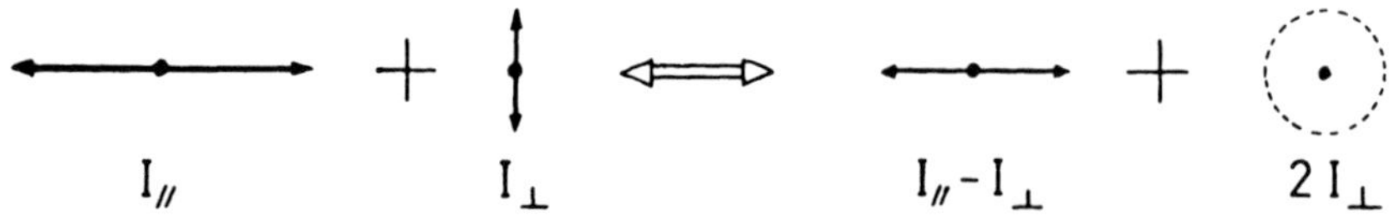

Figure 3.31 Decompositions of a partially unpolarized source.

polarization, because unpolarized light remains unpolarized when it propagates in "ordinary" single-mode fiber.

Note that there are some similarities between a naturally unpolarized source and a wave depolarized with a Lyot depolarizer, but care must be taken in this second case, since the actual depolarization is not perfect in practice (this is described in Chapter 5).

REFERENCES

[1] Graindorge, Ph., H. J. Arditty, M. Papuchon, J. P. Huignard, and Ch. Bordé, "Forced Reciprocity Using Phase Conjugation," Fiber-Optic Rotation Sensors, Springer Series in Optical Sciences, Vol. 32, 1982, pp. 368–374.

[2] Sagnac, G., "L'éther lumineux démontré par l'effet du vent relatif d'éther dans un interféromètre en rotation uniforme," Compte-rendus de l'Académie des Sciences, Vol. 95, 1913, pp. 708–710. Sagnac, G., "Sur la preuve de la réalité de l'éther lumineux par l'expérience de l'interférographe tournant," Comptes rendus de l'Académie des Sciences, Vol. 95, 1913, pp. 1410–1413.

[3] Vali, V., and R.W. Shorthill, "Fiber Ring Interferometer," Applied Optics, Vol. 15, 1976, pp. 1099–1100 (SPIE MS 8, pp. 135–136).

[4] Ulrich, R., "Fiber-Optic Rotation Sensing With Low Drift," Optics Letters, Vol. 5, 1980, pp. 173–175 (SPIE MS 8, pp. 170–172).

[5] Ulrich, R., and M. Johnson, "Fiber-Ring Interferometer: Polarization Analysis," Optics Letters, Vol. 4, 1979, pp. 152–154 (SPIE MS 8, pp. 233–235).

[6] Ulrich, R., "Polarization and Depolarization in the Fiber-Optic Gyroscope," Fiber-Optic Rotation Sensors, Springer Series in Optical Sciences, Vol. 32, 1982, pp. 52–77 (SPIE MS 8, pp. 239–264).

[7] Pavlath, G. A., and H. J. Shaw, "Multimode fiber gyroscope," Fiber-Optic Rotation Sensors, Springer Series in Optical Sciences, Vol. 32, 1982, pp. 364–367.

[8] Pavlath, G. A., and H. J. Shaw, "Birefringence and Polarization Effects in Fiber Gyroscopes," Applied Optics, Vol. 21, 1982, pp. 1752–1757 (SPIE MS 8, pp. 265–270).

[9] Davis, W. C., W. L. Pondrom, and D. E. Thompson, "Fiberoptic Gyro Using Magneto-Optic Phase-Nulling Feedback," Fiber-Optic Rotation Sensors, Springer Series in Optical Sciences, Vol. 32, 1982, pp. 308–315.

[10] Martin, J. M., and J. T. Winkler, "Fiber-Optic Laser Gyro Signal Detection and Processing Technique," SPIE Proceedings, Vol. 139, 1978, pp. 98–102.

[11] Bergh, R. A., H. C. Lefèvre, and H. J. Shaw, "All-Single-Mode Fiber-Optic Gyroscope With Long-Term Stability," Optics Letters, Vol. 6, 1981, pp. 502–504 (SPIE MS 8, pp. 178–180).

[12] Lefèvre, H. C., "Comments About Fiber-Optic Gyroscopes," SPIE Proceedings, Vol. 838, 1987, pp. 86–97 (SPIE MS 8, pp. 56–67).

[13] Kiesel, E., "Impact of Modulation Induced Signal Instabilities on Fiber-Gyro Performances," SPIE Proceedings, Vol. 838, 1987, pp. 129–139 (SPIE MS 8, pp. 399–409).

[14] Ezekiel, S., and H. J. Arditty, "Fiber-Optic Rotation Sensors," Fiber-Optic Rotation Sensors, Springer Series in Optical Sciences, Vol. 32, 1982, pp. 2–26 (SPIE MS 8, pp. 3–27).

[15] Youngquist, R. C., L. F. Stokes, and H. J. Shaw, "Effects of Normal Mode Loss in Dielectric Waveguide Directional Couplers and Interferometers," Journal of Quantum Electronics, Vol. QE 19, 1983, pp. 1888–1896 (SPIE MS 8, pp. 352–360).

[16] Yajima, H., "Dielectric thin-film optical branching waveguide," Applied Physics Letters, Vol. 22, 1973, pp. 647–649.

[17] Arditty, H. J., M. Papuchon, and C. Puech, "Reciprocity Properties of a Branching Waveguide," Fiber-Optic Rotation Sensors, Springer Series in Optical Sciences, Vol. 32, 1982, pp. 102–110.

[18] Izutzu, M., Y. Nakai, and T. Sueta, "Operating Mechanism of the Single-Mode Optical Waveguide Y Junction," Optics Letters, Vol. 7, 1982, pp. 136–138.

[19] Lefèvre, H. C., R. A. Bergh, and H. J. Shaw, "All-Fiber Gyroscope With Inertial-Navigation Short-Term Sensitivity," Optics Letters, Vol. 7, 1982, pp. 454–456 (SPIE MS 8, pp. 197–199).

[20] Bergh, R. A., H. C. Lefèvre, and H. J. Shaw, "Single-Mode Fiber-Optic Polarizer," Optics Letters, Vol. 5, 1980, pp. 479–481.

[21] Varnham, M. P., D. N. Payne, and E. J. Tarbox, "Coiled Birefringent Fiber Polarizer," Optics Letters, Vol. 9, 1984, pp. 306–308.

[22] Lefèvre, H. C., "Single-Mode Fibre Fractional Wave Devices and Polarization Controllers," Electronics Letters, Vol. 16, 1980, pp. 778–780.

[23] Davies, D. E. N., and S. A. Kingsley, "Method of Phase Modulating Signals in Optical Fibres: Application to Optical Telemetry Systems," Electronics Letters, Vol. 10, 1974, pp. 21–22.

[24] Papuchon, M., and C. Puech, "Integrated Optics: A Possible Solution for the Fiber Gyroscope," SPIE Proceedings, Vol. 157, 1978, pp. 218–222.

[25] Arditty, H. J., J. P. Bettini, Y. Bourbin, Ph. Graindorge, H. C. Lefévre, M. Papuchon, and S. Vatoux, "Integrated-Optic Fiber Gyroscope: Progresses Towards a Tactical Application," Proceedings of OFS 2'84, Stuttgard, VDE Verlag, 1984, pp. 321–325.

[26] Lefèvre, H. C., J. P. Bettini, S. Vatoux, and M. Papuchon, "Progress in Optical Fiber Gyroscopes Using Integrated Optics," AGARD-NATO Proceedings, Vol. CCP-383, 1985, pp. 9A1–9A13 (SPIE MS 8, pp. 216–227).

[27] Lefèvre, H. C., S. Vatoux, M. Papuchon, and C. Puech, "Integrated Optics: A Practical Solution for the Fiber-Optic Gyroscope," SPIE Proceedings, Vol. 719, 1986, pp. 101–112 (SPIE MS 8, pp. 562–573).

[28] See, for example, "Fiber Optic Gyros: 15th Anniversary Conference," SPIE Proceedings, Vol. 1585, 1991.

[29] Suchosky, P. G., T. K. Findakly, and F. L. Leonberger, "LiNbO3 Integrated Optical Components for Fiber-Optic Gyroscopes," SPIE Proceedings, Vol. 993, 1988, pp. 240–243.

[30] Kintner, E. C., "Polarization Control in Optical-Fiber Gyroscopes," Optics Letters, Vol. 6, 1981, pp. 154–156 (SPIE MS 8, pp. 236–238).

[31] Fredricks, R. J., and R. Ulrich, "Phase-Error Bounds of Fibre Gyro With Imperfect Polarizer/Depolarizer," Electronics Letters, Vol. 20, 1984, pp. 330–332 (SPIE MS8, pp. 277–278).

[32] Böhm, K., P. Marten, K. Petermann, E. Weidel, and R. Ulrich, "Low-Drift Fibre Gyro Using a Superluminescent Diode," Electronics Letters, Vol. 17, 1981, pp. 352–353 (SPIE MS 8, pp. 181–182).

Chapter 4
Backreflection and Backscattering

4.1 PROBLEM OF BACKREFLECTION

4.1.1 Reduction of Backreflection With Slant Interfaces

Reciprocity applies only to transmitted waves, and in the early days of fiber gyro research, the first problem encountered was the 4% Fresnel backreflection (see Appendix 1) at the air-silica interface of the fiber coil ends, which superimposes a parasitic Michelson interferometer. This was solved by polishing the ends at a sufficient slant angle to avoid backward reflection and aligning them according to refraction laws so as not to degrade the throughput coupling [1] (Figure 4.1).

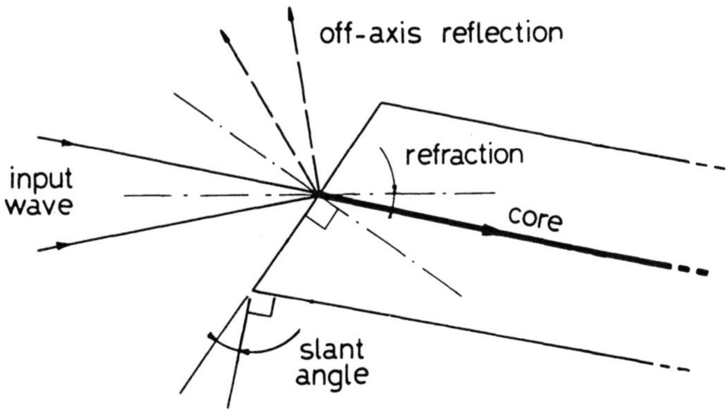

Figure 4.1 Suppression of backreflection at the fiber ends with a slant angle.

Antireflection coating cannot provide a low enough level of backreflected light. With an all-fiber approach, this problem of index mismatch is avoided, but this method is still used at the interface between silica fiber and the $LiNbO_3$ integrated optic circuit [2] (Figure 4.2), which has been found to be the optimal technological approach, as seen earlier.

To evaluate the reduction of backreflection as a function of the slant angle, it is necessary to consider that it is equivalent to the coupling between the real waveguide and its mirror image through the interface plane (Figure 4.3). The throughput coupling loss Γ_θ between two identical waveguides with an angular

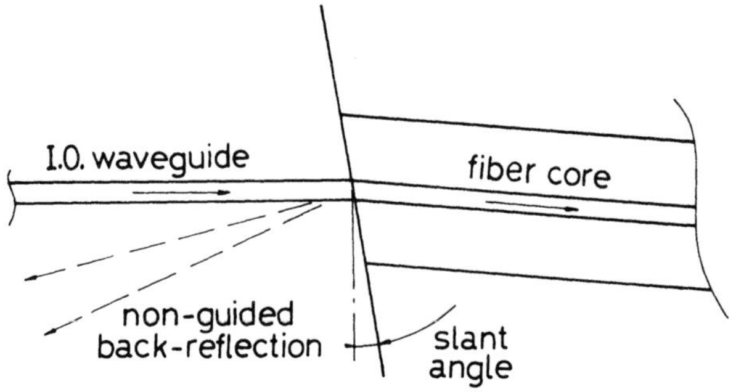

Figure 4.2 Suppression of backreflection between a fiber and an integrated optic waveguide.

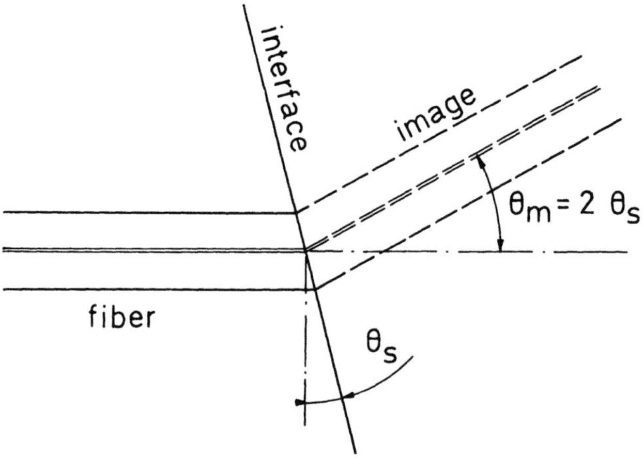

Figure 4.3 Estimation of the backreflection with an "image" fiber.

misalignment θ_m is analyzed in Appendix 2, assuming that the mode is pseudo-Gaussian:

$$\Gamma_\theta = 4.34\left[\frac{\theta_m}{\theta_D'/2}\right]^2 \text{ (in dB)} \tag{4.1}$$

where θ_D' is the full divergence angle at $1/e^2$ of the mode in an index-matched medium:

$$\begin{aligned} \theta_D' &= \theta_D/n \\ \theta_D &\approx 1.5 \text{ NA} \end{aligned} \tag{4.2}$$

and where θ_D is the divergence in a vacuum, n the medium index, and NA the fiber numerical aperture. Since θ_m is twice the slant angle θ_s, the additional loss of backreflection may be estimated as

$$\Gamma_\theta(\theta_s) \approx 70\left[\frac{\theta_s}{\text{NA}}\right]^2 \text{ (in dB)} \tag{4.3}$$

The almost "standard" slant angle of the integrated optic circuit is 10 deg, which implies that the slant angle of the fiber is 15 deg to align both waveguides according to refraction law:

$$n_{SiO_2} \sin 15 \text{ deg} = n_{LiNbO_3} \sin 10 \text{ deg} \tag{4.4}$$

with $n_{SiO_2} \approx 1.45$ and $n_{LiNbO_3} \approx 2.2$. Single-mode fibers used in the fiber gyro have a typical numerical aperture NA $= 0.15$; therefore, the attenuation Γ_{br} of the backreflection is

$$\Gamma_{br} = \Gamma_F + \Gamma_\theta(15 \text{ deg}) \tag{4.5}$$

With 4% between $LiNbO_3$ and SiO_2, the Fresnel backreflection Γ_F is 14 dB, and computing Γ_θ, it is found that $\Gamma_\theta(15 \text{ deg}) = 200$ dB!

However, the validity of the pseudo-Gaussian approximation of the mode shape is lost in the tails, and the formula of Γ_θ applies only for small angular misalignments. Furthermore, spurious scattering at the interface increases the backreflection, but, in practice, the "standard" 10/15-deg combination at the integrated optic interface is perfectly adequate to reduce the backreflection below 60 to 70 dB without decreasing the throughput coupling, compared to the case of perpendicular interfaces.

4.1.2 Influence of Source Coherence

Assuming two backreflection points at the ends of the fiber coil (Figure 4.4), this actually yields six waves at the output of the ring interferometer:

- The two primary reciprocal counterpropagating waves that have exactly the same amplitude A and A' (same modulus and same phase);
- Two waves backreflected at the inputs with different amplitudes A_1 and A_1';
- Two waves backreflected at the outputs with different amplitudes A_2 and A_2'.

If these waves are coherent, there is interference at the output between $(A + A'_1 + A'_2)$ and $(A' + A_1 + A_2)$ instead of only A and A'. Using once again a phasor diagram, it is shown that the phase difference error may depend on the amplitude ratio between the primary waves and the backreflected waves. To limit this error below 10^{-7} rad would require an attenuation of at least 140 dB of the backreflected waves. The path difference between the primary waves and these spurious backreflections is considerable, because it corresponds to the whole optical length of the coil (i.e., several hundreds of meters), and most sources are not coherent enough. In practice, there is a superposition of three independent interferometers:

- The main Sagnac interferometer with A and A';
- Two spurious Michelson interferometers with, respectively, (A_1 , A'_1) and (A_2 , A'_2).

The spurious Michelson signals are simply added in optical intensity to the Sagnac signal, instead of amplitude, and an attenuation of 70 dB is then good enough to get 10^{-7} rad instead of 140 dB.

Furthermore, if a source of particularly short coherence length is used, it becomes possible to reduce the contrast of these parasitic Michelson interfero-

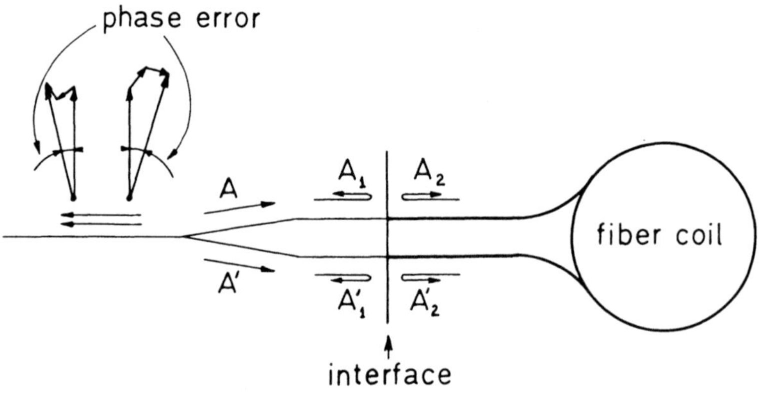

Figure 4.4 Explanation of phase error induced by two backreflections.

meters by unbalancing their arms. In particular, it is advantageous to get the slant angle of the integrated optic interface with a parallelogram substrate shape [3]. With a typical separation of 200 μm between the branch waveguides of the Y-junction and a slant angle of 10 deg, the optical path length unbalance ΔL_{unb} of the spurious Michelson interferometer is (Figure 4.5)

$$\Delta L_{unb} = 2n_{LiNbO_3} 200 \ \mu m \tan 10 \ deg = 150 \ \mu m \tag{4.6}$$

while the coherence length of broadband sources like superluminescent diodes is on the order of 20 to 30 μm.

This analysis of backreflection is a good example of the way parasitic effects related to coherence are suppressed in an interferometric fiber gyro, which involves three steps:

- Firstly, the spurious waves have to be reduced as much as possible with a reasonably simple technological approach. The slant interface is a good solution in the case of backreflection.
- Secondly, the coherence between the primary waves and the spurious waves has to be destroyed to avoid direct coherent detection that is very harmful.
- Thirdly, some further improvements may be obtained if the coherence between the spurious waves is also destroyed.

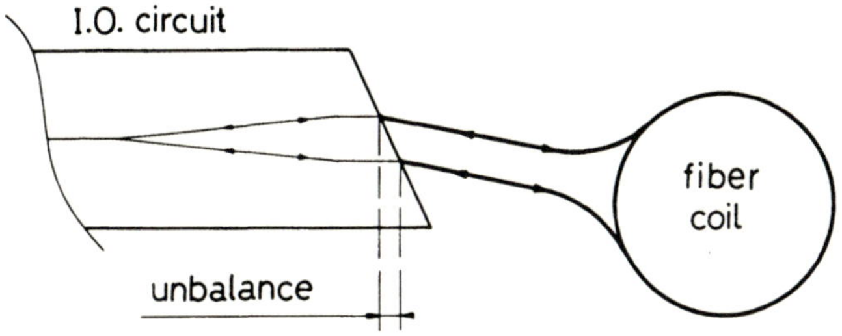

Figure 4.5 Unbalancing of the parasitic Michelson interferometers with a parallelogram-shaped circuit.

4.2 PROBLEM OF BACKSCATTERING

4.2.1 Coherent Backscattering

Propagation in a material suffers various kinds of scattering processes. In a fiber, the light that is scattered inside the numerical aperture of the fiber remains guided in the forward and backward directions. In a fiber gyro, problems arise with Rayleigh backscattering, which preserves the same optical frequency as the throughput wave and can be considered as a randomly distributed backreflection. In a single-mode fiber, there is typically 40 dB of backscattering light (this value is only a gross order of magnitude because backscatter depends on fiber length and the operating wavelength, as will be discussed in Section 4.2.2). With a CW coherent source, this may yield a spurious phase difference error as high as 10^{-2} rad because of the coherent detection process, with the primary waves acting as a local oscillator [4]. Furthermore, since the path differences are not stable, this phase error fluctuates inducing noise. To reduce this effect, it was proposed that additional frequency modulation [5] or phase modulation [6,7] be used to displace this noise outside of the detection bandwidth.

4.2.2 Use of a Broadband Source

This reduction, however, may be obtained more efficiently with an adequate source. It was first proposed to use a pulsed source: in that case, only the light backscattered from the middle of the coil (within the pulse length L_p) will interfere with the primary pulses (Figure 4.6) [4]. However, a broadband source of a short coherence length yields a similar reduction [6,7], considering that the pulses are replaced by wave trains. All these spurious signals arise from parasitic interferometers with various path differences, and, as detailed in Appendix 1, the interference contrast is proportional to the autocorrelation function of the source, which is the Fourier transform of the power (or intensity) spectrum. A pulse has an amplitude spectrum with all the frequency components in phase, while an incoherent broadband source has an amplitude spectrum with all the frequency components in a random phase; but the phase information is lost in the autocorrelation process involved in the interferences, and both pulsed and incoherent sources reduce the contrast of the spurious interferometers because of their narrow autocorrelation function.

However, a broadband source is more advantageous, since its autocorrelation width, equal to the coherence length, is much narrower in practice (20 to 50 μm)

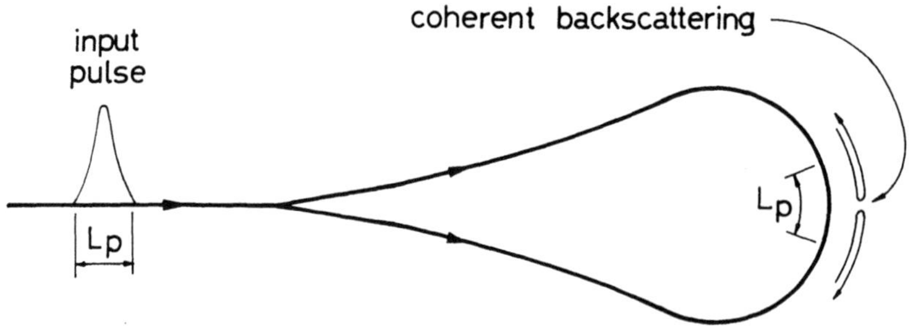

Figure 4.6 Limitation of coherent backscattering with a pulsed source.

than that of a pulse. An equivalent pulse duration should be in the range of 100 femtoseconds! Furthermore, for the same mean power, a pulse has a much higher peak power, which induces nonlinearities, as will be seen later in Section 7.2.

4.2.3 Evaluation of the Residual Rayleigh Backscattering Noise

Rayleigh scattering is caused by dipolar antenna radiation of the material atomic bindings excited by the incoming wave. It has a λ^{-4} dependence in wavelength, and there is a cumulative effect only in fluids or amorphous glasses that have a random structure. Inversely, with an ordered crystal matrix, the various scattered waves would interfere destructively except in the forward direction.

In a fiber, part of the scattered light remains guided in the forward and backward directions. The forward scattering is reciprocal and does not yield any spurious effect in the fiber gyro. On the contrary, the backward scattering behaves like a randomly distributed backreflection. With a Rayleigh scattering attenuation α_R and considering a source with a coherence length L_c, the coherent backscattered intensity I_{cb} coming from the middle of the coil is

$$I_{cb} = I_m \left(1 - 10^{-\alpha_R L_c/10}\right) S \tag{4.7}$$

where I_m is the intensity of the main wave and S is the recapture factor, which may be grossly approximated as the ratio of the acceptance solid angle of the fiber, about equal to NA^2, to a full solid angle of 4π steradian. In practice S is about 10^{-3}.

Taking practical numerical values, $\alpha_R = 2$ dB/km (for $\lambda = 850$ nm) and $L_c = 20\ \mu$m, giving

$$\frac{I_{cb}}{I_m} \approx 10^{-11}$$

which still yields an amplitude ratio:

$$\frac{A_{cb}}{A_m} = \sqrt{\frac{I_{cb}}{I_m}} \approx 3 \times 10^{-6} \tag{4.8}$$

Assuming two main counterpropagating amplitudes A_m and A_m', there are interferences between $(A_m + A'_{cb})$ and $(A'_m + A_{cb})$ at the output. It seems at first that the phase error could be as high as 6×10^{-6} rad, even with a coherence length as small as 20 μm. However, this is not taking into account the benefit brought by the biasing modulation at the proper frequency [8] and the fact that the phase difference between A_m and A'_{cb} is correlated with the one between A'_m and A_{cb} [9].

As a matter of fact, when the phase modulator is operated at the proper frequency f_p of the coil, a wave backscattered from the middle of the coil experiences the phase modulation twice, but with a half-period delay which cancels out the total modulation (Figure 4.7). Furthermore, the main interference wave resulting from $A_m + A'_m$ is modulated in intensity but is not modulated in phase; therefore, the coherent detection process between $(A_m + A'_m)$ as the local oscillator and A_{cb} and A'_{cb} remains at low frequencies and does not yield any spurious signal in the demodulation at f_p [8].

Some benefit also comes from the correlation between the phases of both backscattered waves. Let us consider a scattering point: if the phase accumulated in the propagation by the main primary waves is ϕ_p, the phases ϕ_{cb} and ϕ'_{cb}, due to the propagation of the two waves backscattered at a given point, are such as:

$$\phi_{cb} + \phi'_{cb} = 2\phi_p \tag{4.9}$$

and therefore the phase differences due to propagation between the main and backscattered waves are opposite:

$$\Delta\phi_{pcb} = \phi_p - \phi_{cb} = -\Delta\phi'_{pcb} = -(\phi_p - \phi'_{cb}) \tag{4.10}$$

However, the scattering process adds a $\pi/2$ phase lag on each spurious wave, and the actual phase differences are [9]

$$\begin{aligned} \Delta\phi_{cb} &= \Delta\phi_{pcb} - \pi/2 \\ \Delta\phi'_{cb} &= \Delta\phi'_{pcb} - \pi/2 = -\Delta\phi_{cb} - \pi \end{aligned} \tag{4.11}$$

Using again a phasor diagram (Figure 4.8), it can be seen that if only the propagation is taken into account, the phase errors carried in each direction are opposite,

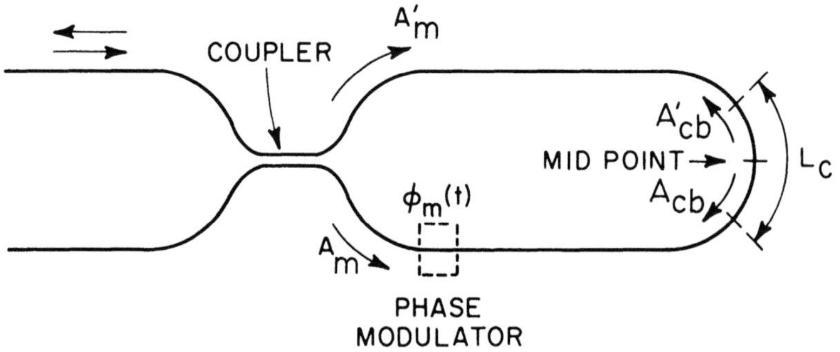

Figure 4.7 Effect of a phase modulator on the waves backscattered in the middle of the coil.

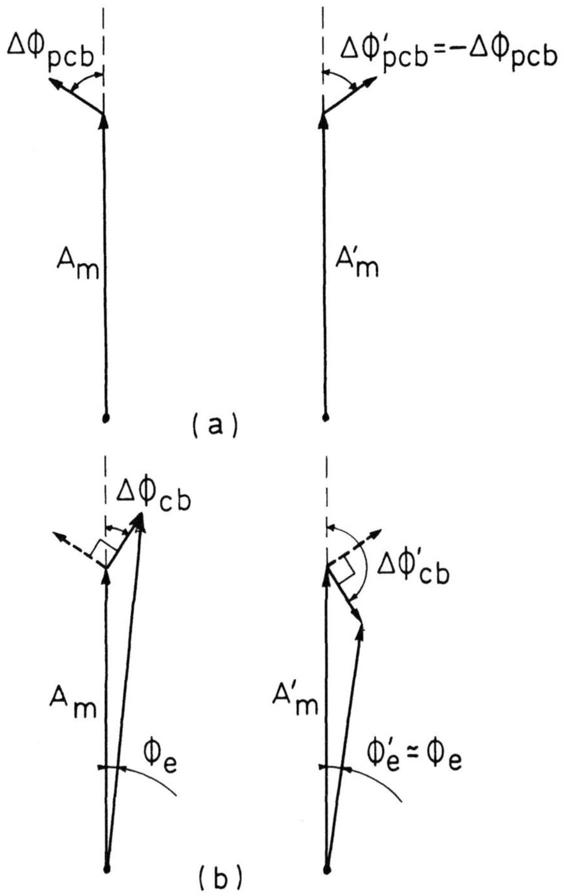

Figure 4.8 Phasor diagram of the main primary amplitudes A_m and A'_m, and the amplitudes A_{cb} and A'_{cb} backscattered in the middle of the coil: (a) effect of propagation; (b) additional effect of the $\pi/2$ phase lag.

and therefore the error on the phase difference would be doubled. However, with the additional $\pi/2$ phase lag, the phase errors ϕ_e and ϕ_e' are actually

$$\phi_e = \frac{|A'_{cb}|}{|A_m|} \sin\Delta\phi_{cb} \tag{4.12}$$

$$\phi_e' = \frac{|A_{cb}|}{|A'_m|} \sin\Delta\phi'_{cb} \tag{4.13}$$

Since $\sin\Delta\phi'_{cb} = \sin(-\pi - \Delta\phi_{cb}) = \sin\Delta\phi_{cb}$, both phase errors become equal when the moduli $|A'_{cb}|$ and $|A_{cb}|$ are equal. Since on the beamsplitter one backscattered wave is transmitted twice while the other is reflected (or coupled) twice, these moduli are equal when the splitter is precisely 50-50 (assuming that the loss is symmetrical in the coil), which eliminates the effect of Rayleigh backscatter.

To summarize, the low coherence length of the broadband source preferably used in an interferometric fiber gyro decreases considerably the Rayleigh backscattering noise, and a further reduction is obtained with a biasing modulation at the proper frequency and with a precise 50-50 separation in the coil splitter. Other reduction techniques remain possible if a longer coherence length is used [10], but they are difficult to reduce to practice.

In any case, the reduction of Rayleigh backscattering noise is reasonably demanding in terms of short coherence length, compared to other effects, and a multimode laser diode is sufficient to get low noise. This spurious effect does not require a coherence length as short as that of a superluminescent diode (see Section 9.2.1).

REFERENCES

[1] Arditty, H. J., H. J. Shaw, M. Chodorow, and R. K. Kompfner, "Re-entrant Fiber-Optic Approach to Rotation Sensing," SPIE Proceedings, Vol. 157, 1978, pp. 138–148.

[2] Lefèvre, H. C., J. P. Bettini, S. Vatoux, and M. Papuchon, "Progress in Optical Fiber Gyroscopes Using Integrated Optics," AGARD-NATO Proceedings, Vol. CPP-383, 1985, pp. 9A1–9A13 (SPIE MS 8, pp. 216–227).

[3] Lefèvre, H. C., "Comments About Fiber-Optic Gyroscopes," SPIE Proceedings, Vol. 838, 1987, pp. 86–97 (SPIE MS 8, pp. 56–67).

[4] Cutler, C. C., S. A. Newton, and H. J. Shaw, "Limitation of Rotating Sensing bby Scattering," Optics Letters, Vol. 5, 1980, pp. 488–490 (SPIE MS 8, pp. 343–345).

[5] Davis, J. L., and S. Ezekiel, "Closed-Loop, Low-Noise Fiber-Optic Rotation Sensor," Optics Letters, Vol. 6, 1981, pp. 505–507 (SPIE MS8, pp. 186–188).

[6] Böhm, K., P. Russer, E. Weidel, and R. Ulrich, "Low-Noise Fiber-Optic Rotation Sensing," Optics Letters, Vol. 6, 1981, pp. 64–66 (SPIE MS 8, pp. 183–185).

[7] Bergh, R. A., H. C. Lefèvre, and H. J. Shaw, "All-Single-Mode Fiber-Optic Gyroscope," Optics Letters, Vol. 6, 1981, pp. 198–200 (SPIE MS8, pp. 175–177).

[8] Lefèvre, H. C., R. A. Bergh, and H. J. Shaw, "All-Fiber Gyroscope With Inertial-Navigation Short-Term Sensitivity," Optics Letters, Vol. 7, 1982, pp. 454–456 (SPIE MS 8, pp. 197–199).

[9] Takada, K., "Calculation of Rayleigh Backscattering Noise in Fiber-Optic Gyroscopes," J.O.S.A., Vol. A2, 1985, pp. 872–877 (SPIE MS 8, pp. 361–366).

[10] Mackintosh, J. M., and B. Culshaw, "Analysis and Observation of Coupling Ratio Dependence of Rayleigh Backscattering Noise in a Fiber-Optic Gyroscope," Journal of Lightwave Technology, Vol. 7, 1989, pp. 1323–1328.

Chapter 5

Analysis of Polarization Nonreciprocities With Broadband Source and High-Birefringence Fiber

5.1 DEPOLARIZATION EFFECT IN HIGH-BIREFRINGENCE POLARIZATION-PRESERVING FIBERS

The principle of polarization preserving fibers (see Appendix 2) is to create a strong birefringence that splits the polarization-mode degeneracy of ordinary single-mode fiber. When light is coupled in one eigen polarization state, it will remain mostly in this input state, since the induced birefringence is much higher than the spurious defects.

The behavior of polarization-preserving fibers may be explained with two orthogonal polarization modes and random coupling points, which transfer a small amount of light from one mode to the other. The mean rate of intensity (or power) transfer is called the h-parameter of the fiber [1], with

$$dI_c/dL = h \cdot I_p \tag{5.1}$$

where I_p is the intensity of the primary input polarization mode and I_c is the intensity of the crossed mode. A typical value of h is 10^{-5} m^{-1} (a polarization conservation of 20 dB/km).

The most popular polarization-preserving fibers used in a fiber gyro are based on linear birefringence induced by an additional stressing structure in the cladding. They conserve a linear state of polarization when this state is coupled at the input along one principal axis of fiber birefringence.

In a fiber gyro, this polarization conservation maintains most of the power in the primary reciprocal waves, avoiding signal fading. Only a small part is transferred in the crossed nonreciprocal wave, and, in addition, when a broadband source is used, it gets depolarized because it propagates at a different velocity compared to the main mode [2]. The propagation of a broad-spectrum wave may be analyzed by considering that it is composed of wave trains with a length equal to the decoherence length L_{dc}. (As described in Appendix 1, the coherence function is a bell-shaped function without strict width boundary. The definition of the coherence length varies depending on publications. In this book we use the classical definition for the coherence length L_c [3], which is actually an rms half-width or half-width at 1σ: this is the length that preserves a good interference contrast. We prefer to call "decoherence" length L_{dc}, the other usual definition based on the simple formula $\lambda^2/\Delta\lambda_{FWHM}$, where $\Delta\lambda_{FWHM}$ is the full width at half maximum of the source intensity spectrum. L_{dc} is actually a half-width at about 4σ: this is the length that destroys the contrast almost completely. Note that some authors are using a third definition for a Gaussian coherence function: the half-width at $1/e$, which is actually a half-width at 2σ.) When there is a crossed-polarization coupling point, the main and crossed wave trains happen to lose their overlap because of their velocity difference (Figure 5.1), which yields statistical decorrelation between both crossed polarizations, called depolarization. This is obtained after a propagation length, called the depolarization length L_d [2] and that is defined with

$$L_d/L_{dc} = 1/\Delta n_b = \Lambda/\lambda \tag{5.2}$$

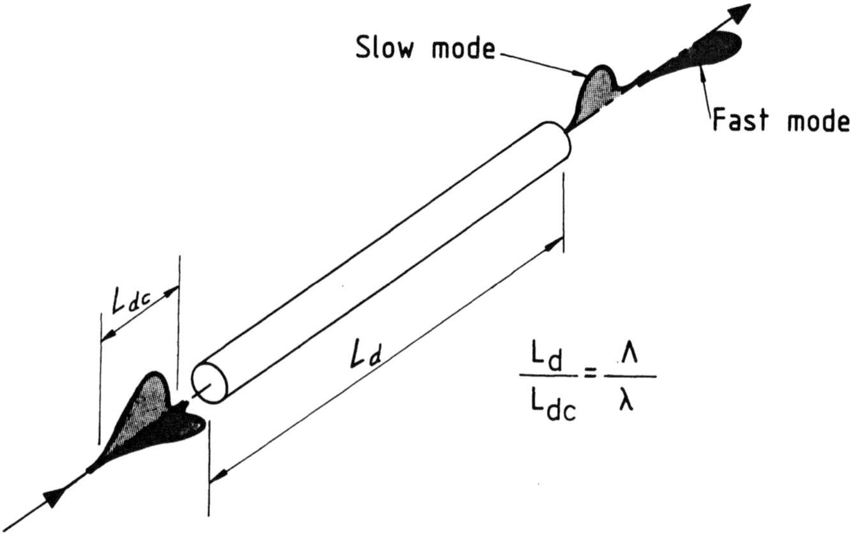

Figure 5.1 Propagation of a wave train in a high-birefringence fiber.

where Δn_b is the birefringence index difference and Λ the fiber birefringence beat length. As explained in Appendix 2, this simple formula applies only if the birefringence index difference Δn_b is wavelength-independent, which is the case with stress-induced birefringence fiber commonly used in the fiber gyro. In the most general case, dispersion has to be taken into account and there is $L_d/L_{dc} = 1/[\Delta n_b + \omega(d(\Delta n_b)/d\omega)]$ [2]. For example, an SLD at 850 nm has a typical spectrum width $\Delta\lambda_{FWHM}$ of 20 nm, which gives

$$L_{dc} = 36\ \mu\text{m}$$

and a typical Δn_b value is 5×10^{-4}; therefore,

$$L_d = 7\ \text{cm}$$

However, this depolarization is not completely equivalent to a naturally unpolarized light. Unpolarized light remains perfectly unpolarized independently of the birefringence of the propagation medium (if the attenuation is not polarization-dependent): two orthogonal polarizations are always statistically decorrelated. With depolarization of a broad spectrum with a birefringent propagation, on the other hand, a good correlation may be restored by compensating for the delay. For example, if, after a propagation in a birefringent medium, light propagates along the same length of birefringent medium, but the principal axes have been rotated by 90 deg, both polarization modes propagate half the way in the fast mode and half the way in the slow mode: they are depolarized in the middle, but they get polarized again at the end, since the first delay is compensated for by the second one.

The whole analysis of the polarization problem in the fiber gyro aims at avoiding such compensation in order to take full advantage of this depolarization effect which drastically reduces nonreciprocities.

5.2 ANALYSIS OF POLARIZATION NONRECIPROCITIES IN A FIBER GYROSCOPE USING AN ALL-POLARIZATION-PRESERVING WAVEGUIDE CONFIGURATION

5.2.1 Intensity-Type Effects

Let us first consider a ring interferometer using a polarization-preserving fiber or waveguide with a perfect input; that is, the state of polarization of the light source is maintained at the input in the same mode, mode 1, and, in particular, is aligned on the transmission mode of the polarizer. When there is one polarization coupling point M in the coil, the input wave train of the broadband source yields at the output two primary wave trains that are still in mode 1 and that are perfectly in phase because of reciprocity. There are also two coupled wave trains in mode 2 that have propagated along different paths (Figure 5.2(a)). This should induce a phase difference, but because of their short coherence length, they do not overlap and these spurious interferences lose their contrast (except for a coupling in the middle of the coil, which may be disregarded for the moment) [4].

If there are two coupling points M and M′ placed at about the same distance from the splitter (Figure 5.2(b)), the cross-coupled wave trains overlap and interfere [5,6]. This is similar to the problem of backreflection, in which one reflecting point is not harmful but two symmetrical reflection points yield a spurious Michelson interferometer. There are actually six wave trains:

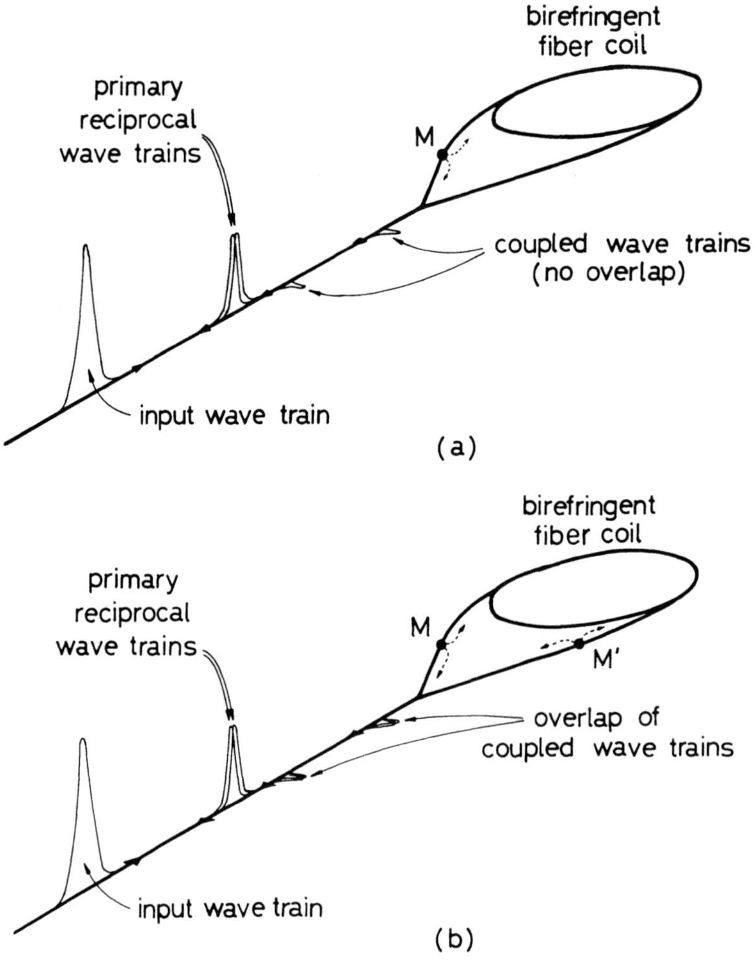

Figure 5.2 Propagation of a wave train in a polarization-preserving ring interferometer: (a) with one crossed-coupling point M; (b) with two symmetrical crossed-coupling points M and M′.

- Two primary reciprocal wave trains with the same phase accumulated in the propagation along the length L of the coil:

$$\phi_{111} = \phi'_{111} = \beta_1 L$$

where β_1 is the propagation constant of mode 1.

- Two secondary wave trains that have propagated in mode 1 between the splitter and the coupling points and that have been coupled in mode 2 for the rest of the propagation. Their respective phases are

$$\phi_{122} = \beta_1 L_M + \beta_2 (L_{MM'} + L_{M'}) \tag{5.3}$$

$$\phi'_{122} = \beta_1 L_{M'} + \beta_2 (L_{MM'} + L_M) \tag{5.4}$$

where β_2 is the propagation constant of mode 2, L_M is the length between the splitter and M, $L_{MM'}$ is the length between M and M′, and $L_{M'}$ is the length between the splitter and M′.

- Two other secondary wave trains that have propagated in mode 1 between the splitter and the coupling points and also between the two coupling points, and that have been coupled in mode 2 between the coupling points and the splitter at the output. Their respective phases are

$$\phi_{112} = \beta_1 (L_M + L_{MM'}) + \beta_2 L_{M'} \tag{5.5}$$

$$\phi'_{112} = \beta_1 (L_{M'} + L_{MM'}) + \beta_2 L_M \tag{5.6}$$

Both pairs of secondary wave trains have the same spurious phase difference:

$$\Delta\phi_s = (\beta_1 - \beta_2) (L_M - L_{M'}) \tag{5.7}$$

Note that the propagation between the two coupling points does not yield any nonreciprocal phase difference. The nonreciprocity is only related to the difference in accumulated birefringence $(\beta_1 - \beta_2)(L_M - L_{M'})$ between the splitter and both symmetrical coupling points.

These two pairs of coherent wave trains are in the crossed-mode 2 at the output and are attenuated by the polarizer. The phase error signal is therefore coming from an intensity-type effect, and it is reduced by the intensity rejection ratio of the polarizer:

$$\Delta\phi_e = 2\epsilon^2\sqrt{\rho_M \rho_{M'}}\, \sin\Delta\phi_s \tag{5.8}$$

where ρ_M and $\rho_{M'}$ are the intensity crossed-coupling ratio at M and M′. Since $\Delta\phi_s$ may vary over 2π, the mean value of this phase error is zero, and its rms deviation is

$$\sigma_{\Delta\phi_e} = \epsilon^2\sqrt{\rho_M \rho_{M'}} \cdot \sqrt{2} \tag{5.9}$$

Note that there are additional wave trains that have experienced a crossed coupling twice, at the input and at the output. They come back in the same mode 1 and have the same phase because of reciprocity:

$$\phi_{121} = \phi'_{121} = \beta_1 L_M + \beta_2 L_{MM'} + \beta_1 L_{M'} \tag{5.10}$$

When there are several pairs of symmetrical coupling points, each pair creates two parasitic interference waves, and since these waves are not coherent with each other, the spurious intensity signals are simply added.

It is possible to evaluate the effect of a fiber coil with a simple model [6]. The random coupling along the coil may be regarded as created by a discrete coupling point M_i for each depolarization length L_d (Figure 5.3), with an intensity coupling $h \cdot L_d$. With a coil length L, there are $N = L/L_d$ coupling points and $N/2$ pairs of symmetrical points. Each pair generates two pairs of secondary wave trains (in each direction) that yield the same phase-bias error; therefore,

$$\Delta\phi_{ei} = 2\epsilon^2[h\ L_d \sin\Delta\phi_{si}] \tag{5.11}$$

since hL_d is the intensity ratio between the main wave trains and the secondary cross-coupled wave trains. The phase differences $\Delta\phi_{si}$ are randomly spread over 2π and the total effect is the summation of independent random variables $\Delta\phi_{ei}$, which are centered about zero. Therefore, the mean value of the total phase error $\Delta\phi_e$ is also zero and the square of its rms deviation is equal to the sum of the square of the rms deviations of each variable:

$$\sigma^2_{\Delta\phi_e} = \frac{N}{2}(\sigma_{\Delta\phi_{ei}})^2 \tag{5.12}$$

We have ${\sigma_{\Delta\phi_{ei}}}^2 = 2\epsilon^4 h^2 {L_d}^2$ because the mean value of $\sin^2\Delta\phi_{si}$ is 1/2; therefore,

$$\sigma^2_{\Delta\phi_e} = N\epsilon^4 h^2 {L_d}^2 = \epsilon^4 h^2 L^2/N$$

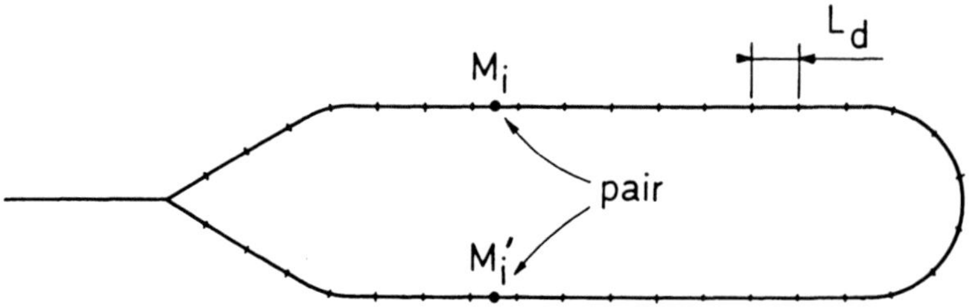

Figure 5.3 Simple model of pairs of symmetrical discrete crossed-coupling points M_i and M'_i for each depolarization length L_d.

since $L = N \cdot L_d$, and

$$\sigma_{\Delta\phi_e} = \epsilon^2 h L_d \sqrt{N} = \frac{\epsilon^2 h L}{\sqrt{N}}$$

where $h \cdot L$ is the total coupling of polarization along the coil. The rms deviation of the birefringence phase error of a fiber coil is not proportional to the polarization coupling ratio $h \cdot L$ in the coil, as might at first be thought. It is reduced proportionally to the square root of the number N of polarization lengths along the coil [6].

For example, with a typical depolarization length of 10 cm, N is equal to 10^4 in a 1-km coil. Since the conservation is typically 20 dB/km, there are 10^4 couping points of 60 dB which yield before filtering by the polarizer a total phase error (rms value) of only 10^{-4} rad, not 10^{-2} rad. A polarizer of 30 dB of rejection is good enough to reduce this intensity type bias error below 10^{-7} rad. This 30-dB value must be compared with the 140 dB that would be required without polarization conservation and depolarization effect!

This result applies only to couplings that are continuously distributed along the fiber coil, the random coupling being regarded as a stationary stochastic process. However, in practice, the main limiting factor is not the coil itself but the localized couplings at the two ends of the coil. As a matter of fact, in the previous example, the effect of the fiber coil is equivalent to a pair of symmetrical points with 40 dB of polarization cross-coupling, even if the total conservation of the coil is only 20 dB. To preserve this benefit, it is desirable to control the polarization cross-couplings of the coil ends at better than this 40-dB value.

With an all-fiber configuration, problems arise with the polarization conservation of the coil coupler that is typically in the 20 to 30 dB range. There are other parasitic couplings at the splices between the coupler leads and the coil ends, but it is possible to unbalance the lengths of both coupler leads. If their dissymmetry is larger than the depolarization length (i.e., typically 10 cm), the coherence between these spurious waves is supressed.

With an integrated optic circuit, the polarization coupling at the splitter (with a Y-junction in particular) is negligible, since the strong birefringence of $LiNbO_3$ provides excellent polarization conservation. Parasitic couplings come from the coil ends butt-coupled on the circuit. A 40-dB coupling is only a 0.5-deg misalignment of the principal axes; however, in practice, it is difficult to have a rugged connection of a fiber without inducing stress that yields locally an additional birefringence, and thus a residual polarization cross-coupling. The depolarization length in $LiNbO_3$ is much shorter than in a fiber (about 500 μm instead of 10 cm, since the birefringence index difference is 8×10^{-2} instead of 5×10^{-4}), but it is still too long to get a sufficient imbalance between both branches of the Y-junction. As a matter of fact, typical values are a branch separation of 200 μm and a slant angle of 10 deg for the parallelogram-shape circuit; that is, an imbalance of only 200 μm × tan10 deg = 35 μm compared to a depolarization length of 500 μm.

This shows that it is very important to accurately control the magnitude and the position of the various localized polarization cross-couplings, even if they seem buried in the total system cross-coupling. Note, however, that these intensity-type effects may be further reduced with a second polarizer placed at the detector, since they are carried by waves that are crossed-polarized with respect to the main primary waves.

5.2.2 Comment About Length of Depolarization L_d vs. Length of "Polarization Correlation" L_{pc}

As is explained in Appendix 1, the contrast or visibility of interference fringes is equal to the autocorrelation function γ_c of the centered spectrum. Most problems are usually analyzed, independently of the exact "bell-shape" of the function, with a half-width of the function called the coherence time τ_c or the corresponding coherence length $L_c = c\tau_c$. As outlined in Section 5.1, their exact definition is not always precisely defined and it deserves to be recalled more accurately.

The classical definition [3] is similar to normalized quadratic averaging, using γ_c as a probability distribution:

$$L_c^2 \int_{-\infty}^{+\infty} \gamma_c^2(\Delta L_{op}) d(\Delta L_{op}) = \int_{-\infty}^{+\infty} \Delta L_{op}^2 \gamma_c^2(\Delta L_{op}) d(\Delta L_{op}) \tag{5.13}$$

where ΔL_{op} is the optical path length difference. The coherence length is thus the rms half-width of γ_c or its half-width at 1σ, σ being the standard deviation. This definition is the most significant in classical interferometry, since the problem is usually to define the length that preserves a good interference contrast.

As we have already seen, the length that suppresses the contrast is also very important: it may be called the decoherence length L_{dc}. Since, in theory, the bell-shaped autocorrelation function reaches zero asymptotically at infinity, it should be infinite; but, in practice, a half-width at 4σ is significant. This decoherence length L_{dc} is defined simply as

$$L_{dc} = \frac{\bar{\lambda}^2}{\Delta\lambda_{FWHM}} \approx 4L_c \tag{5.14}$$

where $\bar{\lambda}$ is the mean wavelength of the broad spectrum and $\Delta\lambda_{FWHM}$ is its full width at half maximum.

When wave trains are used to analyze the effect, it has to be considered that L_{dc} is their full width, since the full width of the autocorrelation of a pulse is double the full width of this pulse.

Applying these concepts to the problem of depolarization in a birefringent medium, it seems more suitable to relate the depolarization length L_d to the decoherence length L_{dc}, with $L_d = L_{dc}/\Delta n_b$. In particular, to ensure that two crossed-polarization coupling points do not yield any spurious signal, their imbalance has to be equal or larger than L_d with this last definition. If $L_c/\Delta n_b$ is used, the signal is still 80% of the perfectly contrasted case. It seems more suitable to call $L_c/\Delta n_b$ the length of polarization correlation L_{pc}, with

$$L_{pc} \approx L_d/4$$

This is the length of propagation in the birefringent medium, which preserves a good degree of polarization (i.e., a good statistical correlation between both orthogonal polarization components).

The case of random coupling along the coil is more delicate. In Section 5.2.1 we considered the number N of depolarization lengths L_d along the fiber, which yields a "safe" evaluation of the upper limit of the phase error. However, we may want to reach a more precise quantification of the effect, and, in particular, to define the actual half-width that has to be used. In any case, if L_{pc} happens to be more suitable than L_d, this multiplies by four the number of averaged elements, which gives a phase error that is only $\sqrt{4} = 2$ times lower than previously found.

It is possible to use a refined model by dividing the fiber coil in elementary length segments L_e that are shorter than the depolarization length L_d, but that remain longer than the autocorrelation width of the stochastic crossed-coupling process in the fiber, which is assumed to be very short (Figure 5.4). Taking two symmetrical elementary segments, they yield a phase-bias error:

$$\Delta\phi_{eii} = 2\epsilon^2 h L_e \sin\Delta\phi_{sii} \tag{5.15}$$

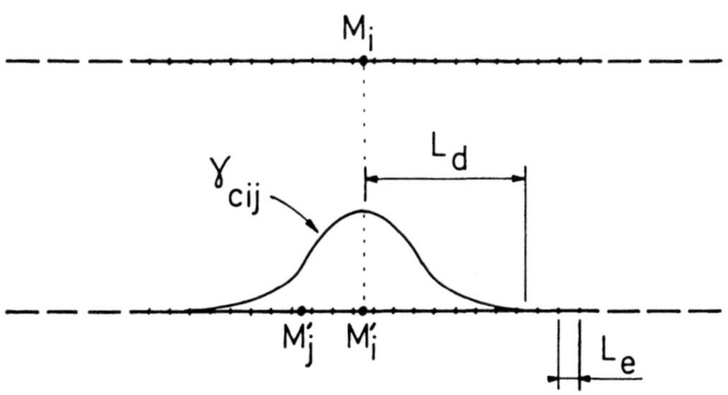

Figure 5.4 Refined model with elementary segments L_e.

The factor of 2 takes into account the input and ouput crossed-couplings, and $h \cdot L_e$ is the intensity crossed-coupling ratio along L_e. However, an elementary segment i yields a spurious signal with its symmetrical segment i, but also with the other segments j, which are around this symmetrical segment. Compared to the symmetrical case, the error is reduced proportionately to the autocorrelation function γ_c of the source:

$$\Delta\phi_{eij} = 2\epsilon^2 h L_e \gamma_{cij} \sin\Delta\phi_{sij} \tag{5.16}$$

where γ_{cij} is the value of γ_c for the actual path difference between the two slightly unbalanced segments. Similar to the analysis of Section 5.2.1, the total phase error $\Delta\phi_e$ is a random variable that is the sum of the independent variables $\Delta\phi_{ij}$. Since $\Delta\phi_{sij}$ are uniformly spread over 2π, they are all centered about a zero mean value, and the square of the rms deviation of $\Delta\phi_e$ is the sum of the square of the rms deviation of all these independent variables:

$$\sigma^2_{\Delta\phi_e} = \sum_i \sum_j \sigma^2_{\Delta\phi_{eij}} \tag{5.17}$$

For a given element i, we first have to sum over j. As a matter of fact, the various $\Delta\phi_{eij}$ are independent, since L_e is assumed to be longer than the autocorrelation width of the stochastic cross-coupling process in the fiber, and

$$\sum_j \sigma^2_{\Delta\phi_{eij}} = 2\epsilon^4 h^2 L_e{}^2 \left[\sum_j \gamma^2_{cij}\right] \tag{5.18}$$

Replacing discrete summation with an integral summation, we have

$$\sum_j \gamma^2_{cij} = \frac{\Sigma \gamma^2_{cij} L_e}{L_e} \approx \frac{1}{L_e} \frac{\int_{-\infty}^{+\infty} \gamma_c^2 dL}{\Delta n_b} \tag{5.19}$$

We have to introduce a new length of polarization correlation L'_{pc} with

$$L'_{pc} = \frac{\int_{-\infty}^{+\infty} \gamma_c^2 dL}{\Delta n_b} \tag{5.20}$$

and

$$\sum_j \sigma^2_{\Delta\phi_{ij}} = 2\epsilon^4 h^2 L_e L'_{pc} \tag{5.21}$$

We have now to sum over i (i.e., the $L/2L_e$ segments over half the coil), and

$$\sigma^2_{\Delta\phi_e} = \epsilon^4 h^2 L L'_{pc} \quad (5.22)$$

Compared to the analysis of Section 5.2.1, L'_{pc} replaces L_d. In the particular case of a Gaussian autocorrelation, L'_{pc} may be computed easily, since $\gamma_c(L) = e^{-L^2/4L^2_c}$ and $\int_{-\infty}^{+\infty} e^{-x^2}dx = \sqrt{\pi}$. It is found that

$$L'_{pc} = L_{pc}\sqrt{2\pi} \approx 2.5\ L_{pc} \approx L_d/1.5 \quad (5.23)$$

This refined analysis shows that the length L'_{pc} which yields the exact number of averaged elements along the coil has an intermediate value between the depolarization length L_d and the length of polarization correlation L_{pc}; but in any case the phase bias error induced by a fiber coil with a stationary random cross-coupling remains very similar to the simple analysis of Section 5.2.1.

5.2.3 Amplitude-Type Effects

As described in Section 3.4.1, the main source of birefringence nonreciprocity with ordinary fiber is caused by an amplitude-type effect where light that is misaligned at the input with respect to the transmission mode of the polarizer is coupled back in the main mode at the output. It carries a nonreciprocal phase error, but also interferes with the primary waves, which yields an amplification of the spurious error term with a coherent detection process [7].

The use of birefringent fiber for the coil and also for the couplers and the fiber pigtail of the source suppresses this effect. As a matter of fact, the primary waves always remain in the same fast (or slow) mode, and light that is misaligned at the input accumulates some path difference with respect to the primary waves in the birefringent input lead of the ring interferometer. When it is cross-coupled in the coil, it cannot compensate for this previous path difference. The birefringence of the input fiber lead destroys the correlation between the primary input wave that has a polarization aligned with the polarizer and the crossed input wave: this suppresses (to first order) the amplitude-type effect [6,8].

On the other hand, if there is no birefringence in the input lead, it is found that cross-coupling in the first depolarization length of the coil yields an amplitude-type effect, since the crossed wave remains coherent with the primary wave in this case [4,9]. Even if the coupling is typically only 60 dB along a depolarization length, this would require a rejection of 60 dB, assuming an input misalignment of 20 dB, to limit the amplitude-type phase error below 10^{-7} rad.

It is seen that the birefringence of the input lead is as important as that of the coil to get full advantage of the depolarization effect. The need of a birefringent input lead makes integrated optics very desirable because of the strong natural birefringence of $LiNbO_3$ [6]. It is necessary to be careful, however, to use the same fast (or slow) mode in the circuit and in the fiber to avoid path compensation that could restore correlation between the crossed-polarization modes.

However, this simple analysis assumes that the primary waves are composed of a single wave train that propagates in front of all the parasitic wave trains if the fast mode is used, or behind if the slow mode is used, avoiding any coherent detection. A real system is actually more complex because multiple crossed couplings have to be taken into account. The primary reciprocal waves are composed of a first main wave train, but there are also additional smaller wave trains coming from an even number of crossed-couplings along the system. They do not carry any birefringence-induced phase error, but, contrary to the main primary wave trains, they may be coherent with a spurious signal and serve as a local oscillator [6]. This effect is much smaller than if it were the first main wave train, but it should be addressed when the best performance is looked for. For similar reasons, the exact autocorrelation function of the broadband source must be known, since a residual multimode laser structure also yields additional wave trains delayed by the optical length of the cavity.

In practice, a careful analysis has to be performed of the amount of crossed-coupling at the various defect points (source pigtailing, fiber splices, coupler, integrated optic circuit pigtailing), the birefringence delay between these points, and the actual coherence of the source to ensure that any second-order amplitude-type error is sufficiently reduced by the polarizer.

Note that it is possible to further improve the decorrelation effect of the birefringence of the input lead with an additionnal modulation of this birefringence [10]. However, such a method increases the complexity of the device, and present polarizer performance allows very low phase error to be obtained without this additional modulator.

5.3 USE OF A DEPOLARIZER

Polarization-preserving fiber is the optimal technical choice for high performance, but it remains possible to use standard nonpolarization-preserving fiber without polarization control and without getting signal fading if a Lyot-type depolarizer is inserted into the system [11]. Such a depolarizer is based on the loss of polarization correlation with propagation of a broadband source in a birefringent medium. This is similar to the depolarization of parasitic crossed waves that we saw in Section 5.2, but it is now done on purpose. As a matter of fact, when a linearly polarized broadband source is aligned at 45 deg of the principal axes of the birefringent medium, the input wave train is split evenly along both axes and that, at the output, the two wave trains do not overlap if the path is longer than the depolarization length L_d. To get depolarization with any input state of polarization, a Lyot depolarizer is actually composed of a first birefringent medium with a length L_L and an index difference Δn_b, and a second one rotated by 45 deg and with a length $2L_L$, twice as large. The wave train is first split unevenly as a function of the input state of polarization, and each secondary wave train is split again, but now evenly, along the rotated axis of the second birefringent medium. At the output, the four wave trains do not overlap, and there is the same decorrelated power along each axis (Figure 5.5) if L_L is larger than L_d. When such a depolarized light is sent through an ordinary fiber, it remains depolarized if the residual fiber birefringence (particularly bending-induced birefringence, described in Appendix 2) does not compensate for the delay given by the depolarizer. Such depolarizers may be fabricated with short lengths of high-birefringence polarization-preserving fiber or with integrated optic waveguides that have a natural crystal birefringence.

A first depolarizer has to be used between the source and the polarizer and a second one in the coil, but the immediate drawback of this approach is an additional loss of 3 dB at the input through the polarizer, and a second loss at the output. Furthermore, depolarization is not perfect, especially when it is obtained with fibers that always have residual spurious cross-coupling, and extreme care must be taken to avoid path compensation that could restore the contrast of amplitude-type error signals.

In pratice, the use of depolarizers does not yield results that are as good as with polarization-preserving fibers. Its main benefit is in terms of the cost of the

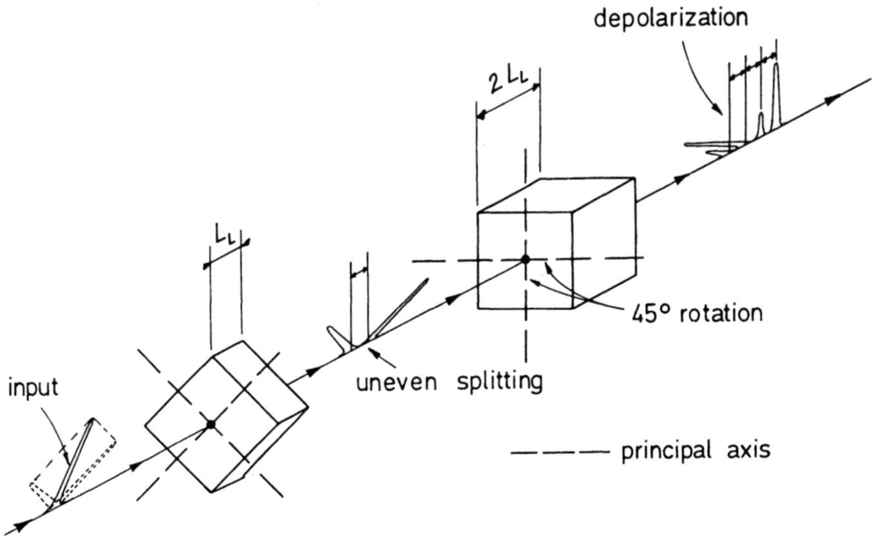

Figure 5.5 Principle of a Lyot depolarizer.

fiber coil, but, for the whole system, the cost saving is not obvious because of the increased complexity brought by the additional depolarizers.

5.4 TESTING WITH OPTICAL COHERENCE DOMAIN POLARIMETRY (OCDP) BASED ON PATH-MATCHED WHITE-LIGHT INTERFEROMETRY

This analysis of polarization nonreciprocities with polarization-preserving fibers or with depolarizers has shown the importance of depolarization effects brought about by birefringence when a low temporal coherence source is used. However, the location and the magnitude of the various cross-coupling points have to be determined precisely, even if they seem buried in the background cross-coupling of the system, to evaluate the residual bias error.

The usual polarization measurement techniques cannot differentiate a series of defects, but this problem has been solved by applying techniques of path-matched white-light interferometry [12,13,14]. By analogy with the well-known method of optical time domain reflectometry, (OTDR), [15] used to test the attenuation of ordinary fibers and its coherence domain counterpart, or OCDR [16], this technique can be called optical coherence domain polarimetry (OCDP).

The principle of path-matched white-light interferometry is based on the use of a broadband source (white light being historically the first available) in an unbalanced interferometer. When the path imbalance is much longer than the coherence length, the interference contrast vanishes (see Appendix 1). With wave trains, the path difference suppresses their overlap at the output (Figure 5.6). However, passing through a second readout interferometer that has the same path imbalance, the interference contrast is restored to one-half, because among the four output wave trains, two overlap, since they have both propagated along one short and one long path in each interferometer.

Now if one splitter in the first interferometer has a very low power-splitting ratio (ϵ^2 and $1 - \epsilon^2$ with $\epsilon^2 << 1/2$), the contrast of the restored interference is ϵ, the amplitude ratio, and not ϵ^2, the intensity ratio (Figure 5.7). The powerful wave train serves actually as a "local oscillator" for the coherent detection process

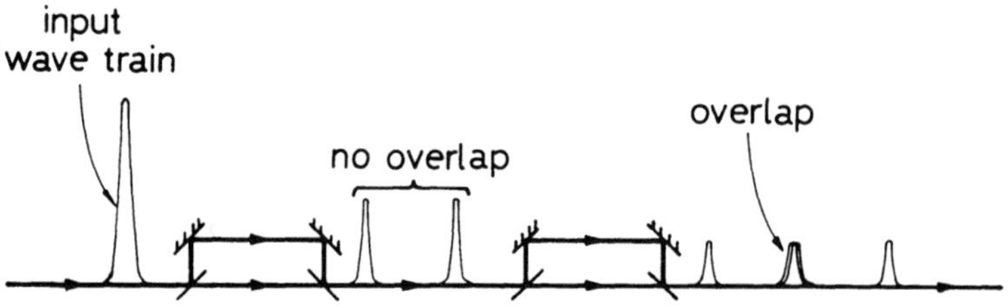

Figure 5.6 Principle of white-light interferometry: propagation of a wave train in a pair of path-matched interferometers.

(see Appendix 1) of the low-power wave train. For example, a splitting ratio of 60 dB thus yields a contrast of 10^{-3} and not 10^{-6}.

A polarization cross-coupling point in a high-birefringence fiber is equivalent to this weak coupling. The main wave train propagates along the primary mode, and the cross-coupled wave train propagates along the other, the fast mode being equivalent to the short path and the slow mode being equivalent to the long path. An output polarizer aligned at 45 deg of the principal axes of the birefringent fiber is equivalent to the 3-dB recombiner (Figure 5.8). Using a readout interferometer, particularly a Michelson interferometer, the birefringence path imbalance in the fiber may be compensated for, yielding a contrast recovery equal to the amplitude ratio of the polarization cross-coupling, which makes this method of OCDP very sensitive. When several coupling points are located in series along the fiber, they

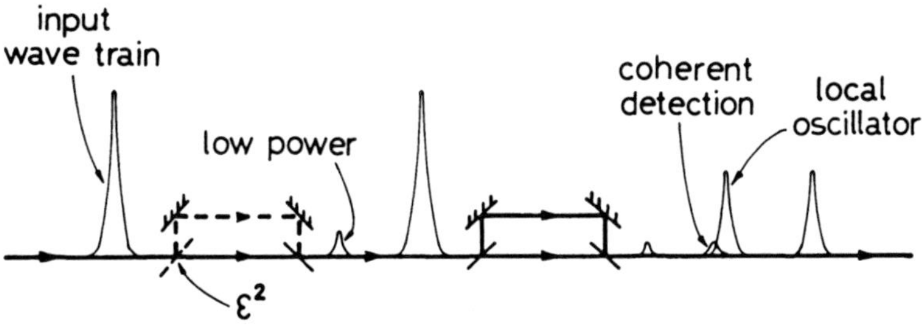

Figure 5.7 White-light interferometry with a very low-power splitting ratio ϵ^2 of one splitter.

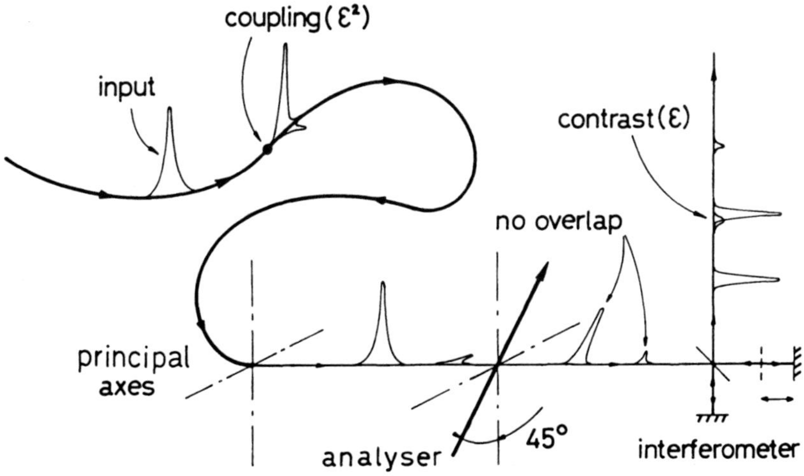

Figure 5.8 Principle of optical coherence domain polarimetry (OCDP).

can be measured separately by scanning one mirror of the Michelson interferometer.

The path difference ΔL_r in the readout interferometer gives the distance L_f between the point that is analyzed and the fiber end:

$$\Delta L_r = \Delta n_b L_f \tag{5.24}$$

where Δn_b is the birefringence index difference of the fiber. The spatial resolution is equal to the depolarization length L_d, about 10 cm in a fiber, but, as we have already seen, 0.5 mm in a $LiNbO_3$ integrated optic circuit.

This technique was first proposed to test the spatial distribution of the random coupling along the fiber [12], which is very useful, since it is important to control the degradation of the polarization conservation of the coil by the winding process, and in particular to check that it is not due to a few localized cross-coupling points. However, the main interest of this technique is the in situ control of the in-line components, like couplers, and of the connections [13,14], particularly the fiber connections on the integrated optic circuit, which, as we have seen, are crucial for limiting the bias error, since they are symmetrical with respect to the splitter. It may also be applied to the test of depolarizers.

This OCDP testing is perfectly adapted to the case of the fiber gyro, because it measures spurious polarization defects in exactly the same way as they induce phase error in a ring interferometer. As an example, it can be used for nondestructive control of the fiber pigtail alignment of a packaged superluminescent diode, despite its low degree of polarization. A partially polarized source is usually considered to be the superposition of two orthogonal, incoherent polarizations; but when tested by white-light interferometry, it should actually be considered to be the superposition of a perfectly unpolarized source and a perfectly linearly polarized source (Figure 3.31). For example, with a 4-to-1 power ratio between the two axes, the superluminescent diode is actually the superposition of a perfectly unpolarized source with two-fifths of the power and a perfectly polarized source with three-fifths of the power. The unpolarized component does not give any contrast recovery in the readout interferometer in the same way that unpolarized light eliminates birefringence bias error in a fiber gyro (see Section 3.4.4), while the polarized component behaves like an ordinary polarized source: a misalignment of the principal axes of the pigtail produces a polarization coupling that can be measured with the value of the contrast that is recovered when the path difference of the readout interferometer compensates for the birefringence of the fiber pigtail. Notice that the Michelson interferometer used in this setup is also a suitable system for precisely measuring the coherence function of the broadband source and, in particular, for evaluating its residual multimode laser structure [17] (Figure 5.9).

Another interesting feature of this method is the measurement of polarizer rejection [13]. On entering a linearly polarized wave train at 45 deg of the principal axes of an integrated optic polarizer on a birefringent substrate, there are, at the output, a main transmitted wave train and a secondary wave train attenuated by

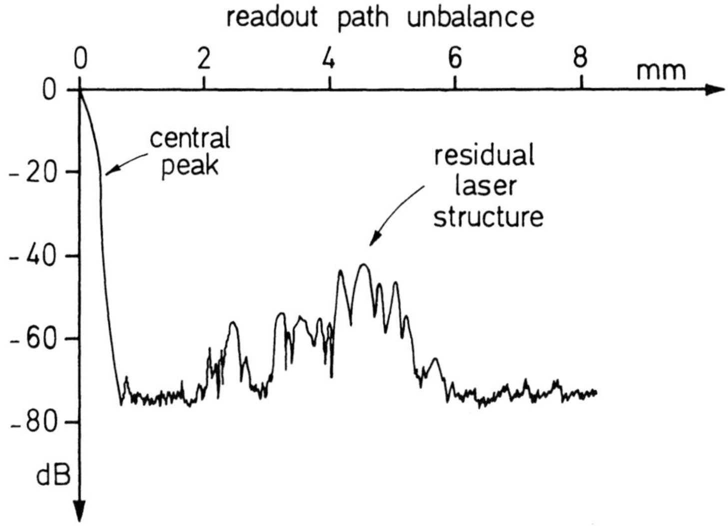

Figure 5.9 Measurement of the effect of the residual multimode laser structure of a superluminescent diode with its coherence function (log scale).

the polarizer (Figure 5.10). An output polarizer at 45 deg recombines both wave trains in the same polarization state, and a readout interferometer yields a measurement of the amplitude extinction ratio of the integrated optic polarizer! Note that a high rejection of the input and output polarizers of the setup is not required.

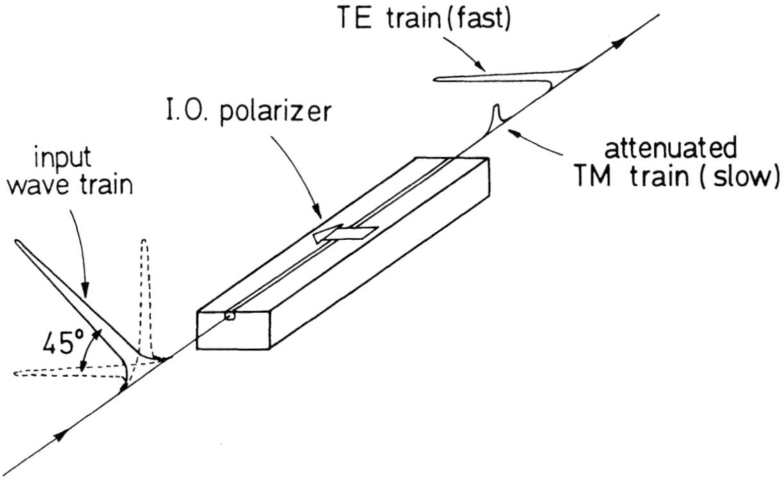

Figure 5.10 Propagation of a wave train in an integrated-optic polarizer fabricated on a birefringent substrate (case of a fast transmitted TE mode, as with x-cut y-propagating $LiNbO_3$ circuit).

The additional polarizers are used only to split the input light into the two modes of the polarizer under test and to recombine them at the output. It is even possible to differentiate the rejection and the cross-coupling of the fiber connections when the circuit is pigtailed. As a matter of fact, polarized light has to be entered at 45 deg of the axes of the input birefringent fiber lead of the IO circuit, and a polarizer has to be placed at 45 deg of the axes of the output lead. The polarizer rejection is measured with a readout imbalance that corresponds to the total birefringence of the two leads and the circuit, while the cross-coupling at each connection is measured with an imbalance that relates to the birefringence of the corresponding lead (Figure 5.11). This also applies to all-fiber polarizers such as "coiled" polarizers.

Finally, this technique can be extended to spatial modes and, in particular, to test the quality of spatial filtering in the short length of single-mode fiber used at the input-output port of the ring interferometer [13]. Light coupled in the fiber propagates in the fundamental LP_{01} mode, but if the fiber is not perfectly single-mode, there is also some light in the higher-order LP_{11} mode. As with polarization modes, there is a velocity difference and the main LP_{01} wave train does not overlap in time with the attenuated LP_{11} wave train at the output. They must be recombined and sent into a readout interferometer to measure the amplitude extinction ratio of the spurious LP_{11} mode. It can be clearly observed with optical coherence domain reflectometry (OCDR) [18] that path-matched white-light interferometry

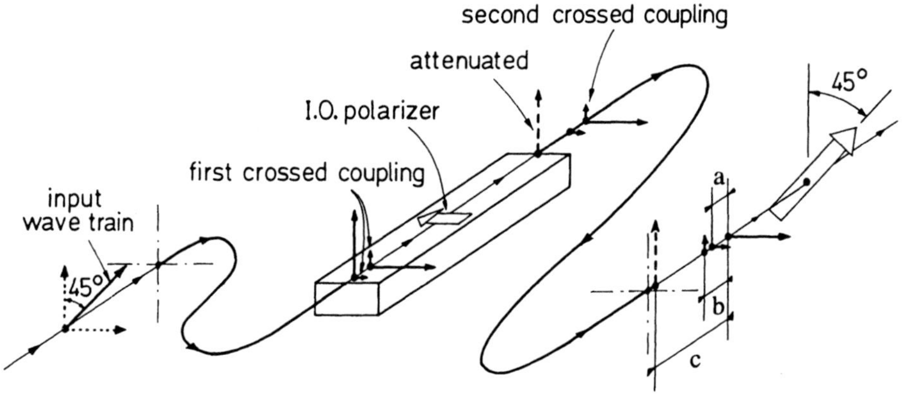

Figure 5.11 OCDP testing of a pigtailed integrated optic polarizer with (a) the path difference of the wave train corresponding to the first crossed coupling; (b) the path difference of the wave train corresponding to the second crossed coupling; (c) the path difference corresponding to the wave train attenuated by the IO polarizer (it is assumed that the IO polarizer transmits the fast TE mode).

equalizes group transit time difference and not that of phase transit time. The discrepancy is negligible with stress-induced birefringence, but it may be very different with spatial modes where there are dispersion effects.

OCDP testing does provide a very powerful tool for controlling and understanding the various problems of polarization and birefringence. It allows one to optimize the compromise among polarization rejection, polarization couplings, and depolarization, and to get a very low bias error despite the limited performance of the components. In the early days of fiber gyros, intensity coupling ratios were measured when testing the selectivity of components and assembling the system, while the spurious signals were dependent on amplitude ratios because of the high coherence of the laser source. Broadband sources and OCDP testing put the problem back in the right perspective: amplitude coupling ratios are measured (and localized!) during mounting, which means high sensitivity for test and control, while the spurious effects are now dependent on intensity ratios, which means reduced parasitic signals in the final setup. Note finally that OCDR techniques, which are, as we have seen, very similar to OCDP, would be a useful tool to more accurately evaluate residual backreflections at splices and connections between fiber and integrated optic circuit.

REFERENCES

[1] Kaminow, I. P., "Polarization in Optical Fibers," IEEE Journal of Quantum Electronics, Vol. QE-17, 1981, pp. 15–21.

[2] Rashleigh, S. C., W. K. Burns, R. P. Moeller, and R. Ulrich, "Polarization Holding in Birefringent Single-Mode Fibers," Optics Letters, Vol. 7, 1982, pp. 40–42.

[3] Born, M., and E. Wolf, Principles of Optics, Pergamon Press, 1975.

[4] Burns, W. K., and R. P. Moeller, "Polarizer Requirements for Fiber Gyroscopes With High-Birefringence Fiber and Broad-Band Source," Journal of Lightwave Technology, Vol. LT2, 1984, pp. 430–435 (SPIE MS 8, pp. 271–276).

[5] Arditty, H. J., J. P. Bettini, Y. Bourbin, Ph. Graindorge, H. C. Lefèvre, M. Papuchon, and S. Vatoux, "Integrated-Optic Fiber Gyroscope, Progresses Towards a Tactical Application," Proceedings of OFS 2'84, Stuttgart, VDE-Verlag, 1984, pp. 321–325.

[6] Lefèvre, H. C., J. P. Bettini, S. Vatoux, and M. Papuchon, "Progress in Optical Fiber Gyroscopes Using Integrated Optics," Proceedings of AGARD-NATO, Vol. CPP-383, 1985, pp. 9A1–9A13 (SPIE MS 8, pp. 216–227).

[7] Fredricks, R. J., and R. Ulrich, "Phase-Error Bounds of Fibre Gyro With Imperfect Polariser/Depolarizer," Electronics Letters, Vol. 20, 1984, pp. 330–332 (SPIE MS 8, pp. 277–278).

[8] Jones, E., and J. W. Parker, "Bias Reduction by Polarization Dispersion in the Fibre-Optic Gyroscope," Electronics Letters, Vol. 22, 1986, pp. 54–56 (SPIE MS 8, pp. 286–287).

[9] Burns, W. K., "Phase-Error Bounds of Fiber Gyro With Polarization-Holding Fiber," Journal of Lightwave Technology, Vol. LT4, 1986, pp. 8–14 (SPIE MS 8, pp. 279–285).

[10] Carrara, S. L. A., B. Y. Kim, and H. J. Shaw, "Bias Drift Reduction in Polarization-Maintaining Fiber Gyroscope," Optics Letters, Vol. 12, 1984, pp. 214–216 (SPIE MS 8, pp. 291–293).

[11] Böhm, K., P. Marten, K. Petermann, E. Weidel, and R. Ulrich, "Low-Drift Gyro Using a Superluminescent Diode," Electronics Letters, Vol. 17, 1981, pp. 352–353 (SPIE MS 8, pp. 181–182).

[12] Takada, K., J. Noda, and K. Okamoto, "Measurement of Spatial Distribution of Mode Coupling in Birefringent Polarization-Maintening Fiber With New Detection Scheme," Optics Letters, Vol. 11, 1986, pp. 680–682.

[13] Lefèvre, H. C., "Comments About the Fiber-Optic Gyroscope," SPIE Proceedings, Vol. 838, 1987, pp. 86–97 (SPIE MS 8, pp. 56–67).

[14] Takada, K., K. Chida, and J. Noda, "Precise Method for Angular Alignment of Birefringent Fiber Based on an Interferometric Technique," Applied Optics, Vol. 26, 1987, pp. 2979–2987.

[15] Barnoski, M. K., and M. S. Jensen, "Fiber Waveguides, a Novel Technique for Investigating Attenuation Characteristics," Applied Optics, Vol. 15, 1976, pp. 2112–2115.

[16] Youngquist, R. C., S. Carr, and D. E. N. Davies, "Optical Coherence Domain Reflectometry: A New Evaluation Technique," Optics Letters, Vol. 12, 1987, pp. 158–160.

[17] Martin, P., G. Le Boudec, and H. C. Lefèvre, "Test Apparatus of Distributed Polarization Coupling in Fiber Gyro Coils Using White Light Interferometry," SPIE Proceedings, Vol. 1585, 1991.

[18] Kohlhaas, A., C. Frömchen, and E. Brinkmeyer, "High-Resolution OCDR for Testing Integrated-Optical Waveguides: Dispersion-Corrupted Experimental Data Corrected by a Numerical Algorithm," Journal of Lightwave Technology, Vol. 9, 1991, pp. 1493–1502.

Chapter 6
Transience-Related Drift and Noise

6.1 EFFECT OF TEMPERATURE TRANSIENCE

Because of reciprocity, the two counterpropagating paths are equalized in a ring interferometer; but this is strictly valid only if the system is time-invariant. Actually, we have already seen (Section 3.2.2) that biasing phase modulation can be generated with a reciprocal modulator placed at one end of the coil acting as a delay line, which provides a filtering behavior with a sinusoidal transfer function: $\sin[(\pi/2) \cdot (f/f_p)]$. This also applies to parasitic phase shifts generated by the environment, particularly with nonuniform temperature change [1]. The two interfering waves do not see the perturbation at exactly the same time, unless it is applied in the middle of the coil.

Considering that the phase perturbation ϕ_{pert} has low frequencies compared to the proper frequency of the coil, the gyro behaves like a dc rejection filter with the usual 6 dB per octave roll-off and differentiation properties (Figure 6.1): the phase error $\Delta\phi_e$, and therefore the drift of the rotation rate measurement, is proportional to the temporal derivative of the temperature, which may be very harmful, particularly during the warmup.

An elementary fiber segment δz, placed at a distance z along the coil of length L (Figure 6.2), yields a phase error:

$$\Delta\phi_e(z) = \frac{2\pi}{\lambda}\frac{dn}{dT}\frac{dT}{dt}(z)\frac{L-2z}{v}\,\delta z \qquad (6.1)$$

considering that the main perturbation is coming from the index change dn/dT equal to 10^{-5}/C in silica (the length change less than 10^{-6}/C is negligible in com-

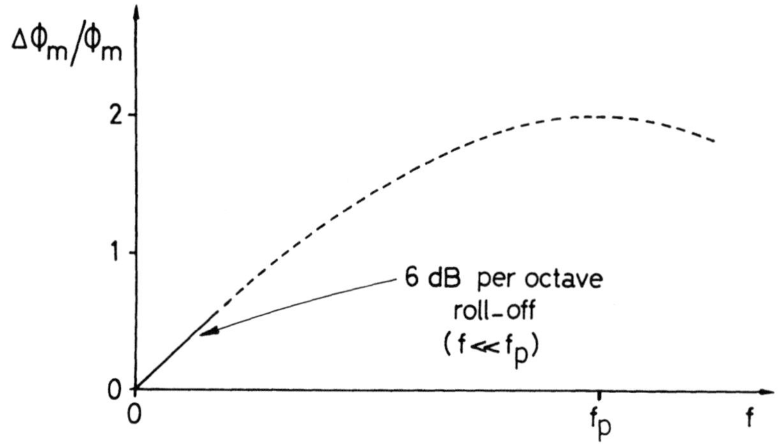

Figure 6.1 Transfer function for phase perturbation about dc.

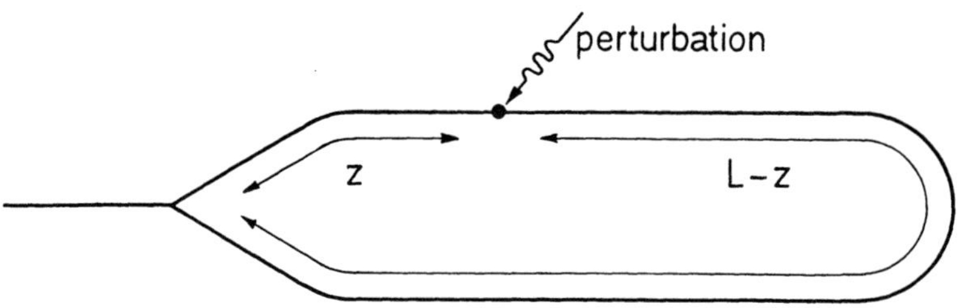

Figure 6.2 Effect of an asymmetrical perturbation.

parison), and where v is the wave velocity and dT/dt is the rate of temperature change of δz. Taking, for example, a coil of 200m composed of 40 layers of 50 turns, and assuming that the first external layer is heated solely at a moderate rate of 0.1C/s, a phase error as high as 5×10^{-5} rad is yielded.

However, if the same temperature change is experienced by segments at equal distance from the middle of the coil, the effect is canceled out, since $\Delta\phi_e(z)$ becomes opposite to $\Delta\phi_e(L - z)$. This compensation is obtained with symmetrical winding [1]: the fiber is wound from the middle, and by alternating layers coming from each half-coil length, this places symmetrical segments in proximity (Figure 6.3(a)). Compared to ordinary winding, this so-called dipolar winding reduces the effect of thermal transience by a factor approximately equal to the number of layers [2]. An even better compensation, approximately equal to the square of the number of layers, is obtained with a quadrupolar winding [2] (Figure 6.3(b)). With a pair of symmetrical layers in a dipolar winding, a radial heat propagation always first

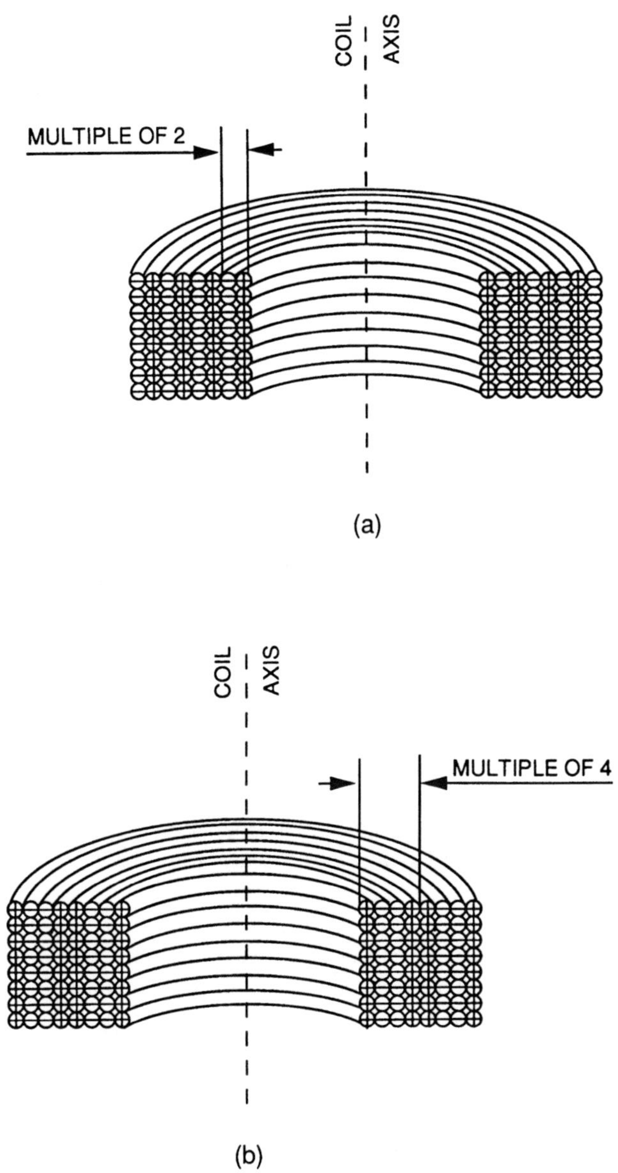

Figure 6.3 Section view of symmetrical windings (one-half coil length is represented by + and the other by −): (a) dipolar; (b) quadrupolar.

reaches the layer of the same half-coil length, which yields a residual transient sensitivity. With the quadrupolar winding, the layer order is reversed from pair to pair, which further improves the compensation. To obtain the best results, the number of layers must be a multiple of four and all the layers must have the same number of turns. Some additional benefit is also obtained by thermally insulating the coil to slow down its rate of temperature change.

With an ideal winding, the asymmetry between the locations of two fiber segments in proximity is limited by the length of a single fiber loop. This problem of thermal transience is easily solved with the compact and short coil of medium sensitivity gyros, but a good thermal design of the system is required for high-sensitivity gyros using a coil with a long length and large diameter.

6.2 EFFECT OF ACOUSTIC NOISE AND VIBRATION

Acoustic noise and vibrations are also a potential source of parasitic effects related to transient phenomena. As a matter of fact, fiber optics are used to make a very sensitive hydrophone, with acoustic pressure yielding phase change through the photoelastic effect [3]; but with the ring configuration of the interferometric fiber gyro, the absolute effect is greatly reduced.

In any case, a perturbation at a given frequency may yield a spurious effect only at this frequency and its harmonics: it does not affect the mean bias offset and does not reduce the sensitivity to rotation if the spurious phase variation remains in the linear range of the sine response, which is always the case in practice. Therefore, it is necessary to consider only perturbations within the detection bandwidth of the system (i.e., a few kilohertz). In this frequency range, the acoustic wavelength in the fiber is equal to several meters (sound velocity is 7 km/s in silica), which makes the perturbation quasi-uniform over the coil dimension: symmetrical fiber segments experience the same phase modulation, which makes the residual effect negligible. This applies to external perturbations and to the proper fluctuations due to phonons [4].

Vibrations are a more severe problem, and an adequate potting has to be used to ensure a good coil ruggedness. An adequate mechanical design is also required to avoid specific resonances within the system, but this is eased because of the solid-state configuration of the fiber gyro.

REFERENCES

[1] Schupe, D. M., "Thermally Induced Nonreciprocity in the Fiber-Optic Interferometer", Applied Optics, Vol. 9, 1980, pp. 654–655 (SPIE MS 8, pp. 294–295).

[2] Frigo, N. J., "Compensation of Linear Sources of Non-Reciprocity in Sagnac Interferometers", SPIE Proceedings, Vol. 412, 1983, pp. 268–271 (SPIE MS 8, pp. 302–305).

[3] Yurek, A. M., "Status of Fiber-Optic Acoustic Sensing", Proceedings of 8th Optical Fiber Sensors Conference, Monterey, 1992, pp. 338–341.

[4] Logozinskii, V. N., "Fluctuations of the Phase Shift Between Opposite Waves in a Fiber Ring Interferometer", Soviet Journal of Quantum Electronics, Vol. 11, 1981, pp. 536–538 (SPIE MS8, pp. 296–298).

Chapter 7
Truly Nonreciprocal Effects

7.1 MAGNETO-OPTIC FARADAY EFFECT

Even in a perfectly reciprocal system, the Sagnac phase shift is not the only effect to be truly nonreciprocal. In particular, due to the magneto-optic Faraday effect, a longitudinal magnetic field **B** modifies the phase of a circularly polarized wave by an amount determined by the Verdet coefficient V of the medium. The sign of this phase shift depends on the left- or right-handed character of the circular polarization and also on the relative direction of the field and light propagation vectors. It is well known that this phase shift may manifest itself as a change θ_F in the orientation of linearly polarized light resulting from the opposite phase shifts of its copropagating left- and right-handed circularly polarized components: $\theta_F = V \cdot B \cdot L$, where L is the medium length. It may also be detected as a phase difference $\Delta\phi_F$ in a ring fiber interferometer where identical circularly polarized waves counterpropagate around the coil (Figure 7.1). As seen in Appendix 1, this phase difference is the double of the angle of Faraday rotation θ_F:

$$\Delta\phi_F = 2\, V \cdot B \cdot L \tag{7.1}$$

At first it seems that the total Faraday effect along a given path is proportional to the line integral of **B** along this path. For a closed path, the result should be different from zero according to Ampère law only if the path encloses an electrical current. A toroidal closed path configuration has been used to demonstrate an electrical current fiber sensor [1], but a fiber-optic gyroscope should not be sensitive to environmental magnetic fields, because of the absence of traversing electrical current. However, this is actually true only if the same state of polarization is

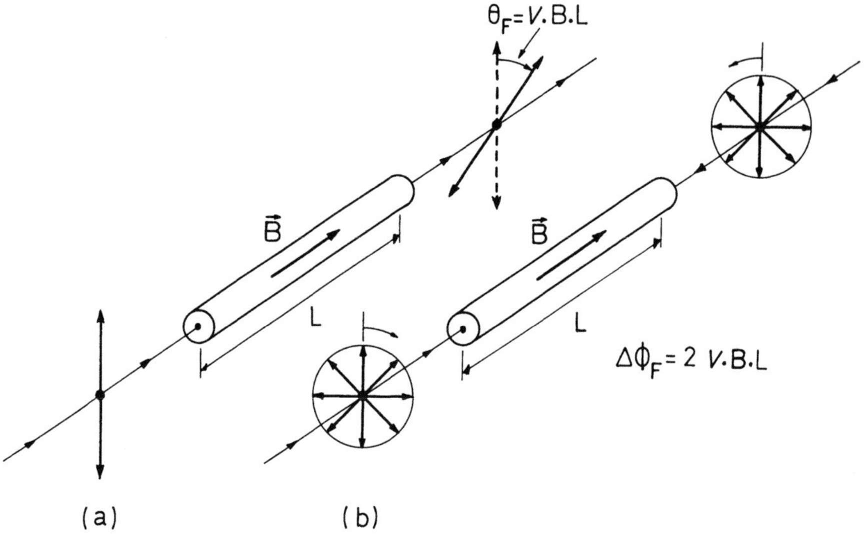

Figure 7.1 Faraday effect: (a) rotation of a linear polarization; (b) phase difference between two counterpropagating circular polarizations.

preserved along the fiber. The Faraday phase shift accumulated along an elementary length vector **dz** is

$$d\phi_F = \alpha_p \cdot V \cdot \mathbf{B} \cdot \mathbf{dz} \tag{7.2}$$

where α_p is a coefficient that depends on the state of polarization. It is equal to zero for linear polarization and to ± 1 for circular polarizations. It has an intermediate value for elliptical polarization. The total phase difference between both counterpropagating waves is represented by the relationship

$$\Delta\phi_F = 2\int_{\text{closed path}} \alpha_p \cdot V \cdot \mathbf{B} \cdot \mathbf{dz} \tag{7.3}$$

which may be different from zero even if the line integral $\int \mathbf{B} \cdot \mathbf{dz}$ is equal to zero, because α_p is not constant. This is due to polarization changes along the fiber arising from residual birefringence [2,3,4]. A configuration using bending-induced birefringence to enhance sensitivity to an external magnetic field and make a magnetometer with a ring interferometer has even been demonstrated [5].

Assuming that the influence of the earth's magnetic field B_{earth} were integrated constructively along the whole fiber length L, the maximum nonreciprocal phase difference would be

$$\Delta\phi_{F\max} = 2 \cdot V \cdot B_{\text{earth}} \cdot L \tag{7.4}$$

The Verdet constant V has a λ^{-2} wavelength dependence and is equal to 2 rad $m^{-1}T^{-1}$ at 0.85 μm, and since B_{earth} is typically 0.5G (or 5×10^{-5} tesla), $\Delta\phi_{Fmax}$ would be as high as 0.2 rad with a 1-km coil length. Experimentally, it was observed [2–4] that there is a compensation factor of about 10^3 in a gyroscope using an ordinary fiber, which yields a measurement error equivalent to approximatively the earth rotation rate (i.e., 15 deg/h).

Notice that the Faraday effect is also given in scientific textbooks as a function of the H field. Since in diamagnetic material like silica, **B** and **H** are proportional and the relative magnetic permeability is very close to unity, the unit change of the Verdet constant V is obtained by multiplying its "B-value" by $\mu_0 = 4\pi \times 10^{-7}\ Hm^{-1}$ (or $T \cdot A^{-1} \cdot m$); that is, the "H-value" of V is 2.5×10^{-6} rad A^{-1} for a wavelength of 0.85 μm.

The use of polarization-preserving fiber, which has been seen to be very helpful to reduce birefringence-induced nonreciprocities, is also reducing the magnetic dependence, and, in practice, the residual Faraday phase error becomes on the order of 1 μrad for 1G (10^{-4} tesla). However, the effect is not completely nulled out because of the residual rotation of the birefringence axes of practical fibers [6]. Their preforms experience very high stress, which tends to give a helicoidal shape to the stressing rods, and stress-induced high-birefringence fibers used to preserve polarization are drawn with a slowly changing orientation of their principal axes [7].

When the principal axes of a linear-birefringence fiber are rotated, the eigen polarization mode does not have a perfectly linear state of polarization. This can be viewed simply by considering a Poincaré sphere (see Appendix 2) defined with respect to a "rest" reference frame that rotates with the principal axes at the twist rate t_w (in rad/m). In this rest frame, the linear birefringence is represented with a stable equatorial vector $\mathbf{\Delta\beta}_l$, but there is an additional circular birefringence vector $\mathbf{\Delta\beta}_c$ aligned along the polar axis to take into account the change of frame of reference (Figure 7.2). The magnitude of $\mathbf{\Delta\beta}_c$ is equal to t_w, but it corresponds to the opposite direction of rotation. The total birefringence $\mathbf{\Delta\beta}_t$ is obtained simply with the vectorial sum $\mathbf{\Delta\beta}_l + \mathbf{\Delta\beta}_c$. The magnitude $\Delta\beta_c$ is much smaller than $\Delta\beta_l$, otherwise the polarization would not be preserved at all; therefore, the two stable orthogonal states of polarization are slightly elliptical to correspond to the intersection of $\mathbf{\Delta\beta}_t$ with the Poincaré sphere. Going back to the "laboratory" frame, the two states preserve the same constant ellipticity, but their minor and major axes follow the rotation of the principal axes of the birefringence of the fiber. The polarization is "dragged" by the twist of the birefringence axes and becomes slightly elliptical.

In a ring interferometer using such a polarization preserving fiber, it is possible to consider that a magnetic field has a negligible dependence on the state of polarization that is the same in both opposite directions. However, it modifies the phase of the counterpropagating waves as a function of the coefficient α_p equal to the ellipticity of the state; that is, the ratio $\Delta\beta_c/\Delta\beta_l = -t_w/\Delta\beta_l$. The accumulated Faraday phase difference is, therefore,

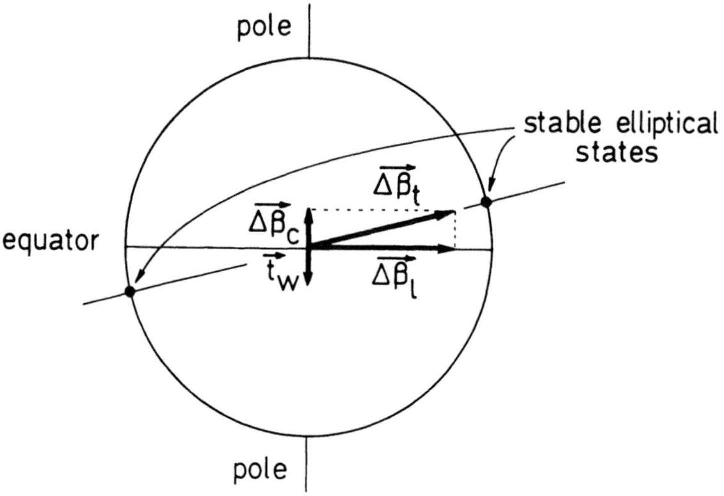

Figure 7.2 Poincaré sphere describing the stable elliptical states of polarization in a high-birefringence fiber with rotated principal axes.

$$\Delta\phi_F = \frac{2V}{\Delta\beta_l}\int_{\text{closed path}} t_w(z)\,\mathbf{B}\cdot\mathbf{dz} \tag{7.5}$$

Assuming a circular coil of radius R, this yields

$$\Delta\phi_F = \frac{2VB}{\Delta\beta_l}\int_{\text{closed path}} t_w(z)\cos\left(\frac{z}{R} - \theta_B\right)dz \tag{7.6}$$

where θ_B is the angle of the **B** field with respect to the reference axes. This formula is equivalent to a "synchronous demodulation" of the twist rate $t_w(z)$ at a "frequency" $(2\pi R)^{-1}$ and with an integration "time" L.

The residual magnetic dependence is therefore coming from the spatial frequency components of $t_w(z)$ equal to the inverse of a turn perimeter $2\pi R$ within a bandwidth equal to the inverse of the total length of the coil. Assuming that $t_w(z)$ is a random function with a constant power density, the usual result of detection of a white noise with a lock-in amplifier may be applied.

It is difficult to evaluate directly this power density, but it is possible to predict the effect of a length increase of a known fiber: it is similar to a random walk process, and the magnetic dependence of the phase error should increase as the square root of the length (which is equivalent to an integration time) increases. On the other hand, the phase error should be independent of the coil diameter and the turn perimeter if the noise power density of the parasitic twist is uniform (i.e., frequency-independent). A good order of magnitude of the practical phase error induced by the earth's magnetic field (i.e., 0.5G) is 0.5 to 2 $\mu rad/\sqrt{km}$ (at $\lambda = 0.85$ μm), depending on the quality of the fiber. Since the Sagnac sensitivity is proportional to the length and the diameter of the coil, the equivalent rotation rate error due to an external magnetic field should decrease proportionally to the coil diameter and to the square root of its length.

This phase error is coming from a random process, the parasitic twist rate of the high-birefringence fiber; but for a given gyro coil, this rate is stable over time, and the magnetic dependence does not yield long-term drift or noise if the gyro keeps the same orientation with respect to the magnetic field. The gyro has an axis and a coefficient of magnetic sensitivity that are stable and can be modeled to compensate for the predictable bias offset. This sensitive axis lies approximately in the plane of the coil, since the Faraday effect is nulled out for a magnetic field orthogonal to the direction of propagation.

If the application requires a very low magnetic dependence, it is possible to get a further improvement of one to two orders of magnitude by shielding the sensing coil with a high permeability material such as μ-metal. Notice that because of the λ^{-2} dependence of the Faraday effect, the use of a long wavelength (i.e., 1.3 or 1.55 μm) reduces the phase error by a factor of 3 to 4 compared to 0.85 μm for similar fiber defects.

As we have seen, polarization-preserving fibers provide a better reduction of Faraday nonreciprocity than ordinary fibers. However, it was shown that if an additional depolarizer is placed between the polarizer and the coil coupler in addition to the coil depolarizer, the Faraday nonreciprocity is also greatly reduced, even with an ordinary fiber coil [8].

7.2 NONLINEAR KERR EFFECT

Another important case of truly nonreciprocal effect may arise due to the nonlinear optical Kerr effect [9]. Reciprocity is indeed based on the linearity of propagation equation (see Section 3.1), but an imbalance in the power levels of the counter-propagating waves can produce a small nonreciprocal phase difference because of propagation nonlinearity induced by the high optical power density in the very small silica fiber core. Slow variations in the splitting ratio of the power divider feeding the sensing coil may therefore translate directly into bias drift. Experimentally, a power difference of 1 μW (e.g., arising from 10^{-3} splitting imbalance of a 1-mW source) gives a nonreciprocal index difference as small as 10^{-15}; but when integrated along a few hundred meters of fiber, this produces a phase difference of a few 10^{-5} rad, at least two orders of magnitude above the theoretical sensitivity limit. It could be reduced by simply reducing the power in the fiber, but this would increase the relative influence of detection noise.

The Kerr-induced rotation-rate error results in fact from a complex four-wave mixing process and not simply from an intensity self-dependence of the propagation constant of each counterpropagating wave. It depends also on the intensity of the opposite wave [9,10]. In a linear medium, the electric polarization vector **P** is defined as (see Appendix 1)

$$\mathbf{P} = \chi_e \epsilon_0 \mathbf{E} \tag{7.7}$$

but when the wave has a high energy density (i.e., a large **E** field), an additional nonlinear term depending on the third order susceptibility $\chi_e^{(3)}$ and the scalar square $|\mathbf{E}|^2$ of the electric field has to be taken into account and **P** becomes

$$\mathbf{P} = \chi_e \epsilon_o \mathbf{E} + \chi_e^{(3)} \epsilon_0 |\mathbf{E}|^2 \mathbf{E} \tag{7.8}$$

The relative dielectric permittivity $\epsilon_r = 1 + \chi_e$ is changed to

$$\epsilon_r + \delta\epsilon_r = 1 + \chi_e + \chi_e^{(3)} |\mathbf{E}|^2 \tag{7.9}$$

and the actual index of refraction $n = \sqrt{\epsilon_r}$ has an additional nonlinearity term δn_{NL}:

$$\delta n_{NL} = \frac{\chi_e^{(3)} |\mathbf{E}|^2}{2n} \tag{7.10}$$

In a ring interferometer where two fields $\mathbf{E}_1$ and $\mathbf{E}_2$ propagate in opposite directions, two polarization vectors $\mathbf{P}_1$ and $\mathbf{P}_2$ have to be considered for each propagation direction. The former relationship between the **P** and **E** vectors of a single wave applies, but now each counterpropagating wave cannot be considered independently. The total polarization vector $\mathbf{P}_1 + \mathbf{P}_2$ has to be related to the total field $\mathbf{E}_1 + \mathbf{E}_2$, and therefore

$$\mathbf{P}_1 + \mathbf{P}_2 = \chi_e \epsilon_0 (\mathbf{E}_1 + \mathbf{E}_2) + \chi_e^{(3)} \epsilon_0 |\mathbf{E}_1 + \mathbf{E}_2|^2 (\mathbf{E}_1 + \mathbf{E}_2) \tag{7.11}$$

A potential source of nonreciprocity comes from the $|\mathbf{E}_1 + \mathbf{E}_2|^2$ term, which represents the intensity of the standing wave resulting from the interference between both counterpropagating fields $\mathbf{E}_1$ and $\mathbf{E}_2$.

Assuming continuous monochromatic waves with the same linear state of polarization and the same frequency ω and opposite propagation constant β and $-\beta$, we have

$$\mathbf{E}_1 = \mathbf{E}_{10} e^{i(\omega t - \beta z)} \qquad \mathbf{E}_2 = \mathbf{E}_{20} e^{i(\omega t + \beta z)} \tag{7.12}$$

where z is the spatial longitudinal coordinate along the fiber coil. Since $|\mathbf{E}_1 + \mathbf{E}_2|^2 = |\mathbf{E}_1{}^2| + |\mathbf{E}_2|^2 + \mathbf{E}_1\mathbf{E}_2{}^* + \mathbf{E}_2\mathbf{E}_1{}^*$, this yields

$$|\mathbf{E}_1 + \mathbf{E}_2|^2(\mathbf{E}_1 + \mathbf{E}_2) = + (|\mathbf{E}_{10}|^2 + |\mathbf{E}_{20}|^2)\mathbf{E}_1 + (|\mathbf{E}_{10}|^2 + |\mathbf{E}_{20}|^2)\mathbf{E}_2 + (\mathbf{E}_{10}\mathbf{E}_{20}{}^*e^{-2i\beta z} + \mathbf{E}_{20}\mathbf{E}_{10}{}^*e^{2i\beta z})\mathbf{E}_{10}e^{i(\omega t - \beta z)} + (\mathbf{E}_{10}\mathbf{E}_{20}{}^*e^{-2i\beta z} + \mathbf{E}_{20}\mathbf{E}_{10}{}^*e^{2i\beta z})\mathbf{E}_{20}e^{i(\omega t + \beta z)} \quad (7.13)$$

The first two terms of this relationship depend both on the sum of the field squares (i.e., intensities) of the two waves and therefore yield the same nonlinear index change for $\mathbf{E}_1$ and $\mathbf{E}_2$ in each opposite direction. On the other hand, the two last terms induce a nonreciprocity, since

$$(\mathbf{E}_{10}\mathbf{E}_{20}{}^*e^{-2i\beta z} + \mathbf{E}_{20}\mathbf{E}_{10}{}^*e^{2i\beta z})\mathbf{E}_{10}e^{i(\omega t - \beta z)} = \mathbf{E}_{10}{}^2\mathbf{E}_{20}{}^*e^{i(\omega t - 3\beta z)} + \mathbf{E}_{20}|\mathbf{E}_{10}|^2e^{i(\omega t + \beta z)} = |\mathbf{E}_{10}|^2\mathbf{E}_2 + \mathbf{E}_{10}{}^2\mathbf{E}_{20}{}^*e^{i(\omega t - 3\beta z)} \quad (7.14)$$

and similarly,

$$(\mathbf{E}_{10}\mathbf{E}_{20}{}^*e^{-2i\beta z} + \mathbf{E}_{20}\mathbf{E}_{10}{}^*e^{2i\beta z})\mathbf{E}_{20}e^{i(\omega t + \beta z)} = |\mathbf{E}_{20}|^2\mathbf{E}_1 + \mathbf{E}_{20}{}^2\mathbf{E}_{10}{}^*e^{i(\omega t - 3\beta z)} \quad (7.15)$$

The effect of the terms at a spatial frequency of 3β or -3β is averaged out in the propagation, but the two other terms at β and $-\beta$ are phase-matched and yield a constant susceptibility change as the waves propagate. Each polarization vector is actually

$$\mathbf{P}_1 = \chi_e\epsilon_0\mathbf{E}_1 + \chi_e^{(3)}\epsilon_0(|\mathbf{E}_1|^2 + 2|\mathbf{E}_2|^2)\mathbf{E}_1$$
$$\mathbf{P}_2 = \chi_e\epsilon_0\mathbf{E}_2 + \chi_e^{(3)}\epsilon_0(2|\mathbf{E}_1|^2 + |\mathbf{E}_2|^2)\mathbf{E}_2 \quad (7.16)$$

This gives a different nonlinear change of the index of refraction for each opposite direction:

$$\delta n_{NL1} = \frac{\chi_e^{(3)}(|\mathbf{E}_1|^2 + 2|\mathbf{E}_2|^2)}{2n}$$
$$\delta n_{NL2} = \frac{\chi_e^{(3)}(|\mathbf{E}_2|^2 + 2|\mathbf{E}_1|^2)}{2n} \quad (7.17)$$

and a nonreciprocal index difference:

$$\Delta n_K = \delta n_{NL1} - \delta n_{NL2} = \frac{\chi_e^{(3)}(|\mathbf{E}_2|^2 - |\mathbf{E}_1|^2)}{2n} \quad (7.18)$$

Assuming a uniform intensity distribution over the core area and a diameter of about 5 μm, this Kerr-induced index difference may be evaluated from the value of $\chi_e^{(3)}$ in silica as a function of the power difference ΔP (proportional to $|\mathbf{E}_2|^2 - |\mathbf{E}_1|^2$) between both directions as [9]:

$$\Delta n_K/\Delta P \approx 2 \times 10^{-15}\ \mu\text{W}^{-1}$$

This difference is very small, but as the Sagnac effect, it is integrated along the whole length L of the fiber coil and gives rise to a significant phase difference $\Delta\phi_K$. For a wavelength of 0.633 μm [9]:

$$\Delta\phi_K/L \cdot \Delta P \approx 2 \times 10^{-5}\ \text{rad} \cdot \mu\text{W}^{-1} \cdot \text{km}^{-1} \tag{7.19}$$

This analysis shows that the Kerr-effect nonreciprocity results solely from the formation of a nonlinear index grating, due to the interference between the two counterpropagating waves within the fiber, which yields a standing wave. As stated early on in [10], if the contrast of this standing wave is washed out by some process, then nonreciprocity should decrease. This important point explains why the use of a broadband source with a short coherence length greatly reduces the Kerr nonreciprocity: the standing wave is contrasted only over a distance equal to the coherence length L_c in the middle of the fiber coil (Figure 7.3), and therefore the effect of the nonreciprocal index difference is integrated only along L_c and not along the whole fiber length L!

The cancellation of the Kerr nonreciprocity with a broadband source was originally explained with the statistics of the light intensity fluctuation [11,12]. As a matter of fact, this original explanation considered the case of an intensity modulated wave which yields a nonlinear index perturbation that depends on time t and position z in the fiber:

$$\begin{aligned} \delta n_{NL1} &\propto [I_1(z,t) + 2I_2(z,t)] \\ \delta n_{NL2} &\propto [I_2(z,t) + 2I_1(z,t)] \end{aligned} \tag{7.20}$$

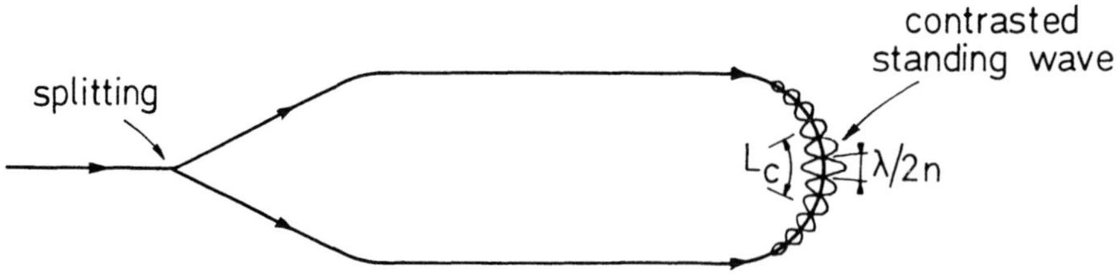

Figure 7.3 Contrasted standing wave in the middle of the fiber coil.

An important feature of these equations, as we have already seen, is that the cross-effect of the power of one wave is twice its self-effect. The use a square-wave intensity modulation of a monochromatic source was first proposed to reduce the Kerr nonreciprocity [13]. In this case, the crossed effect is present only when both counterpropagating intensities are coincident (Figure 7.4) (i.e., half the time), while the self-effect is present all the time. Therefore, the factor of 2 of the cross-effect is reduced to an averaged value of unity, which effectively cancels out the nonreciprocity, since the mean phase perturbation becomes identical in both directions.

Such a kind of compensation is not restricted to square waves, and it applies if the mean value $<I>$ of the modulated intensity is equal to its standard deviation $\sigma_I = \sqrt{<I^2> - <I>^2}$ [11,12]. By virtue of the central limit theorem, a polarized broadband source has a random intensity with an exponential probability distribution:

$$p(I) = \begin{cases} \dfrac{1}{<I>} e^{-I/<I>} & \text{if } I \geq 0 \\ 0 & \text{if } I < 0 \end{cases} \tag{7.21}$$

and this fulfills the requirement $<I> = \sigma_I$, which ensures the absence of Kerr-induced nonreciprocity.

This last explanation, though, gives the misleading feeling that the reduction of Kerr-effect nonreciprocity with a broadband source that "happens" to have the adequate statistics is only very "fortunate." On the other hand, the direct analysis of the nonlinear process shows clearly that it is related to an interference phenomenon of standing wave and that a low-coherence source reduces the parasitic effect in the same way that is already very beneficial with backreflection, backscattering,

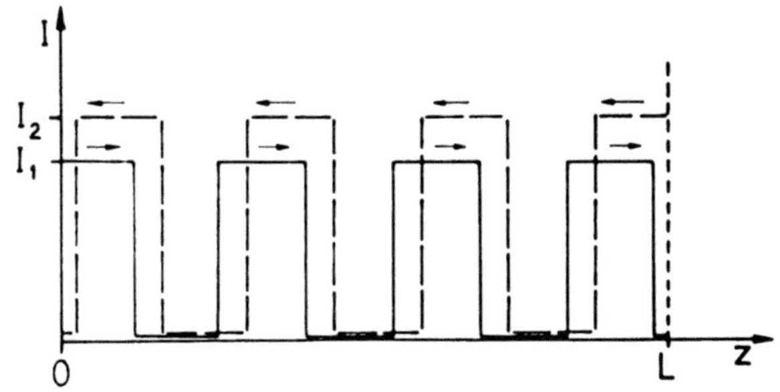

Figure 7.4 Square-wave intensity modulation of the counterpropagating waves to reduce Kerr-induced nonreciprocity.

or polarization nonreciprocities, which also depend on parasitic interference phenomena.

However, the similarity in terms of coherence between this nonlinear effect and the other coherence-related linear effects is limited to the use of broadband souce with continuous light emission that destroys the contrast of the standing wave, but ensures that both counterpropagating intensities are permanent in the fiber. A very short pulse can also limit the coherent effect of backreflection, backscattering, and polarization nonreciprocities, but for the nonlinearity problem each counterpropagating pulse would experience mostly the self-effect, which would yield a nonreciprocity with a power imbalance. Furthermore, for the same mean power the nonlinearity would be further increased, since it depends on the peak power, which is much higher in the case of a pulse.

Note that it would be interesting to study the effect of an additional phase modulation, particularly in the middle of the loop, to see if it could also be possible with this mean to reduce the contrast of the standing wave and the related Kerr nonreciprocity despite a high-coherence source.

REFERENCES

[1] Arditty, H. J., Y. Bourbin, M. Papuchon, and C. Puech, "Current Sensor Using State-of-the-Art Fiber-Optic Interferometric Techniques," Proceedings of IOOC, Paper WL3, 1981.

[2] Böhm, K., K. Petermann, and E. Weidel, "Sensitivity of a Fiber-Gyroscope to Environmental Magnetic Fields," Optics Letters, Vol. 7, 1982, pp. 180–182 (SPIE MS 8, pp. 328–330).

[3] Schiffner, G., B. Nottbeck, and G. Schröner, "Fiber-Optic Rotation Sensor: Analysis of Effects Limiting Sensitivity and Accuracy," Springer Series in Optical Sciences, Vol. 32, 1982, pp. 266–274.

[4] Bergh, R. A., H. C. Lefèvre, and H. J. Shaw, "All Single-Mode Fiber Optic Gyroscope," Springer Series in Optical Sciences, Vol. 32, 1982, pp. 252–255.

[5] Bergh, R. A., H. C. Lefèvre, and H. J. Shaw, "Geometrical Fiber Configuration for Isolators And Magnetometers," Springer Series in Optical Sciences, Vol. 32, 1982, pp. 400–405.

[6] Hotate, K., and K. Tabe, "Drift of an Optical Fiber Gyroscope Caused by the Faraday Effect: Influence of the Earth's Magnetic Field," Applied Optics, Vol. 25, 1986, pp. 1086–1092 (SPIE MS 8, pp. 331–337).

[7] Marrone, M. J., C. A. Villaruel, N. J. Frigo, and A. Dandridge, "Internal Rotation of the Birefringence Axes in Polarization-Holding Fibers," Optics Letters, Vol. 12, 1987, pp. 60–62.

[8] Blake, J., "Magnetic Field Sensitivity of Depolarized Fiber Optic Gyros," SPIE Proceedings, Vol. 1367, 1990, pp. 81–86.

[9] Ezekiel, S., J. L. Davis, and R. W. Hellwarth, "Intensity Dependent Nonreciprocal Phase Shift in a Fiberoptic Gyroscope," Springer Series in Optical Sciences, Vol. 32, 1982, pp. 332–336 (SPIE MS 8, pp. 308–312).

[10] Kaplan, A. E., and P. Meystre, "Large Enhancement of the Sagnac Effect in a Nonlinear Ring Resonator and Related Effects," Springer Series in Optical Sciences, Vol. 32, 1982, pp. 375–385.

[11] Bergh, R. A., B. Culshaw, C. C. Cutler, H. C. Lefèvre, and H. J. Shaw, "Source Statistics and the Kerr Effect in Fiber-Optic Gyroscopes," Optics Letters, Vol. 7, 1982, pp. 563–565 (SPIE MS 8, pp. 313–315).

[12] Petermann, K., "Intensity-Dependent Nonreciprocal Phase Shift in Fiber-Optic Gyroscopes for Light Sources With Low Coherence," Optics Letters, Vol. 7, 1982, pp. 623–625 (SPIE MS 8, pp. 322–323).

[13] Bergh, R. A., H. C. Lefèvre, and H. J. Shaw, "Compensation of the Optical Kerr Effect in Fiber-Optic Gyroscopes," Optics Letters, Vol. 7, 1982, pp. 282–284 (SPIE MS 8, pp. 316–318).

Chapter 8

Scale Factor Accuracy

8.1 PROBLEM OF SCALE FACTOR IN THE INTERFEROMETRIC FIBER GYROSCOPE

The modulation-demodulation detection scheme described in Section 3.2.2 provides a very good bias, since it preserves the reciprocity of the ring interferometer. However, if a high-performance gyroscope must have a stable and low-noise bias, it also requires good accuracy over the whole dynamic range and not only about zero. The measurement of interest is the integrated angle of rotation and not simply the rate. Any past error will affect the future information. This constraint implies the need for an accurate measurement at any rate (i.e., an accurate scale factor). Furthermore, the intrinsic response of an interferometer is sinusoidal, while the desired rate signal of a gyroscope should be linear.

This problem is solved with a closed-loop (or phase-nulling) signal processing approach [1,2]. The demodulated biased signal (or open-loop signal) is used as an error signal that is fed back into the system to generate an additional feedback phase difference $\Delta\phi_{FB}$ that is opposite to the rotation-induced phase difference $\Delta\phi_R$ (Figure 8.1). The total phase difference $\Delta\phi_T = \Delta\phi_R + \Delta\phi_{FB}$ is servo-controlled on zero, which provides good sensitivity, since the system is always operated about a high-slope point. With such a closed-loop scheme, the new measurement signal is the value of $\Delta\phi_{FB}$. This yields a linear response with good stability, since this feedback value $\Delta\phi_{FB}$ is independent of the returning optical power and of the gain of the detection chain. The measured value of the rotation rate becomes

$$\Omega = -\frac{\lambda c}{2\pi LD}\Delta\phi_{FB} \tag{8.1}$$

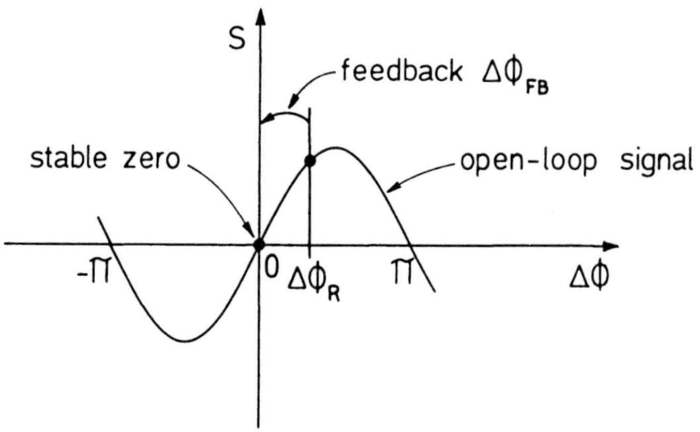

Figure 8.1 Principle of closed-loop operation of an interferometric fiber gyro with a feedback phase difference $\Delta\phi_{FB}$.

A closed-loop operation provides a stable and linear measurement of the optical phase difference in the interferometer, but the accuracy of the scale factor of the rotation rate measurement also depends on the stability of the source wavelength λ and on the geometrical stability of the coil through the product length $\times$ diameter (LD).

8.2 CLOSED-LOOP OPERATION

8.2.1 Use of a Frequency Shift

The first proposed arrangements of closed-loop operation used a frequency shift generated with acousto-optic modulators (AOM), also called Bragg cells [1,2]. As a matter of fact, the Sagnac effect may be interpreted as a Doppler effect on the coil beam splitter (see Section 2.1.1). Therefore, a frequency shifter placed at the beginning of the coil may null out the Doppler shift of the Sagnac effect. The absolute phase ϕ_{abs} accumulated in the propagation is

$$\phi_{abs} = \frac{2\pi nL}{\lambda} = \frac{2\pi nL}{c} \cdot f \tag{8.2}$$

where f is the temporal frequency of the wave. Then a feedback frequency shift Δf_{FB} generates a feedback phase difference $\Delta\phi_{FB}$:

$$\Delta\phi_{FB} = \frac{2\pi nL}{c} \cdot \Delta f_{FB} \tag{8.3}$$

The new measurement is now Δf_{FB}, which is linearly dependent on the rotation rate Ω:

$$\Delta f_{FB} = -\frac{D}{n\lambda}\Omega \tag{8.4}$$

To get high sensitivity, the error signal is obtained with an additional phase modulator and the usual modulation-demodulation scheme of the open-loop scheme [3].

However, to cover the whole dynamic range, this feedback frequency shift $\Delta\phi_{FB}$ has to vary around zero between plus and minus several hundred kilohertz. Since AOMs work only at a high center frequency (typically 50 to 100 MHz) with a relative bandwidth of several percent, this requires the use of two cells to generate the feedback frequency shift $\Delta\phi_{FB}$ by difference. They can be placed in opposition at one end of the coil, the first cell generating an upshift and the second one a downshift (Figure 8.2(a)), or they can be placed at both ends of the coil with the same shift sign (Figure 8.2(b)). One is operated at a constant center frequency f_c, yielding a constant wavelength shift $\Delta\lambda_c$, and the other is operated at a controlled frequency $f_c + \Delta f_{FB}$, yielding a wavelength shift $\Delta\lambda_c + \Delta\lambda_{FB}$. When the gyro is at rest, the feedback frequency shift Δf_{FB} should ideally be zero, but the intrinsic single-mode reciprocity has been destroyed [4]. Considering first the symmetrical case (Figure 8.2(b)), the source emits a wavelength λ, but the actual wavelength of the counterpropagating waves is:

- λ between the splitter and the modulators at the input;
- $\lambda + \Delta\lambda_c$ between both modulators;
- $\lambda + 2\Delta\lambda_c$ between the modulators and the splitter at the output.

Both counterpropagating waves preserve the same wavelength $\lambda + \Delta\lambda_c$ between both modulators, but they have a difference of $2\Delta\lambda_c$ between the splitter and the modulators. This wavelength difference between the opposite waves yields a nonreciprocal phase difference $\Delta\phi_1$ when they propagate between the splitter and the first modulator, and another nonreciprocal phase difference $\Delta\phi_2$ with an opposite sign between the splitter and the second modulator. When the distances between the splitter and each modulator are perfectly equal, the total nonreciprocal phase difference $\Delta\phi_1 + \Delta\phi_2$ is nulled out, but a small imbalance ΔL_{AOM} in the exact symmetry of the two modulators yields a residual phase error:

$$\Delta\phi_e = \Delta\phi_1 + \Delta\phi_2 = \frac{2\pi f_c}{c}\Delta L_{\mathrm{AOM}} \tag{8.5}$$

With a typical center frequency f_c of 100 MHz, this nonreciprocal phase error is as high as 2 μrad/μm.

When the two modulators are placed in opposition at one end of the coil (Figure 8.2(a)), a similar result is obtained where ΔL_{AOM} is now the distance

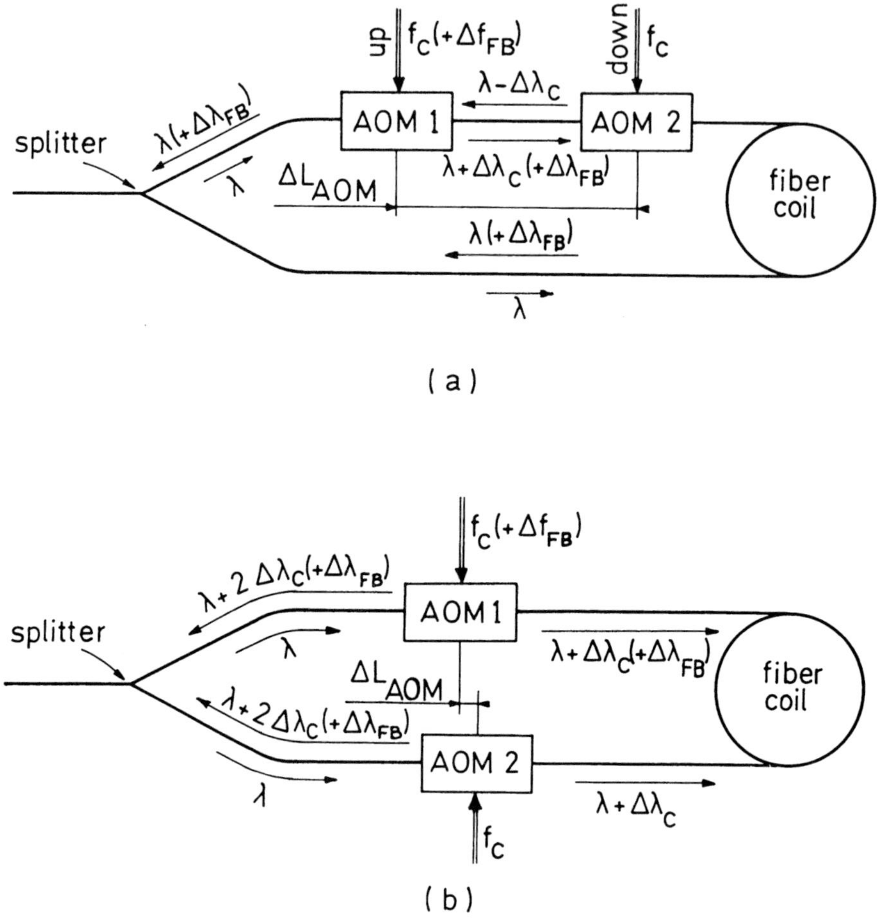

Figure 8.2 Closed-loop scheme with two acousto-optic frequency modulators (AOM): (a) modulators in series; (b) modulators at both ends of the coil.

between both modulators. There is the same problem of stability, but there is also a constant bias offset related to the averaged value of ΔL_{AOM}.

Several solutions have been proposed to improve the mechanical stability of this assembly, including a double Bragg cell where both frequency shifters are combined in the same component [5] and an integrated optic implementation [6], but this closed-loop approach with acousto-optic frequency shifters is complex technologically and the power consumption is relatively high. Furthermore, even with an improved stability, these approaches do not respect reciprocity. Note that this sensitivity to modulator imbalance has been proposed as a sensing method to detect any effect that affects this imbalance [7].

8.2.2 Analog Phase Ramp or Serrodyne Modulation

This stability problem of the acousto-optic frequency shifter is now completely overcome by the use of a linear phase ramp [8–11]. A frequency is the derivative of a phase, and a phase ramp modulation $\phi_{PR}(t) = \dot{\phi}t$ (where $\dot{\phi}$ is the slope) that is applied with a phase modulator instead of a frequency shifter is equivalent to a frequency shift $\Delta f = \dot{\phi}/2\pi$

$$\sin[2\pi ft + \dot{\phi}t] = \sin[2\pi(f + (\dot{\phi}/2\pi))t] \tag{8.6}$$

Such a processing scheme, also called serrodyne modulation, allows one to work positively or negatively about zero, depending on the sign of the ramp slope, thereby eliminating the former need of a high center frequency, which has been shown to destroy the intrinsic reciprocity of the ring interferometer. However, the ramp cannot be infinite, and, in practice, a sawtooth modulation form has to be used with a very fast flyback at the reset (Figure 8.3). This requires a phase modulator with a flat efficiency over a large bandwidth, which is one of the main technical advantages of integrated optics (see Appendix 3).

The effect on a gyro can be viewed directly by considering the group delay difference $\Delta\tau_g$ between the long and short paths that connect the phase modulator and the splitter. Like the case of the biasing modulation (see Section 3.2.2), the same phase ramp feedback modulation $\phi_{PR}(t)$ is applied on the two opposite waves; but because of the delay $\Delta\tau_g$, this generates a feedback phase difference $\Delta\phi_{PR}(t)$ with

$$\Delta\phi_{PR}(t) = \phi_{PR}(t) - \phi_{PR}(t - \Delta\tau_g) \tag{8.7}$$

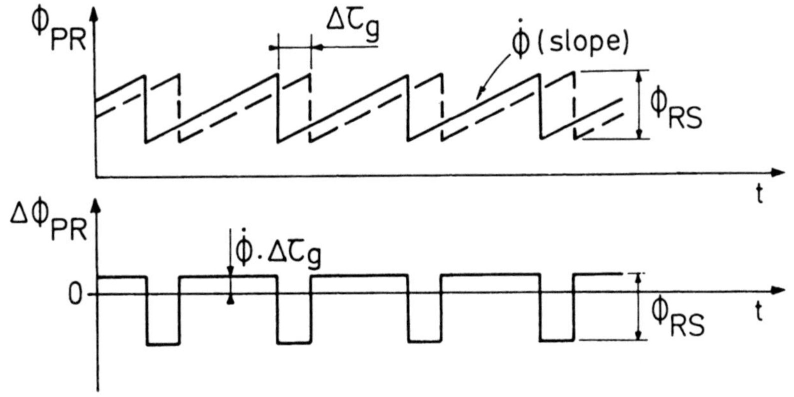

Figure 8.3 Analog sawtooth phase ramp modulation ϕ_{PR} and induced feedback phase difference $\Delta\phi_{PR}$.

It is equal to $\dot{\phi} \cdot \Delta\tau_g$ during the ramp and to $\dot{\phi} \cdot \Delta\tau_g - \phi_{RS}$ during a time $\Delta\tau_g$ after the flyback. The value ϕ_{RS} is the height of the phase reset. This reset induces an error unless ϕ_{RS} is equal to 2π rad (or a multiple of 2π), the period of the interferometer response [9,10].

Let us assume that when the processing loop is closed, the slope $\dot{\phi}$ is adjusted to compensate for the rotation-induced phase difference $\Delta\phi_R$ and that the total phase difference $\Delta\phi_T$ is nulled out:

$$\Delta\phi_T = \Delta\phi_R + \Delta\phi_{PR} = 0 \tag{8.8}$$

that is:

$$\Delta\phi_R = -\dot{\phi} \cdot \Delta\tau_g \tag{8.9}$$

After the reset, $\Delta\phi_T$ becomes equal to ϕ_{RS} instead of zero. Assuming, for simplicity, a biased sine response of the interferometer (Figure 8.4), the signal is zero when $\Delta\phi_T = 0$, but becomes $\sin\phi_{RS}$ during the time $\Delta\tau_g$ after each reset. the gating out of this spurious signal has been proposed [8], but it has been shown that this is actually a very convenient error signal for checking the phase modulator efficiency with a second feedback loop activated at each reset [9].

With such a second processing loop, the reset is precisely controlled on 2π rad, and therefore the counting of the positive and negative resets provides an

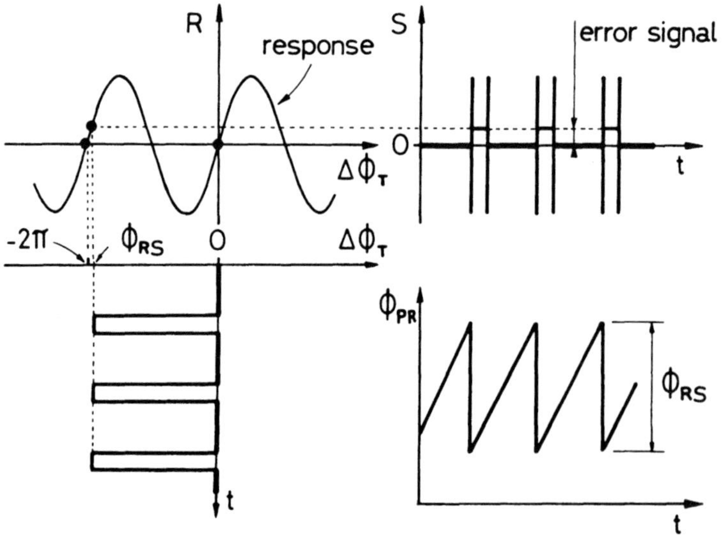

Figure 8.4 Effect of the reset of the phase ramp.

accurate measurement of the angle of rotation. The slope $\dot{\phi}$ is proportional to the rate Ω, since $\Delta\phi_R = -\dot{\phi}\Delta\tau$:

$$\dot{\phi} = -\frac{2\pi LD}{\lambda c \Delta\tau_g}\Omega \tag{8.10}$$

Because $\Delta\tau_g = nL/c$, we have

$$\dot{\phi} = -\frac{2\pi D}{n\lambda} \cdot \Omega \tag{8.11}$$

Each reset corresponds precisely to 2π rad, and also to an angular increment θ_{inc}, defined by

$$\theta_{inc} = \int_0^T \Omega dt = -\frac{n\lambda}{2\pi D}\int_0^T \dot{\phi} dt \tag{8.12}$$

where T is the period of the sawtooth modulation, and since $\int_o^T \dot{\phi} dt = 2\pi$ for any slope, we have

$$\theta_{inc} = \frac{n\lambda}{D} \tag{8.13}$$

This incremental value is independent of the coil length L, but it is inversely proportional to the coil diameter D. For a wavelength $\lambda = 850$ nm, the product $\theta_{inc} \cdot D$ is equal to 25 arcsec · cm. Note that the frequency of these "2π-increments" is the same as the one obtained with the direct frequency shift feedback and as the one naturally generated in a laser gyro (see Section 2.2.1).

An important feature of this phase ramp processing technique is that the control of the 2π reset does not have to be very accurate, since it has only a third-order dependence on the actual scale factor accuracy [10]. Let us consider (Figure 8.5) that, starting from an ideal phase ramp, there is a small decreasing ϵ_m in the gain of the modulation chain, the actual modulation ϕ_{PRa} being related to the ideal modulation ϕ_{PRi} by

$$\phi_{PRa} = (1 - \epsilon_m)\phi_{PRi} \tag{8.14}$$

and the induced differences by

$$\Delta\phi_{PRa} = (1 - \epsilon_m)\Delta\phi_{PRi} \tag{8.15}$$

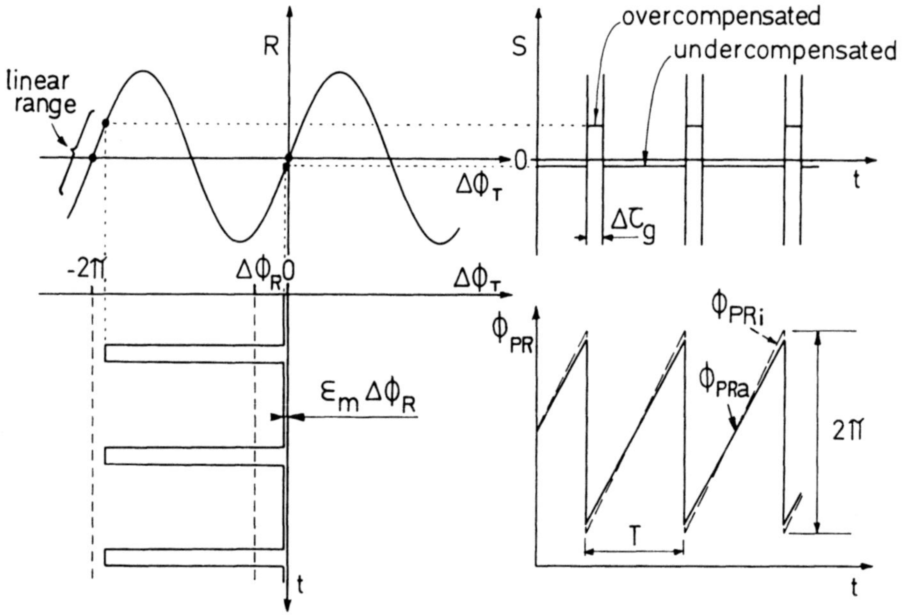

Figure 8.5 Effect of a change of the ideal gain of the modulation chain.

The actual phase ramp now undercompensates for the rotation rate-induced phase difference $\Delta\phi_R$ and the total phase difference becomes

$$\Delta\phi_{Ta} = \Delta\phi_R + \Delta\phi_{PRa} = \epsilon_m \Delta\phi_R \tag{8.16}$$

However, if the ramp slope is decreased, the reset height is also decreased with the same coefficient ϵ_m, and during a time $\Delta\tau_g$ after the reset, the actual phase ramp overcompensates $\Delta\phi_R$:

$$\Delta\phi_{Ta} = 2\pi - \epsilon_m(2\pi - \Delta\phi_R) \tag{8.17}$$

Therefore, over a full period T of the sawtooth modulation, and assuming a biased sine response of the open-loop error signal, the signal is

$$\begin{array}{ll} \sin[\epsilon_m \Delta\phi_R] & \text{during } T - \Delta\tau_g \\ -\sin[\epsilon_m(2\pi - \Delta\phi_R)] & \text{during } \Delta\tau_g \end{array} \tag{8.18}$$

If these phase errors are small, the sine response may be linearized (ignoring third-order terms) and the mean signal becomes

$$<S> = \frac{\epsilon_m \Delta\phi_R (T - \Delta\tau_g) - \epsilon_m (2\pi - \Delta\phi_R)\Delta\tau_g}{T} \tag{8.19}$$

but since

$$\Delta\phi_R = -\dot{\phi}\Delta\tau_g \tag{8.20}$$

$$\Delta\phi_R \int_o^T dt = -\Delta\tau_g \int_o^T \dot{\phi} dt \tag{8.21}$$

$$\Delta\phi_R \cdot T = 2\pi\Delta\tau_g \tag{8.22}$$

the mean error signal $<S>$ remains equal to zero independently of the exact value of the reset if ϵ_m is small enough to keep the operating points in the linear parts of the sine response.

This analog phase ramp feedback scheme looks very attractive, but it requires a very short and stable flyback time to yield high scale factor stability and linearity [10,12]. Typically, a stability of 10 ppm requires a flyback time of less than 1% of the transit time through the coil (i.e., less than a few tens of nanoseconds).

8.2.3 Digital Phase Ramp

This problem of analog phase ramp flyback is very simply solved with a digital approach [4,9,13]. Instead of a continuous ramp, the "digital phase ramp" generates phase steps ϕ_s with a duration equal to $\Delta\tau_g$. Because of the delay through the coil, the induced phase difference $\Delta\phi_{FB}$ is constant and equal to the step value ϕ_s. These phase steps and the resets can be synchronized with a square-wave biasing modulation (see Section 3.2.2), which is preferably operated at the proper frequency: with a half-period equal to this same transit time $\Delta\tau_g$ (Figure 8.6). The amplitude ϕ_s of the phase step is set by the phase-nulling feedback loop to be opposite to the rotation-induced phase difference $\Delta\phi_R$:

$$\phi_s = -\Delta\phi_R$$

and this value ϕ_s provides a linearized readout of the rate.

The real "magic" of the digital phase ramp technique is the fact that the use of digital logic and a D/A (digital/analog) converter naturally yields an adequate synchronized reset with the automatic overflow of the converter for any value of

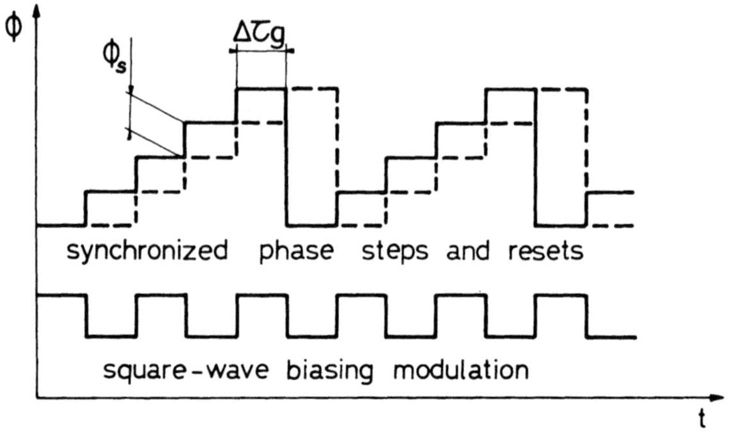

Figure 8.6 Digital phase ramp with steps and resets synchronized with the square-wave biasing modulation.

the step, thus making implementation of this powerful technique very simple (Figure 8.7). A digital register contains the digital value D_s of the phase step with a dynamic range that can be very high (more than 25 bits), and a digital integrator generates the digital value D_R of the staircase ramp. The analog driving voltage of the phase modulator is produced with a D/A converter and a buffer amplifier. With N bits, they transform a digital number D into an analog voltage over a dynamic range between zero and $(2^N - 1)V_{LSB}$, where V_{LSB} is the driving voltage corresponding to the least significant bit (LSB). When D_R becomes higher than $(2^N - 1)$, the automatic overflow yields a voltage equal to $(D_R - 2^N)V_{LSB}$. If the gain of the modulation chain is adjusted such that

$$2^N V_{LSB} = 2V_\pi$$

where V_π is the voltage that generates a π rad phase shift, as detailed in Appendix 3. The overflow automatically generates a reset that is equivalent to the 2π reset of the analog ramp and does not produce any scale-factor error. Note that this automatic overflow may be used with the sole ramp, but also with the digital sum of the ramp and the square-wave modulation (Figure 8.8). This allows the use of a push-pull connection of the two modulators of the Y-junction, which reduces their global nonlinearity (see Section 3.3.4). With the analog approach, the electronic circuitry that generates the sawtooth modulation is so "delicate" that it is preferable to apply this feedback modulation on one modulator and apply the biasing modulation on the second modulator with an independent circuit, which

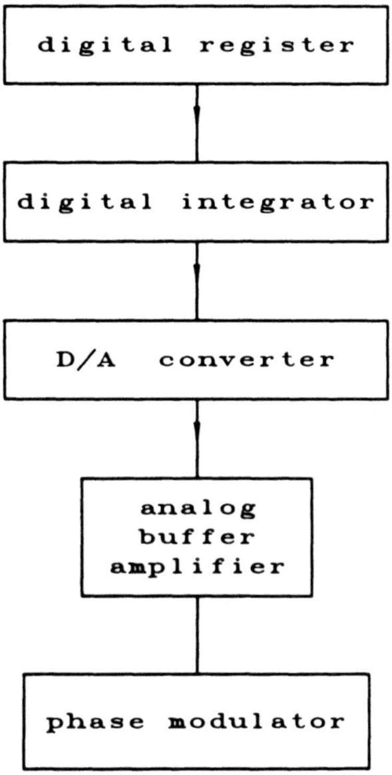

Figure 8.7 Generation of the digital phase ramp.

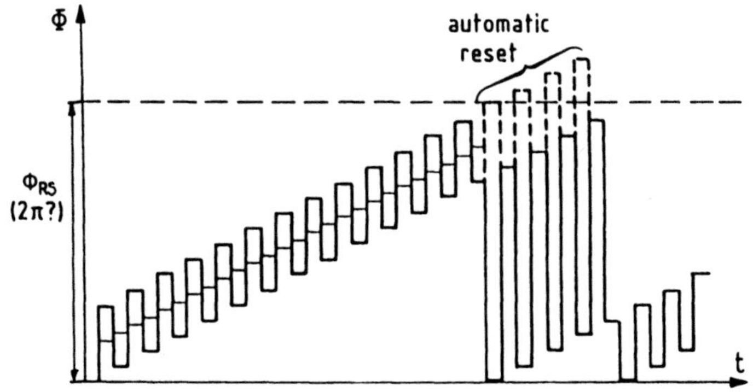

Figure 8.8 D/A overflow when the phase ramp and the phase modulation are added digitally.

makes the setup more complicated. The resets, as the steps, are synchronized with the clock time $\Delta\tau_g$. This eases the control of the exact 2π value of $2^N V_{LSB}$ with a second feedback loop that is activated at each flyback and is not disturbed by the transients of square-wave biasing modulation, since it is also synchronized with $\Delta\tau_g$.

Let us take, for example, the case of a rotation-induced phase difference $\Delta\phi_R$ equal to $-4\pi/5$ rad (Figure 8.9). With an analog ramp, the resets are equal to 2π and have a periodicity of $2.5\Delta\tau_g$, but they overlap with the biasing modulation that has a periodicity of $2\Delta\tau_g$. On the other hand, with the digital phase ramp, the step value is $4\pi/5$ rad and the first reset "waits" for the third clock time $\Delta\tau_g$ while the second reset happens only after $2\Delta\tau_g$. These resets are no longer periodic, but they

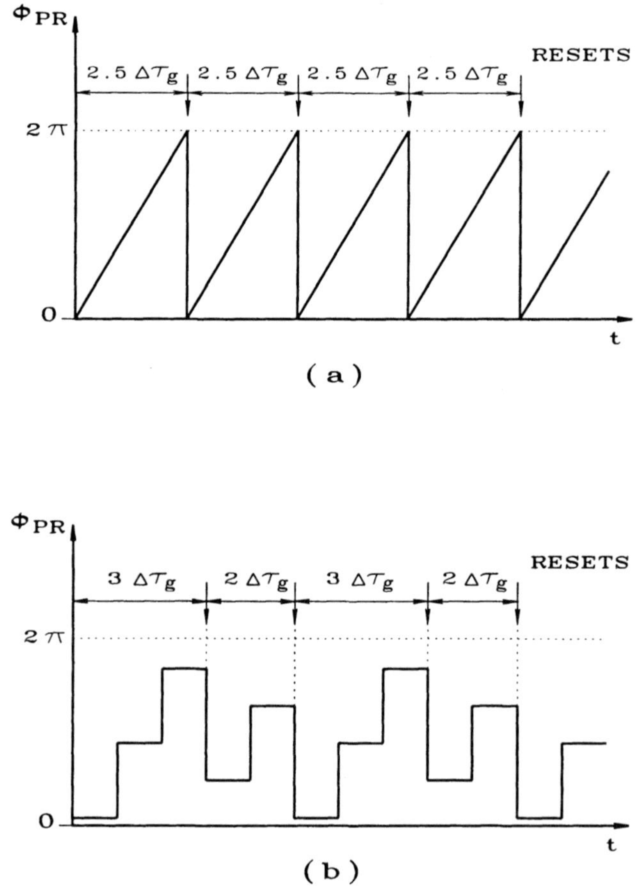

Figure 8.9 Nulling phase ramp for a rotation-induced phase difference of $-4\pi/5$ rad: (a) analog case; (b) digital case.

are now synchronized with the clock time $\Delta\tau_g$ and the biasing modulation. Note that, strictly speaking, the resets of the digital ramp are not equal to 2π. 2π is the difference between the value of where the next step would go if the number of bits were not limited and the value of where it is actually going because of the overflow.

Another very important point is that the digital phase ramp does not require a large number of bits for the D/A converter, though the actual dynamic range between 2π rad and a resolution of 10^{-7} rad is as high as 26 bits! With an N-bit converter, the least significant phase step ϕ_{LSB} is

$$\phi_{LSB} = \frac{2\pi}{2^N} \tag{8.23}$$

The exact value of the required phase step ϕ_s is contained within

$$m\phi_{LSB} \leq \phi_s < (m + 1)\phi_{LSB}$$

where m is an integer smaller that 2^N. The value ϕ_s is stored in the rate register, which must have a large enough number of bits, but only the N most significant bits of the ramp are used in the D/A to drive the phase modulator. Instead of a series of identical steps, the D/A converter generates a ramp composed of m' steps $m\phi_{LSB}$, which undercompensates for the rotation, and m'' steps $(m + 1)\phi_{LSB}$, which overcompensates for the rotation. On average, the feedback phase difference $<\Delta\phi_{FB}>$ is such as

$$<\Delta\phi_{PR}> = \frac{m'm\phi_{LSB} + m''(m + 1)\phi_{LSB}}{m' + m''} \tag{8.24}$$

that is, m' and m'' are such as

$$\phi_s = m\phi_{LSB}\left(1 + \frac{m''}{m' + m''}\right) \tag{8.25}$$

This phase step averaging yields the same averaging of the actual interference signal if the maximum instantaneous error (i.e., ϕ_{LSB}) remains in the linear part of the sine response. With 10 bits, for example, ϕ_{LSB} is equal to $2\pi/2^{10} = 6 \times 10^{-3}$ rad, which corresponds to a rate as high as several thousands of degrees per hour; but the residual nonlinearity of the sine response, $(\phi_{LSB} - \sin\phi_{LSB})/\phi_{LSB} \approx \phi_{LSB}^2/6$, remains below 10 ppm. The averaging also applies if ϕ_s is smaller than ϕ_{LSB}: there is no dead zone. For example, if $\phi_s = [1/(m' + 1)]\phi_{LSB}$, there will be no step during m' clock times and just one step during one clock time (Figure 8.10).

Another averaging process is also very useful for relaxing the requirement of linearity of the D/A converter. A converter defect generates phase errors, and

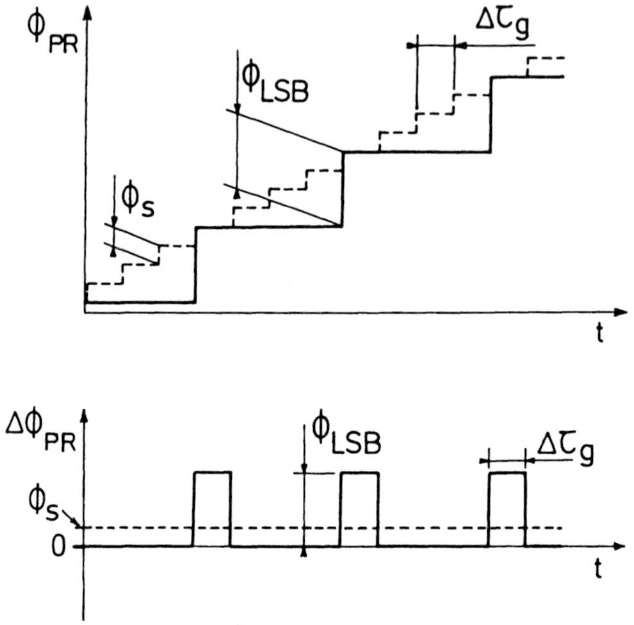

Figure 8.10 Actual phase ramp when the required phase step ϕ_s is smaller than ϕ_{LSB} (case where $\phi_s = \phi_{LSB}/4$).

the value of the phase ramp ϕ_{sj} after j steps may be different from $j\phi_s$; but on average the feedback phase difference is over m steps:

$$<\Delta\phi_{FB}> = [(\phi_{sm} - \phi_{s(m-1)}) + (\phi_{s(m-1)} + \phi_{s(m-2)}) + \ldots + (\phi_{s2} - \phi_{s1}) + (\phi_{s1} - 0)]/m \quad (8.26)$$

$$<\Delta\phi_{FB}> = \frac{\phi_{sm}}{m}$$

The only condition is, therefore, that the linearity error of the converter remain in the linear part of the interferometer sine response. The usual specification of D/A converters is a linearity error of less than one LSB; therefore, the previous condition about ϕ_{LSB} applies also for this second averaging process.

Besides these very useful relaxed constraints on the characteristics of the converter, the digital phase ramp technique also has several basic advantages over its analog counterpart. The true rate measurement is the digital value D_s of the phase step ϕ_s, which is stored in a register of the digital logics. The clock time driving the electronics has to be approximatively matched to the transit time $\Delta\tau_g$

through the coil in order to limit the width of the transient pulses, but the step value is not directly related to $\Delta\tau_g$. When $\Delta\tau_g$ is changing, this slightly modifies the width of the transient spikes that are gated out, but the value of the feedback step ϕ_s remains unchanged. This is a clear advantage over the analog approach, in which $\Delta\tau_g$ is part of the scale factor through the value of the index n. With the digital ramp and a stable electronic clock, the scale factor has only the basic dependence of the Sagnac effect on the geometrical length of the coil (i.e., in particular, a temperature coefficient of 10^{-6}/K with silica fiber), instead of an index dependence, which has a temperature coefficient of 10^{-5}/K.

Furthermore, the usual readout of these ramping techniques is the count of the 2π resets, which yields, as we have already seen, an angular increment value $\theta_{inc} = n\lambda/D$. With the digital approach, in which the rate is stored in a register, it is easy to generate any submultiple value of θ_{inc} without "waiting" for the 2π resets and particularly $\theta_{inc}/2^{N'}$ using N' bits in parallel for the output. This is very important for stabilization and pointing applications where very small increments are required. It is even possible to tailor the value of the increment to avoid any quantification noise within the measurement bandwidth. Assuming, for example, a typical noise of 1 (deg/h)/$\sqrt{\text{Hz}}$ = 1 (arcsec/s)/$\sqrt{\text{Hz}}$ (i.e., a random walk of 1 arcsec/$\sqrt{\text{s}}$ = 0.03 arsec/$\sqrt{\text{ms}}$), the increment may be adjusted to 0.03 arcsec, while θ_{inc} is on the order of few arcseconds to avoid any quantification noise within a 1-kHz bandwidth.

The value of the rate is stored in a digital register and can be used directly with a parallel interface, but this would require a very large number of bits. The I-FOG is intrinsically a rate gyroscope. To generate 2π-increments or subincrements with a logic integration of the rate value simplifies the interface and yields a rate-integrating gyroscope; but the difference between a rate device and a rate integrating device is simply a matter of signal processing logic and interface protocol and is not a basic difference.

Note that the digital ramp approach allows the dynamic range to be easily extended to several fringes. The phase step value is stored in a register that may correspond to more than $\pm\pi$ rad, and the overflow of the D/A converter automatically limits the range of actual phase modulation to less than 2π.

This analysis demonstrates that the digital phase ramp technique is a highly efficient closed-loop method for linearizing and stabilizing the scale factor of an interferometric fiber gyroscope. It has fundamental advantages and does not require stringent performance for the electronic components.

8.2.4 All-Digital Closed-Loop Processing Method

The first implementation of the digital phase ramp technique used an analog demodulator to get the sinusoidal rate signal that serves as the feedback error signal of closed-loop processing (Figure 8.11) [4,13]. Such a method, however, is difficult to implement over 160 dB of dynamic range of an operating FOG because of the intrinsic offset drift of the analog demodulation circuit and the A/D converter that digitalizes the integrated error signal. Typical variations are in the maximum voltage range of 10^{-4} to 10^{-5}/C. Compensation and modeling of this thermal drift are possible, but they increase the complexity of the electronics, yielding higher cost and lower reliability.

This problem can be completely overcome with an all-digital closed-loop approach [14,15]. The value of the modulated output signal is converted directly into digital form at each half-period, and the converter value of the second half-period is simply subtracted digitally from that of the first half-period. Such digital demodulation is intrinsically free of any source of electronic long-term drift. In particular, drift of the analog input offset of the A/D converter is canceled out by the subtraction of the odd and even samples. However, one might think that an A/D converter with a very large number of bits is required to limit the dead zone of the LSB: 160 dB of dynamic range is equivalent to 26 bits, as we have already seen. Fortunately, this crude analysis is not true. Because the sampling frequency is typically in the megahertz range, the bandwidth of the analog output signal is quite large, and thus a great deal of white noise is present (Figure 8.12). Signal-processing theory shows that sampling the analog signal with an LSB just smaller than the σ value of the noise is sufficient. Thus, digital integration yields the same noise reduction as with analog filtering, without any dead zone or spurious offset.

This can be simply understood with a hand-waving argument, by considering an analog signal with a noise extending over a range of several bits of the quantification circuit (Figure 8.13). If the mean value of the analog signal is zero, there

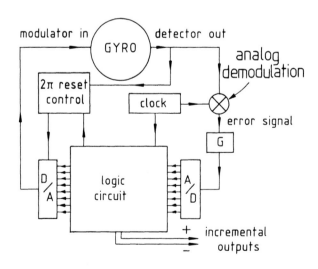

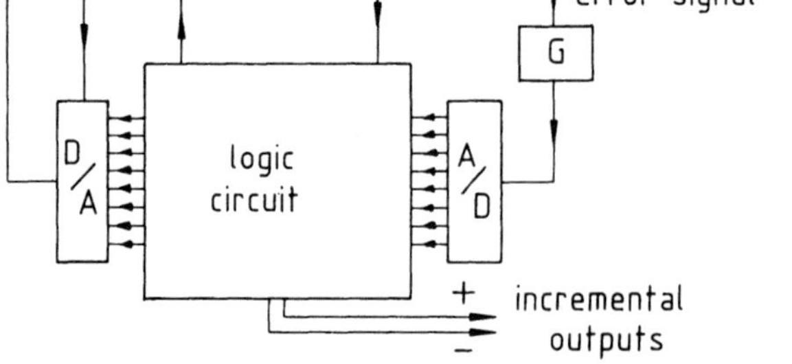

Figure 8.11 Original implementation of digital phase ramp feedback with an analog demodulation.

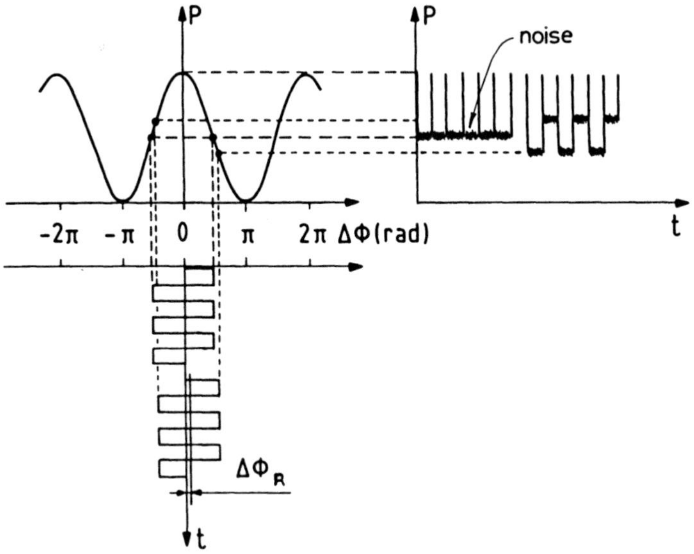

Figure 8.12 Actual signal with its noise.

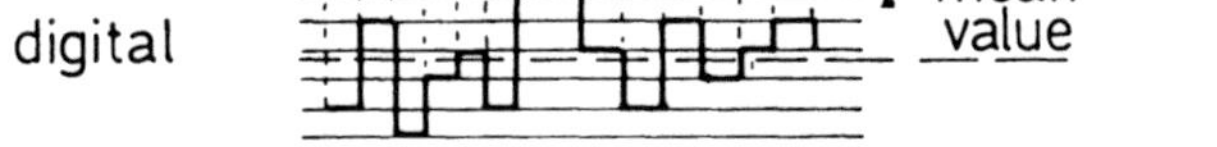

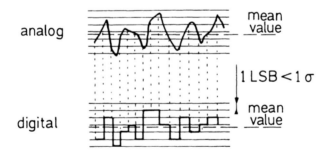

Figure 8.13 Digital quantification of an analog signal with an additional noise.

will be as many digital samples above zero as there are below. If the mean value is slightly positive, although the variation is much smaller than the LSB, it is still possible to measure it because, on average, there will be slightly more digital samples above zero than below. There is no degradation of the mean signal if the σ value of the noise is bigger than the LSB.

In practice, the noise is typically 10^{-6} rad/$\sqrt{\text{Hz}}$, or, in relative optical power value, $10^{-6}/\sqrt{\text{Hz}}$. With an analog bandwidth of 1 MHz, the σ value of the noise becomes 10^{-3} of the $\pi/2$ biasing power; that is, 11 bits are enough to convert the analog signal over the entire dynamic range of power variation, without any dead zone. The use of a digital integrator yields the same noise reduction as low-pass analog filtering, but without any electronic source of long-term drift.

This digital demodulation scheme is naturally compatible with numerical feedback schemes such as the digital phase ramp. This makes it even more efficient, since the error signal is servo-controlled to zero. With its noise extending over $\pm 4\sigma$, it is spread only over three bits, and this avoids the effect of A/D converter nonlinearity which is directly translated into scale-factor nonlinearity when digital demodulation is used with an open-loop scheme [16]. All-digital closed-loop processing yields a simple implementation that will be particularly adequate for integration. The functions are:

- Optical detection;
- Analog gating of the transient pulses and filtering;
- A/D conversion of the gyro signal;
- Digital logic driven with a common clock;
- D/A conversion of the feedback phase modulation.

The entire logic circuitry may be implemented on a single logic circuit. Its functional schematics are (Figure 8.14):

- Subtraction of odd and even samples to demodulate the error signal;
- Digital integration of this demodulated error signal;

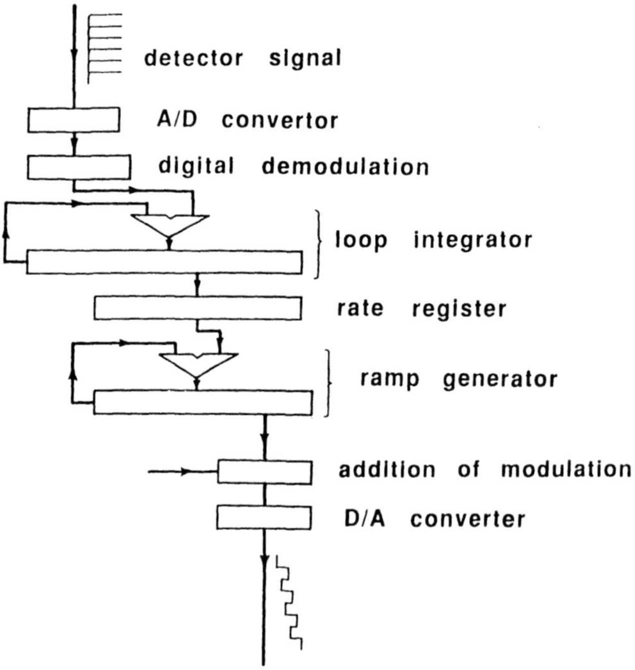

Figure 8.14 Functional schematics of the logic circuit used in the all-digital closed-loop approach.

- Storage of the rotation rate value in a register;
- Second integration of the rate value to generate the digital ramp;
- Digital addition of the square-wave biasing modulation.

This all-digital approach may also be used for the second feedback loop which controls the gain of the phase modulation chain [15], particularly with the 2π reset. The second error signal is obtained by comparing the digital value of the sampled detector signal just before and just after the reset. A digital integrator is also used to close this second loop, and a second D/A converter controls the reference voltage of the first D/A that generates the ramp or the gain of the buffer amplifier (Figure 8.15). This second D/A converter operates about dc to compensate for the long-term drift of the phase modulator response, but there is no stringent requirement of quantification error for this converter either. As we have already seen for the analog ramp (Section 8.2.2), the scale factor error induced by an imperfect 2π reset is only a third-order effect. This also applies to the digital ramp, but with the additional advantage of avoiding the requirement of a very fast flyback, since the transients are synchronized and can be gated out [15]. A control of the 2π reset within 1000 ppm (i.e., 10 bits) is sufficient to ensure a linearity better than 10 ppm!

It is possible to directly view the reason why the digital ramp technique tolerates many defects (such as imperfect 2π reset control, quantification, nonlinearity of the electronic drive, nonlinearity of the modulation response) without degrading the scale factor performance. The actual phase ramp ϕ_{PRa} is the sum of an ideal phase ramp ϕ_{PRi} and a defect ϕ_{PRd}. The induced phase difference is

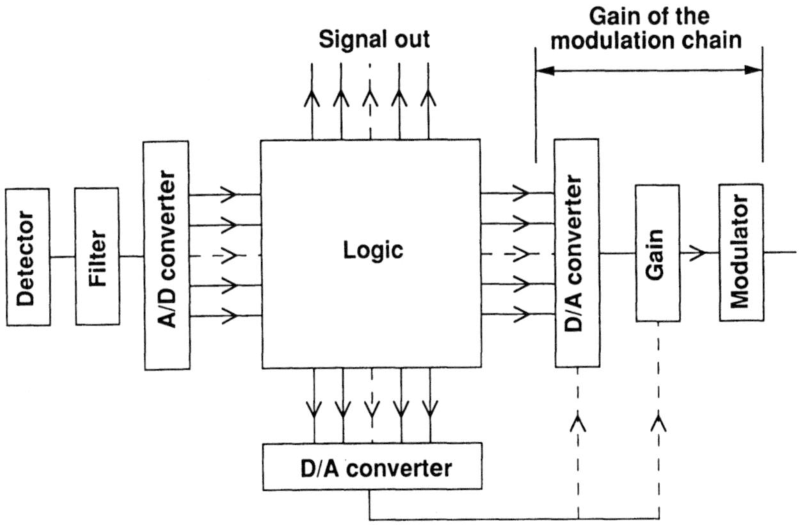

Figure 8.15 Functional schematics of the all-digital approach with a second feedback loop for the control of the modulation amplitude.

$$\Delta\phi_{PRa} = \Delta\phi_{PRi} + \Delta\phi_{PRd} \tag{8.27}$$

As with any phase modulation, the defect $\Delta\phi_{PRd}$ of the phase difference is given by the delay $\Delta\tau_g$ through the coil:

$$\Delta\phi_{PRd} = \phi_{PRd}(t) - \phi_{PRd}(t - \Delta\tau_g) \tag{8.28}$$

Since the mean value of a difference is the difference of the mean values, the mean defect $<\Delta\phi_{PRd}>$ is

$$<\Delta\phi_{PRd}> = <\phi_{PRd}(t)> - <\phi_{PRd}(t - \Delta\tau_g)> \tag{8.29}$$

and both mean values $<\phi_{PRd}(t)>$ and $<\phi_{PRd}(t - \Delta\tau_g)>$ are perfectly equal because of reciprocity; therefore

$$<\Delta\phi_{PRd}> = 0$$

If the defect $\Delta\phi_{PRd}$ remains in the linear part of the biased sine response, the mean error of the interference signal is also averaged out to zero as $\Delta\phi_{PRd}$. This applies to most defects except imperfect transients that yield instantaneous error $\Delta\phi_{PRd}$ much larger than this linear range, but with the digital ramp they can be synchronously gated out.

In summary, the all-digital closed-loop processing method brings a drift-free high-dynamic range without requiring a large number of bits for the various converters, because the quantification errors are canceled out through averaging processes. While reciprocity is a fundamental concept of creating a good optical system, the all-digital closed-loop approach is just as important from the electronic standpoint: it also allows a "perfect" device to be created from imperfect components! It does appear to be the ideal processing technique for the interferometric fiber-optic gyroscope, combining performance, simplicity, and the potential for optimal circuit integration.

8.3 WAVELENGTH CONTROL

8.3.1 Wavelength Dependence of a Ring Interferometer With a Broadband Source

Assuming one has achieved a perfect measurement of the Sagnac phase difference $\Delta\phi_R$ with the phase-nulling closed-loop scheme, the scale factor remains, as we have seen, related to the equivalent area of the coil and to the wavelength of the source. The coil area exhibits a typical variation of less than a ppm per Kelvin because of the silica thermal expansion, which necessitates some control and modeling for high-grade applications. Wavelength stability is more difficult to solve. For example, semiconductor diode sources, which are popular sources for the I-FOG, have a typical wavelength drift of about 400 ppm per Kelvin with temperature and of about 40 ppm per milliampere with driving current, even without taking into account additional factors such as aging or the feedback effect of the light returning to the source. Temperature control of the source and a stabilized driver are sufficient for a medium accuracy in the 100-ppm range, but high-performance applications require a direct wavelength control to be able to reach the ppm range.

A first fundamental question is to precisely define the effective wavelength involved in the scale factor when the broadband source needed for high performance has a relative spectrum width of several percent, which is several orders of magnitude larger than the ppm stability that is looked for. As explained in Section 2.3.1, the rotation-induced phase difference $\Delta\phi_R$ may be expressed with an equivalent path length difference ΔL_R that is perfectly wavelength-independent:

$$\Delta\phi_R = 2\pi\frac{\Delta L_R}{\lambda}$$
$$\Delta L_R = \frac{LD\Omega}{c} \tag{8.30}$$

The intrinsic unbiased interference response is

$$I = \frac{I}{2}\left[1 + \gamma_c(\Delta L_R)\cos\left(2\pi\frac{\Delta L_R}{\lambda_{cent}}\right)\right] \tag{8.31}$$

where γ_c is the coherence function of the source as measured in a scanning interferometer in a vacuum like a Michelson interferometer, and where λ_{cent} is the central wavelength of the spectrum. However, this simple result applies only if the spectrum is symmetrical with respect to the optical spatial frequency σ (i.e., the inverse of the wavelength λ), with the central frequency corresponding to the maximum power. In practice, gyro source spectra have a significant asymmetry, and, as discussed in Appendix 1, the unbiased interference response is, in the most general case,

$$I = \frac{I}{2}\left[1 + \gamma_{ce}(\Delta L_R)\cos\left(2\pi\frac{\Delta L_R}{\lambda}\right) + \gamma_{co}(\Delta L_R)\sin\left(2\pi\frac{\Delta L_R}{\lambda}\right)\right] \tag{8.32}$$

where γ_{ce} is the coherence function of the even component of the asymmetrical spectrum and λ is the mean wavelength. Compared to the simple symmetrical case, there is an additional term $\gamma_{co}(\Delta L_R)\sin(2\pi\Delta LR/\lambda)$, which takes into account the odd component of the spectrum. When the gyro is operated on the central fringe, as is usually the case, this additional term is negligible in practice; however, if it is operated on a wider dynamic range, a shift of the actual mean wavelength is yielded as the phase difference increases, since the zero crossings of the variable term of the interferometer response are not precisely periodic anymore.

This can be seen directly by regarding the interferometer as a filter with a transmission $\mathfrak{T}$ (λ or σ), depending on the wavelength λ (or the spatial frequency σ) for a given path difference ΔL_R

$$\mathfrak{T}(\lambda \text{ or } \sigma) = \frac{1}{2}\left[1 + \cos\left(2\pi\frac{\Delta L_R}{\lambda}\right)\right] = \frac{1}{2}[1 + \cos(2\pi\Delta L_R\sigma)] \qquad (8.33)$$

A broad power spectrum $I(\sigma)$ yields the integrated response,

$$I = \frac{1}{2}\int I(\sigma)[1 + \cos(2\pi\Delta L_R\sigma)]d\sigma \qquad (8.34)$$

For the zero crossing points of the variable cosine part of the response, the mean frequency $\bar{\sigma}$ is such that the product of $I(\sigma)$ with $\cos[2\pi\Delta L_R(\sigma - \bar{\sigma})]$ has a null integral (Figure 8.16). If the spectrum is symmetrical (in terms of frequency), the problem is simple and the mean value is the central value; but when it is asymmetrical, it is more complex and depends on the fringe order. In particular, the mean value $\bar{\sigma}$ is not equal to the value σ_{max}, which corresponds to the maximum intensity. A linear mean, defined with the product of the spectrum with a linear function instead of a sine (Figure 8.17), is a good approximation for gyros working around the zero order between $\pm\pi$ rad [17]. This linear mean is actually equivalent to a center of gravity.

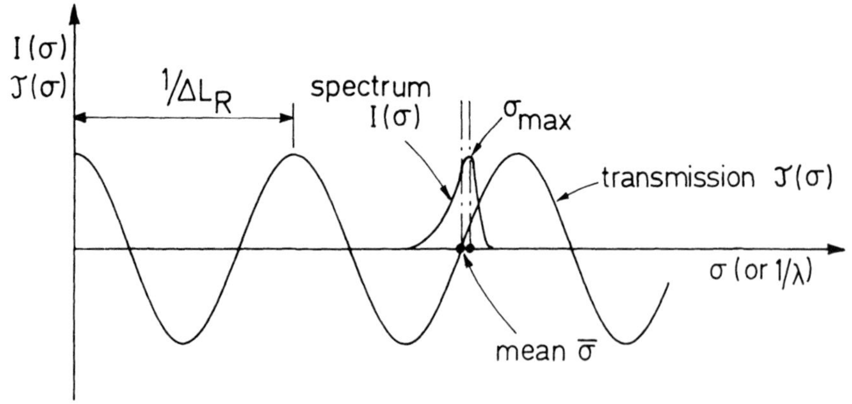

Figure 8.16 Interferometric definition of the mean spatial frequency $\bar{\sigma}$ of a broad spectrum.

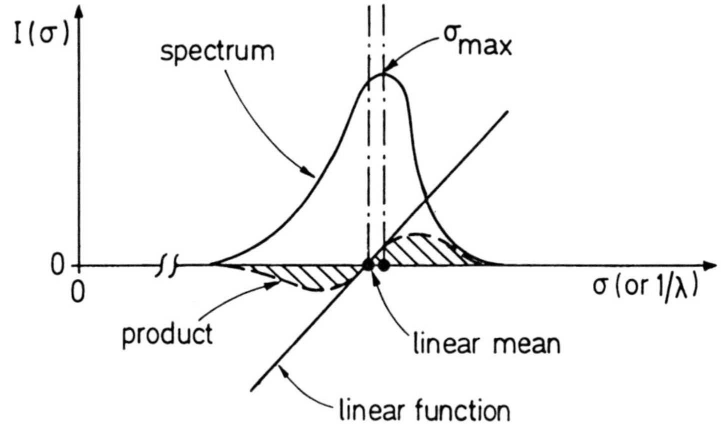

Figure 8.17 Definition of the linear mean of an asymmetrical broad spectrum.

8.3.2 Effect of Phase Modulation

So far, we have considered the unbiased response of the interferometer; however, as we have already seen, an I-FOG has to be operated with a biasing phase modulation and a phase-nulling feedback. Therefore, the wavelength dependence of the modulators may also have to be taken into account.

In the open-loop case, the demodulated biased response at a given wavelength is always a sine independent of the modulation depth; therefore, even with the wavelength dependence of the phase modulator, the total response with a broad spectrum is also a perfect sine. The amplitude of this open-loop response depends slightly on the spectrum, but, in practice, other sources of scale factor error are much more important.

With the phase-nulling scheme using Bragg modulators (see Section 8.2.1), there is a feedback frequency shift Δf_{FB}, which is wavelength-independent; therefore, the scale factor has the same spectrum dependence as in the unbiased case. With analog or digital phase ramps (see Sections 8.2.2 and 8.2.3), the problem is slightly more delicate. A phase modulator based on the elasto-optic effect as a piezoelectric modulator, or on the electro-optic Pockels effect as in integrated optics, is actually a "path-difference modulator." A given driving voltage V_d yields a given path-difference that is, to first order, wavelength-independent. The modulation is mainly due to a geometrical length change or an index change, which are almost wavelength-independent. This yields, for a given voltage ramp or step, a feedback phase difference $\Delta\phi_{PR}$ inversely proportional to the wavelength:

$$\Delta\phi_{PR} = 2\pi\frac{\Delta L_{PR}}{\lambda} \tag{8.35}$$

where ΔL_{PR} is the equivalent path length difference induced by the ramp. Therefore, a rotation yielding

$$\Delta\phi_R = 2\pi\frac{\Delta L_R}{\lambda} \tag{8.36}$$

Both phase differences $\Delta\phi_R$ and $\Delta\phi_{PR}$ have the same wavelength dependence, and phase ramp feedback should be globally wavelength-independent: for any wavelength, the driving voltage yields the same path length difference ΔL_{PR}, which compensates for the rotation-induced difference ΔL_R. However, this would be true only if the effect of the reset is gated out. In this case, other causes of drift of the modulator efficiency or of the gain of the driving electronics (temperature in particular) would then become predominant. We have seen (Section 8.2.2) that a small change of the gain of the phase modulation chain has only a third-order effect on the scale factor accuracy when the reset is close to 2π and when the signal after the reset is taken into account. A wavelength change is equivalent to a gain change, since $\Delta\phi_{FB} = 2\pi\Delta L_{FB}/\lambda$. Therefore, a wavelength change does not modify the mean effect of the feedback ramp while it modifies to first order the rotation-induced phase difference $\Delta\phi_R$. Then the basic wavelength dependence of the Sagnac effect is retrieved with the phase ramp feedback when the effect of the reset is not gated out. Despite this drawback, a controlled 2π reset is preferable, since the wavelength may be controlled independently, while other sources of modulation efficiency drift would not be detectable.

8.3.3 Wavelength Control Schemes

Among the problems to be solved to get a high-performance fiber gyro, wavelength control was the last one to be addressed, and publications on this subject are not very numerous compared to the total literature on the FOG. No definitive answer has yet been given to this problem, in contrast to the problems of optical architecture with reciprocity, those of the various parasitic noises and drifts with a broadband source or those of the signal processing scheme with biasing modulation and phase nulling feedback. This is also more of an engineering problem, the solutions of which are usually kept confidential, rather than a basic theoretical analysis that may be published.

Nevertheless, some design concepts have been described that directly control the linear mean value involved in the scale factor of the gyro. As a matter of fact, it is not desirable to measure the whole spectrum and calculate this averaging. The control device should perform a direct measurement of the mean wavelength.

In addition to temperature control for stabilizing the source spectrum, the simplest approach is using a narrow optical filter in front of the detector at the output. Since each emission wavelength has an independent behavior, this approach is equivalent to using a source with a spectrum equal to the product of the emission spectrum and the transmission of the filter. Assuming that the reference filter has a stable transmission, the stability of the actual spectrum is improved by a factor equal to the square of the ratio ρ_f between the widths of the source spectrum and of the filter response (Figure 8.18). Interference filter, for example, may be as narrow as 5 nm, while superluminescent diodes have a typical width of 15 nm; that is, $\rho_f \approx 3$. This would yield a ten-fold improvement in actual mean wavelength stability. However, there is the drawback of detected power reduced by a factor equal to ρ_f. However, this degrades the theoretical signal-to-noise ratio only by $\sqrt{\rho_f}$.

Some other approaches control the wavelength with the light tapped out at the input by the source splitter. The use of an additional narrow reference source [18] was proposed to stabilize the path difference ΔL_{cont} of a control interferometer on an integral number m of the reference wavelength λ_{ref}, and to adjust the broad spectrum to have the path difference of the stabilized interferometer equal to another integral number m' of the mean wavelength $\bar{\lambda}$ of the broadband source:

$$\Delta L_{\text{cont}} = m\lambda_{\text{ref}} = m'\bar{\lambda}$$

This method is very accurate with a laboratory setup, but it is difficult to make a compact control device for practical applications. Furthermore, note that it measures the mean wavelength on the m'^{th} fringe, which, as we have seen, may be slightly different from the first-order fringe actually involved in the gyro, if the spectrum is asymmetrical.

Another scheme, which looks more practical, is to use a miniaturized grating spectrometer to spread the source spectrum spatially [19]. Two detectors connected with an opposite polarity directly provide the linear mean of the wavelength because they have an additional triangular mask that simulates the product by a linear function.

In any case, all of these schemes require a very stable reference: a stable reference filter, stable reference wavelength, or stable reference spectrometer. A

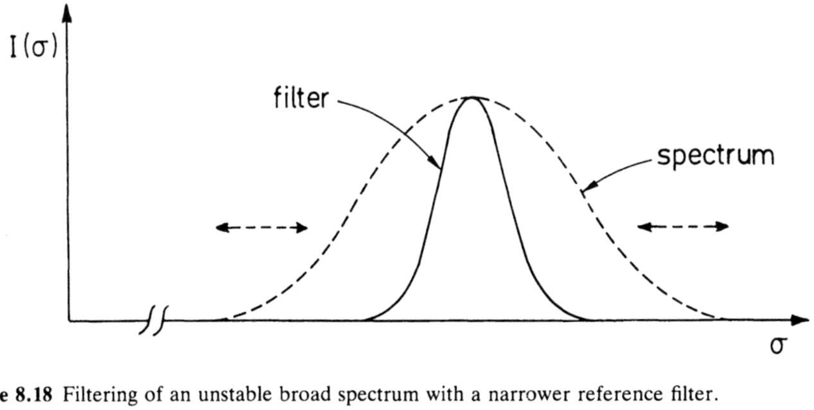

Figure 8.18 Filtering of an unstable broad spectrum with a narrower reference filter.

stability better than 10 ppm in real operation is a difficult but reasonable engineering goal.

Note that another interesting possibility is to use the propagation dispersion in fibers. The group transit time τ_g through the coil has a wavelength dependence (see Section 3.2.3 and Appendix 1), and this transit time variation may be detected with an accurate measurement of the proper (or eigen) frequency f_p of the coil using a square wave modulation with an asymmetrical duty cycle [17] (see Section 3.2.3). However, such a technique does not work at 1.3 μm where silica shows zero dispersion, and it also requires the use of a reference filter to differentiate the transit time variation due to wavelength from the strong temperature dependence.

A similar transit time dependence may be obtained with the frequency shifting feedback scheme [20]; but, as we have seen, this approach does not provide very good performance because it destroys the reciprocity of the interferometer (see Section 8.2.1).

REFERENCES

[1] Davis, J. L., and S. Ezekiel, "Techniques for Shot-Noise-Limited Inertial Rotation Measurement Using a Muti-Turn Fiber Sagnac Interferometer," SPIE Proceedings, Vol. 157, 1978, pp. 131–136 (SPIE MS 8, pp. 138–143).

[2] Cahill, R. F., and E. Udd, "Phase-Nulling Fiber-Optic Laser Gyro," Optics Letters, Vol. 4, 1979, pp. 93–95 (SPIE MS 8, pp. 152–154).

[3] Davis, J. L., and S. Ezekiel, "Closed-Loop, Low-Noise Fiber-Optic Rotation Sensor," Optics Letters, Vol. 6, 1982, pp. 505–507 (SPIE MS 8, pp. 186–188).

[4] Lefèvre, H. C., S. Vatoux, M. Papuchon, and C. Puech, "Integrated Optics: A Practical Solution for the Fiber-Optic Gyroscope," SPIE Proceedings, Vol. 719, 1986, pp. 101–112 (SPIE MS 8, pp. 562–573).

[5] Auch, W., "The Fiber-Optic Gyro—A Device for Laboratory Use Only?" SPIE Proceedings, Vol. 719, 1986, pp. 28–34.

[6] Shimizu, H., R. Ishikawa, and R. Kaede, "Integrated-Optical Frequency Modulator for Fiber-Optic Gyroscope," Electronics Letters, Vol. 22, 1986, pp. 334–335.

[7] Michal, R. J., E. Udd, and J. P. Theriault, "Derivative Fiber-Sensors Based on Phase-Nulling Optical Gyro Development," SPIE Proceedings, Vol. 719, 1986, pp. 150–154.

[8] Kim, B. Y., and H. J. Shaw, "Gated Phase-Modulation Approach to Fiber-Optic Gyroscope With Linearized Scale Factor," Optics Letters, Vol. 9, 1984, pp. 375–377 (SPIE MS 8, pp. 393–395).

[9] Lefèvre, H. C., Ph. Graindorge, H. J. Arditty, S. Vatoux, and M. Papuchon, "Double Closed-Loop Hybrid Fiber Gyroscope Using Digital Phase Ramp," Proceedings of OFS 3/'85, San Diego, OSA/IEEE, Postdeadline Paper 7, 1985 (SPIE MS 8, pp. 444–447).

[10] Kay, C. J., "Serrodyne Modulator in a Fibre-Optic Gyroscope," IEE Proceedings, Part J--Optoelectronics, Vol. 132, 1985, pp. 259–264 (SPIE MS 8, pp. 448–453).

[11] Elberg, A., and G. Schiffner, "Closed-Loop Fiber-Optic Gyroscope With a Sawtooth Phase-Modulated Feedback," Optics Letters, Vol. 10, 1985, pp. 300–302 (SPIE MS 8, pp. 396–398).

[12] Kurokawa, A., K. Kajiwara, N. Usui, Y. Hayakawa, M. Haruna, and H. Nishihara, "Evaluation of a Sawtooth Generator in a Closed-Loop Fiber-Optic Gyroscope," Proceedings of OFS 6/'89, Paris, Springer Proceedings in Physics, Vol. 44, 1989, pp. 107–114.

[13] Lefèvre, H. C., J. P. Bettini, S. Vatoux, and M. Papuchon, "Progress in Optical Fiber Gyroscopes Using Integrated Optics," AGARD-NATO Proceedings, Vol. CPP-383, 1985, pp. 9A1-9A3 (SPIE MS 8, pp. 216–277).

[14] Arditty, H. J., P. Graindorge, H. C. Lefèvre, P. Martin, J. Morisse, and P. Simonpiétri, "Fiber-Optic Gyroscope With All-Digital Processing," Proceedings of OFS 6/'89, Paris, Springer-Verlag Proceedings in Physics, Vol. 44, 1989, pp. 131–136.

[15] Lefèvre, H. C., P. Martin, J. Morisse, P. Simonpiétri, P. Vivenot, and H. J. Arditty, "High Dynamic Range Fiber Gyro With All-Digital Processing," SPIE Proceedings, Vol. 1367, 1990, pp. 72–80.

[16] Auch, W., M. Oswald, and D. Ruppert, "Product Development of a Fiber-Optic Rate Gyro," Proceedings of Symposium Gyro Technology, DGON, Stuttgart, 1987, pp. 3.0–3.19.

[17] Lefèvre, H. C., "Comments About the Fiber-Optic Gyroscope," SPIE Proceedings, Vol. 838, 1987, pp. 86–97 (SPIE MS 8, pp. 56–67).

[18] Chou, H., and S. Ezekiel, "Wavelength Stabilization of Broadband Semi-Conductor Light Sources," Optics Letters, Vol. 10, 1985, pp. 612–614 (SPIE MS 8).

[19] Schuma, R. F., and K. M. Killian, "Superluminescent Diode (SLD) Wavelength Control in High-Performance Fiber-Optic Gyroscopes," SPIE Proceedings, Vol. 719, 1986, pp. 192–193 (SPIE MS 8).

[20] Udd, E., and R. F. Cahill, "From Conception to the Field: Fiber-Optic Gyro Development at McDonnell-Douglas," SPIE Proceedings, Vol. 719, 1986, pp. 17–23.

Chapter 9
Technology of the I-FOG

9.1 RECAPITULATION OF THE OPTIMAL OPERATING CONDITIONS

The interferometric fiber gyroscope, often abbreviated I-FOG, is a ring interferometer that uses a multiturn fiber coil to enhance the Sagnac effect induced by rotation with respect to inertial space. This yields a phase difference $\Delta\phi_R$ between the two counterpropagating waves that is proportional to the rotation rate Ω (see Section 2.3):

$$\Delta\phi_R = \frac{2\pi LD}{\lambda c} \cdot \Omega \tag{9.1}$$

The optimal operation conditions can be summarized as follows:

- Use of a single-mode reciprocal configuration with a truly single-mode filter (single spatial mode and single polarization) at the common input-output port of the interferometer. This ensures that the paths of both opposite waves are perfectly equalized and that only truly nonreciprocal effects as rotation yields a phase difference (see Section 3.2).
- Use of a modulation-demodulation biasing scheme with a reciprocal phase modulator at the end of the fiber coil. The interferometer behaves like a delay line filter because of the transit time through the coil, which yields a high sensitivity operation point without degrading the reciprocity. Best performances are obtained at the proper or eigenfrequency that matches the half-period of the modulation to the coil transit time. The combination of this processing scheme with a reciprocal configuration yields the so-called minimum configuration (see Section 3.2).

- Use of a broadband optical source to take advantage of its short coherence length. This destroys the interference contrast of the various parasitic waves generated in the system by backreflection, backscattering, or polarization cross-coupling (see Chapters 4 and 5). It suppresses the effect of Kerr nonlinearity, which is also related to an interference phenomenon (see Section 7.2).
- Use of polarization-preserving fibers that provide a very beneficial depolarization effect on the crossed-polarized waves because of the intrinsic fiber birefringence and the low coherence of the broadband source (see Chapter 5). This relaxes the constraint of very high polarizer rejection. The accurate analysis of these polarization problems is obtained with optical coherence domain polarimetry (OCDP), based on path-matched white-light differential interferometry (see Section 5.4). These fibers also reduce the influence of the nonreciprocal magneto-optic Faraday effect (see Section 7.1).
- Use of a phase-nulling feedback with a closed-loop processing scheme to linearize and stabilize the Sagnac phase measurement (see Chapter 8). Among the possible techniques, the all-digital closed-loop approach provides fundamental advantages. It combines a digital phase ramp feedback, which yields very good linearity and stability over the whole dynamic range, and a digital demodulation, which is intrinsically free of any source of electronic bias drift.
- Use of wavelength control to stabilize the source spectrum and get an accurate scale factor of the rotation rate measurement (see Section 8.3).

Figure 9.1 summarizes the optimal architecture that combines a Y-coupler configuration, using a multifunction integrated optic circuit, with all-digital closed-loop processing electronics generating a staircase digital phase ramp.

9.2 SOURCE

9.2.1 Superluminescent Diode

The first laboratory experiments were performed with He-Ne gas lasers that can be efficiently coupled into a single-mode fiber, since their Gaussian emission mode is matched to the pseudo-Gaussian fundamental mode of the fiber (see Appendix 2). However, the development of semiconductor emitting diodes, particularly for telecommunications, has made these compact solid-state light sources working with a low driving voltage the ideal choice for practical devices. These diodes are mainly of two kinds: surface light emitting diodes, abbreviated LEDs, and laser diodes, abbreviated LDs [1]. They use III-V semiconductor junctions, AlGaAs-GaAs for a wavelength in the 800- to 850-nm range, or InGaAsP-InP for the 1300- and 1550-nm ranges.

These two sources, though, were not optimal for gyro applications: LEDs based on spontaneous emission have the adequate spectrum width, but they cannot be coupled efficiently into a single-mode fiber, since the emission area is large (50

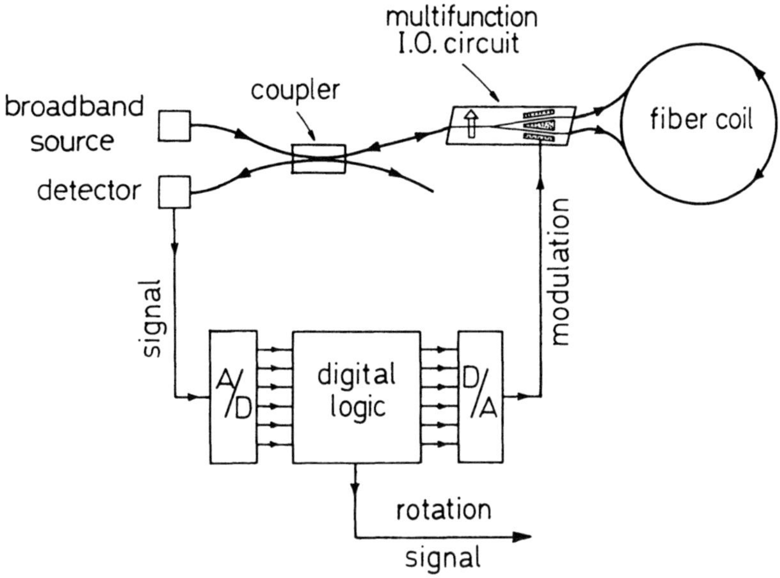

Figure 9.1 Optimal architecture of the I-FOG with a Y-coupler configuration and an all-digital closed-loop processing scheme.

to 100 μm in diameter) compared to the fiber core; LDs, on the other hand, may be coupled efficiently because the wave, generated in a narrow stripe (a few micrometers) by stimulated emission, is spatially coherent, but the spectrum is composed of the narrow emission peaks of the modes of the Fabry-Perot laser cavity (see Appendix 1). To get simultaneously a good spatial coherence and a low temporal coherence, laser diodes have to be modified to make superluminescent (or superradiant) diodes, abbreviated SLDs (or SRDs). The lasing effect is suppressed by decreasing the reflectivity of the mirror facets with an antireflection coating at the emission output and an absorbing region at the other diode end (Figure 9.2) [2–6]. The use of an angled stripe [7], which works on a similar principle as the angled edge of the integrated circuit, has also been demonstrated (see Section 4.1.1).

The gain of a semiconductor diode is very high, and even without cavity feedback, the output power may be almost as high as that of a laser. Along a single pass, the first spontaneous emission photons are amplified by stimulated emission, and the output wave has a spatial coherence similar to that of a laser, since it is also generated in a narrow stripe, yielding an efficient coupling into a single-mode fiber. However, the multimode laser structure of the spectrum is greatly reduced, and SLDs behave like quasi-broadband sources. At 850 nm, the full width at half maximum is on the order of 20 nm.

Practical devices, which are typically 300 to 500 μm long, are hermetically packaged in a rugged casing with a fiber pigtail soldered in front of the emission

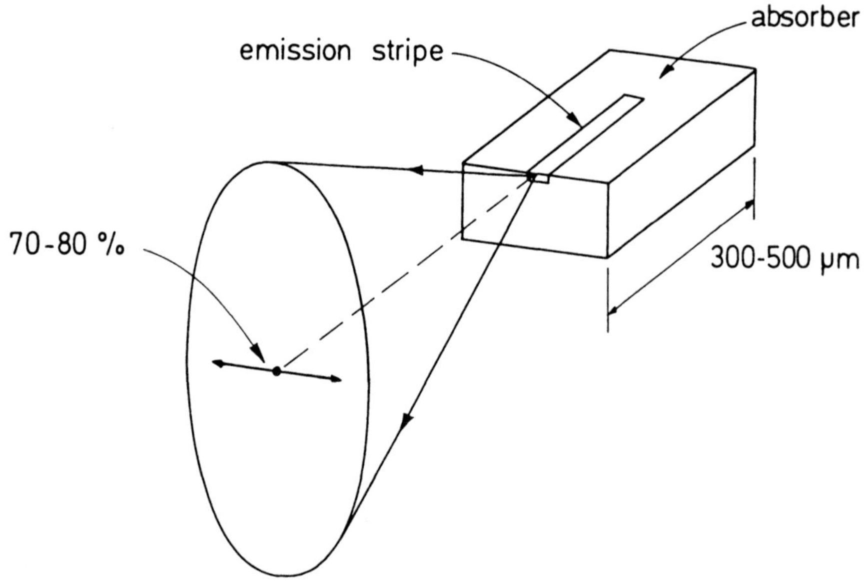

Figure 9.2 Superluminescent diode with a rear absorbing region.

window of the diode. The typical coupled power is 0.5 to 1 mW (the coupling efficiency is 10% to 20%) for a driving current of 50 to 150 mA. A polarization-preserving fiber may be used for pigtailing with adequate orientation of its birefringence axes. SLDs are partially polarized, with 70% to 80% of the power in the "horizontal" polarization parallel to the junction.

The main problem of SLDs is their poor spectrum stability: the mean wavelength has a drift of about 400 ppm/K with temperature and 40 ppm/mA with driving current. A temperature control with a Peltier element and a stable driver yield a stability in the range of 100 ppm; but higher performance requires a direct wavelength control.

9.2.2 Rare-Earth Doped Fiber Sources

To overcome the problem of SLD wavelength stability, work has been devoted over the last years to the development of alternative broadband sources based on rare-earth doped fiber [8]. As with SLDs, rare-earth doping provides a very high amplification gain, and high-power broadband emission may be obtained over a single pass with amplified spontaneous emission (ASE) without requiring a cavity feedback. These wideband fiber sources are often called superfluorescent fiber sources (the distinction between superradiance, superluminescence, superfluorescence, and ASE is not always very clear in the literature. For more information, see Auzel et al. [9]). Their energy levels are much more stable than those of semiconductors and should improve the wavelength stability. They may be pumped with compact high-power laser diodes. Two dopants are particularly efficient: Neodymium (Nd), with an emission around 1060 nm and a pump of 800 nm [8,10], and Erbium (Er), with an emission around 1550 nm and a pump of 980 nm or 1480 nm [11,12].

These sources are very promising for high-performance navigation-grade fiber gyroscopes; however, the subject is not straightforward, and to obtain a very good wavelength stability depends on various parameters, particularly pump wavelength, pump power, and light feedback [13].

Several features of these sources must be pointed out: the emitted light is unpolarized, which is very beneficial for reducing polarization nonreciprocities (see Section 3.4.4). The output power may be very high and is naturally matched to single-mode fiber, which is useful for getting low photon noise or sharing a single source for three gyro axes. The spectrum is very asymmetrical and care must be taken to carefully evaluate the mean wavelength (see Section 8.3.1 and Appendix 1). Note that the demultiplexer required to separate the pump and emission wavelengths may be realized in an all-fiber form with evanescent field couplers that have a significant wavelength dependence.

A last important point is the excess noise induced by the random beating between all the uncorrelated frequency components of a broad spectrum [14]. As seen in Section 2.3.2, the photon shot noise of a flow $\dot{N}$ of uncorrelated "particles" like photons is

$$\sigma_{\dot{N}}^2 = 2\dot{N}\Delta f_{bw} \tag{9.2}$$

where Δf_{bw} is the detection frequency band. However, an optical spectrum with a frequency width Δf yields an excess noise because of the random beating between all its frequency components:

$$\sigma_{\dot{N}_{exc}}^2 = \frac{\dot{N}^2}{\Delta f}\Delta f_{bw} \tag{9.3}$$

Therefore, this excess noise becomes predominant when $\Delta f < \dot{N}/2$. With an SLD at $\lambda = 850$ nm with $\Delta\lambda = 15$ nm, we have $\Delta f = c\Delta\lambda/\lambda^2 = 6 \times 10^{12}$ Hz and the excess noise becomes equivalent to the photon noise for a stream $\dot{N}$ of 1.2×10^{13} photons per second; that is, a power of about 4 μW, which is close to the limit of the returning light, in practice.

However, a superfluorescent fiber source has a "narrower" broad spectrum than an SLD (e.g., $\Delta\lambda = 2$ nm about $\lambda = 1530$ nm; i.e., $\Delta f = 2.5 \times 10^{11}$ Hz), and the excess noise now becomes predominant at only 0.2 μW. Note, however, that this excess noise is an intensity noise, which may be reduced by operating the gyro at a bias point close to a black fringe; that is, a phase bias close to π instead of $\pi/2$ [15]. As a matter of fact, the sensitivity is proportional to the slope of the raised cosine response curve; that is, $\sin\phi_b$ (where ϕ_b is the phase bias) and intensity noise are proportional to the actual power on bias (i.e., the response $(1 + \cos\phi_b)$). Working, for example, at a 0.9π bias instead of $\pi/2$, the sensitivity is reduced by a factor of $\sin(0.9 \times \pi)/\sin(\pi/2) = 0.3$, while intensity noise experiences a reduction six times higher, since $[1 + \cos(0.9 \times \pi)]/[1 + \cos(\pi/2)] = 0.05$. Furthermore, as seen in Section 2.3.2, this does not degrade the theoretical signal-to-photon-noise ratio. Note that it is also an efficient way to reduce the amount of light returning to the source.

9.3 FIBER COIL

As explained previously, the best performances are obtained with a sensing coil made of stress-induced high-birefringence polarization-preserving fiber (Appendix 2). Elliptical core fiber has a higher attenuation and a lower polarization conservation, but it remains an interesting choice for the small coil of medium-grade gyros, since it should lend itself to a simpler fabrication process and a lower cost. Furthermore, a symmetrical quadrupolar winding reduces the effect of temperature transience (see Section 6.1) and a μ-metal shielding may be required to further reduce the magnetic dependence (see Section 7.1). Some other practical specifications are also needed. In particular, high-NA fiber with a highly doped core is suitable for avoiding bending loss in very compact coils: with an NA of 0.16, diameter may be as small as 20 mm without additional attenuation. Careful winding and adequate potting material are important to ensure good pointing accuracy: the sensitivity axis is parallel to the equivalent area vector $\boldsymbol{A}$ defined by the line integral $\int(1/2)\mathbf{r} \times \mathbf{dr}$ along the fiber (see Section 2.1.1). To get the specified performances in a three-axis unit, the axis stability, in radians, must be equal to that of the scale factor; that is, for example, 10^{-5} rad is required with 10 ppm.

The fiber diameter should be as small as possible to limit the volume of the coil. Gyro fibers have a typical cladding diameter of 80 μm instead of the standard 125 μm of telecommunication fibers. The same applies to the protective coating, which should be as thin as possible. However, it must minimize microbending to avoid loss or degradation of the polarization conservation, which actually limits its minimum thickness. The coil characteristics must be conserved over the entire temperature range of the operation. The standard acrylate coating experiences a glass phase transition below −20°C which strongly degrades the polarization conservation, and two-layer coatings with a "soft" inner layer have been developed to solve this problem.

To avoid an increase of attenuation under radiation in military or space applications, the use of long wavelengths (1300 or 1550 nm) is preferable.

A last, very important point is the fiber reliability. This problem is very complex [16,17], but the basic ideas may be outlined. The fiber surface contains very small intrinsic flaws due to the basic structure of silica, and larger extrinsic flaws due to dust or particles included in the fiber during the drawing process. When the fiber is under tensile stress, the size of the flaws is increased, which may eventually cause breakage of the fiber. When wound in coils, the fiber is placed under tensile stress at the outside. The related strain is equal to the ratio between the fiber diameter and the coil diameter. To ensure a good reliability, the whole fiber length has to be proof-tested at a high strain level (typically 0.5% to 2%, while ideal silica may withstand strain up to 10%) for a few seconds to check that the sample does not contain weak points that would otherwise induce breakage. Based on the Weibull model of the weakest link in a chain, the failure probability over the expected lifetime of the gyro may be evaluated as a function of the proof-test level and of the fiber characteristics [16]. The quality of practical fibers ensures a very good reliability for coils in the 10-cm range that experience a strain of about 0.1%. However, for small coils (about 2 to 3 cm), high-strength fiber would have to be used to achieve a very long lifetime. Note that proof-test level is also expressed in terms of stress instead of strain, which are related by the Young modulus of silica (E_{SiO_2} = 70 GPa in SI units). Using pounds per square inch, remember that 1% of strain corresponds to a stress of 100 kpsi. Another useful order of magnitude is that 1% of strain is induced by a force of 10N for a 125-μm fiber, and 4N for an 80-μm fiber.

9.4 "HEART" OF THE INTERFEROMETER

As described in the previous chapters (particularly Chapter 3), the main subject of concern has been the "heart" of the interferometer, composed of a source beam splitter, polarizer, spatial single-mode filter, coil beam splitter, and phase modulators. Because bulk-optic components require delicate alignments to couple light into single-mode fiber, research has been focused on a rugged, all-guided approach, and the users of integrated optics have had the advantage of getting wideband phase modulators. Optimal simplicity is obtained with a hybrid approach: the Y-tap or Y-coupler configuration (see Section 3.3.4) now widely used [18]. A multifunction integrated optic circuit combines a Y-junction for the coil splitter, phase modulators, and preferably the polarizer, while an in-line fiber tap or coupler is used for the source splitter, with its lead acting as the spatial filter.

A very critical component is the polarizer, and a proton-exchanged $LiNbO_3$ circuit (see Appendix 3) that guides only one state of polarization looks like an optimal approach [19]. However, recent experimental results [20] have showed that it is possible to get a similar rejection (typically 60 dB) with a metallic overlay on a Ti-indiffused waveguide, which keeps the competition alive! An in-line fiber polarizer using a metallic overlay also [21] is another interesting alternative. Note that OCDP (see Section 5.4) is an essential technique for evaluating accurately the rejection of the polarizing element.

9.5 DETECTOR

Finally, it is necessary to be careful when choosing the detector so as not to degrade the performance of the system, which would normally be limited by photon shot noise (see Section 2.3.2). Semiconductor PIN photodiodes are ideal because of their very high quantum efficiency: the number of primary electrons generated is very close to the number of input photons and the flow of electrons has about the same shot noise as the theoretical value for the flow of photons. For 850 nm, a silicon (Si) photodiode has to be used, while indium gallium arsenide (InGaAs) is optimal for 1300 and 1550 nm.

The noise of the transimpedance preamplifier of the diode (which converts current into voltage) must not yield a noise-equivalent power (NEP) higher than the shot noise of the primary current. This noise is mainly due to the thermal noise of the charge resistor of the amplifier. The voltage-current conversion is equal to its resistance R, while the thermal noise generated on the output voltage has an rms value σ_v proportional to the square root of this resistance:

$$\sigma_v = \sqrt{4kT_aR\Delta f_{bw}} \tag{9.4}$$

where $k = 1.38 \times 10^{-23}\ J \cdot K^{-1}$ is the Boltzmann constant and T_a the absolute Kelvin temperature.

If the charge resistance R is increased, the conversion factor increases linearly while thermal noise increases only as the square root. This improves the signal-to-thermal-noise ratio. However, the improvement is limited by the gain-bandwidth product of the amplifier, which dictates the maximum resistance that may be used for the required bandwidth.

Photomultiplier tubes are fundamentally not adapted to this application because of their very low quantum efficiency, especially in the near infrared, without considering their size and their high voltage power supply. Avalanche photodiodes (APD) can be useful if the returning power is low, but they will not be ideal in "flying" devices because of the temperature dependence of the gain. The theory of the avalanche effect shows that the fundamental noise of the primary current is degraded by at least a factor of $\sqrt{2}$ compared to a PIN diode. With silicon, an excellent material, the degradation factor is about $\sqrt{3}$ for an avalanche gain of 100, but there is a reduction, equal to this gain, of the relative effect of the preamplifier noise. Figure 9.3 summarizes graphically the relative noises (noise-

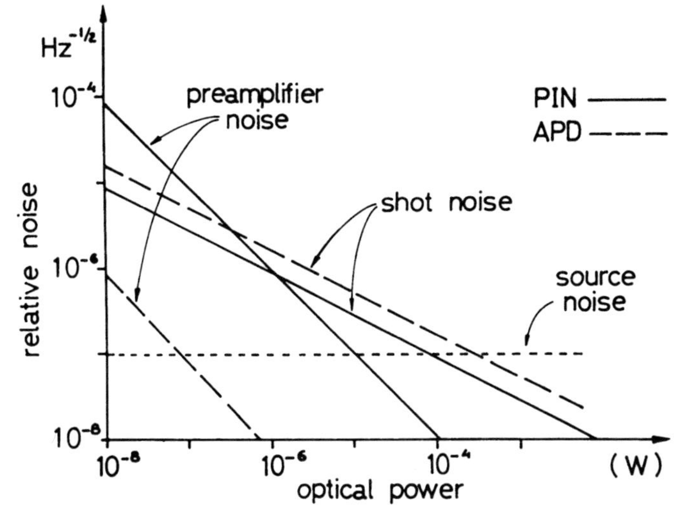

Figure 9.3 Various relative noises as a function of the returning power (silicon diodes at 850 nm).

to-signal ratio!) for various optical powers in a log-log scale considering a Si-PIN detector at 850 nm with a typical NEP of 10^{-12} W/$\sqrt{\text{Hz}}$. In the shot-noise regime, the slope is $-1/2$, and when the system is limited by preamplifier noise the slope is -1. With a typical APD, the shot-noise curve is upshifted by a factor of $\sqrt{3}$ compared to a PIN diode, but the limitation of the preamplifier noise is downshifted by a factor of 100, since there is a direct current amplification with the avalanche effect. In practice, APDs have been useful at a laboratory stage to study the first experimental gyros, which had high loss and hence low returning power (<1μW). However, progress in the loss reduction of the components and coupling efficiency of the source makes 10 μW of returning power a reasonable engineering goal for practical devices, allowing the use of a PIN diode without photon shot-noise degradation.

A last problem to get photon-noise-limited detection could be the proper intensity noise of the source; but at the frequency of operation (100 kHz to 1 MHz) and with superluminescent diodes, it ranges from -130 to -140 dB/Hz (i.e., $3 \times 10^{-7}/\sqrt{\text{Hz}}$ to $10^{-7}/\sqrt{\text{Hz}}$) [22], which would be a limit only with more than 10 μW back at the detector. However, as seen in Section 9.2.2, rare-earth broadband fiber sources may have an excess intensity noise of 110 to 120 dB/Hz because of their "narrow" broad spectrum.

Note finally that the detection unit requires a very careful electronic design to avoid ground loop and electromagnetic circuit coupling problems. The biasing modulation voltage is typically on the order of 1V and the primary current in the detector is typically 1 μA for a few microwatts of returning optical power. To limit bias error due to electronic coupling to below 10^{-7} rad, the coupled current at the modulation frequency has to remain below 10^{-13} A; that is, about a single electron per sampling time (typically, 1 μs for a 200m coil)! This problem of electromagnetic coupling applies also to the driving current of the source.

REFERENCES

[1] See, for example, Fukuda, M., "Reliability and Degradation of Semiconductor Lasers and LEDs," Boston-London: Artech House, 1991.

[2] Lee, T. P., C. A. Burrus, and B. I. Miller, "A Stripe-Geometry Double-Heterostructure Amplified-Spontaneous Emission (Superluminescent) Diode," IEEE Journal of Quantum Electronics, Vol. QE-9, 1973, pp. 820–821.

[3] Wang, C. S., W. H. Cheng, C. J. Hwang, W. K. Burns, and R. P. Moeller, "High Power Low Divergence Superradiant Diode," Applied Physics Letters, Vol. 41, 1982, pp. 587–589.

[4] Wang, C. S., J. S. Fen, R. Fu, V. S. Sunderam, R. Varma, J. Zarrabi, C. Lin, and C. J. Hwang, "High-Power Long-Life Superluminescent Diode," SPIE Proceedings, Vol. 719, 1986, pp. 203–207.

[5] Kwong, N. S. K., K. Y. Lau, N. Bar-Chaim, I. Ury, and K. J. Lee,"High Power, High Efficiency Window Buried Heterostructure GaAlAs Superluminescent Diode With an Integrated Absorber," Applied Physics Letters, Vol. 151, 1987, pp. 1879–1881.

[6] Kwong, N. S. K., N. Bar-Chaim, and T. Chen, "High-Power 1.3 μm Superluminescent Diode," Applied Physics Letters, Vol. 54, 1989, pp. 298–300.

[7] Niesen, J., L. Zinkiewicz, P. H. Payton, and C. Morrison, "Recent Development in 0.83 m Superluminescent Diodes at TRW," SPIE Proceedings, Vol. 719, 1986, pp. 208–215.

[8] Liu, K., M. Digonnet, K. Fesler, B. Y. Kim, and H. J. Shaw, "Superfluorescent Single-Mode Nd: Fiber Source at 1060 nm," Proceedings of OFS'88, New Orleans, 1988, pp. FDD5-1–FDD5-4.

[9] Auzel, F., S. Hubert, and D. Meichenin, "Very Low Threshold CW Excitation of Superfluorescence at 2.72 μm in Er^{3+}," Europhysics Letters, Vol. 7, 1988, pp. 459–462.

[10] Fesler, K. A., R. F. Kalman, M. J. F. Digonnet, B. Y. Kim, and H. J. Shaw, "Behavior of Broadband Fiber Sources in a Fiber Gyroscope," SPIE Proceedings, Vol. 1171, 1989, pp. 346–352.

[11] Morkel, P. R., "Erbium-Doped Fibre Superfluorescent Source for the Fibre Gyroscope," Proceedings of OFS'89, Paris, Springer Proceedings in Physics, Vol. 44, 1989, pp. 143–148.

[12] Wysocki, P. F., R. F. Kalman, M. J. F. Digonnet, B. Y. Kim, "1.55 μm broadband fiber sources pumped near 980 nm," SPIE Proceedings, Vol. 1373, 1990, pp. 66–77.

[13] Wysocki, P. F., K. Fesler, K. Liu, M. J. F. Digonnet, and B. Y. Kim, "Spectrum Thermal Stability of Nd- and Er-Doped Fiber Sources," SPIE Proceedings, Vol. 1373, 1990, pp. 234–245.

[14] Morkel, P. R., R. I. Laming, and D. N. Payne, "Noise Characteristics of High-Power Doped-Fibre Superluminescent Sources," Electronics Letters, Vol. 26, 1990, pp. 96–98.

[15] Lefèvre, H. C., S. Vatoux, M. Papuchon, and C. Puech, "Integrated Optics: A Practical Solution for the Fiber-Optic Gyroscope," SPIE Proceedings, Vol. 719, 1986, pp. 101–112 [SPIE MS 8, pp. 562–573].

[16] Miyajima, Y., "Studies on High-Tensile Proof Tests of Optical Fibers," Journal of Lightwave Technology, Vol. 1, 1983, pp. 340–346.

[17] Kurkjian, C. R., J. T. Krause, and M. J. Matthewson, "Strength and Fatigue of Silica Optical Fibers," Journal of Lightwave Technology, Vol. 7, 1989, pp. 1360–1370.

[18] Ezekiel, S., E. Udd, eds., "Fiber Optic Gyro: 15th Anniversary Conference," SPIE Proceedings, Vol. 1585, 1991.
[19] Suchosky, P. G., T. K. Findakly, and F. L. Leonberger, "$LiNbO_3$ Integrated Optical Components for Fiber-Optic Gyroscopes," SPIE Proceedings, Vol. 993, 1988, 240–243.
[20] Szafraniec, B., C. Laskoskie, and D. Ang, "High-Performance Ti-Indiffused $LiNbO_3$ Multifunction Chip for Use in Fiber Optic Gyros," SPIE Proceedings, Vol. 1585, 1991, pp. 393–404.
[21] Johnstone, W., S. Carey, and B. Culshaw, "Developments in the Characterisation and Performance of Optical Fibre Polarising Devices Using Thin Metal Films," SPIE Proceedings, Vol. 1585, 1991, pp. 365–370.
[22] Dandridge, A., and H. F. Taylor, "Noise and Correlation Effects in GaAlAs Broadband Sources," IEEE Journal of Lightwave Technology, Vol. 5, 1987, pp. 689–693.

Chapter 10
Alternative Approaches for the I-FOG

10.1 ALTERNATIVE OPTICAL CONFIGURATIONS

The minimum configuration (see Section 3.2) using a single-mode filter at the common input-output port of the ring interferometer and a phase modulation-demodulation to bias the response is now almost universally used. In the early days of gyro research, other schemes were proposed, such as the use of an acousto-optic modulator to split or recombine both counterpropagating waves [1,2], or the use of a polarization splitter and a quarter-wave plate to passively bias the signal [3]. These schemes, however, are not significantly simpler than the minimum configuration, and their performances are limited, since they do not fully respect reciprocity.

Among these "nonreciprocal" approaches, one solution, the use of a [3 × 3] coupler [4], is still pursued [5,6] because of its great simplicity. It does not yield very good performances, but it is an interesting scheme for a very low-cost fiber gyro. It is based on the intrinsic phase shift induced in the evanescent wave coupling of a [3 × 3] coupler.

As a matter of fact, we have already seen (Section 3.3.1) that a [2 × 2] coupler yields a $\pi/2$ phase shift for the coupled wave. At the reciprocal port of the interferometer, both waves have experienced the same coupling phase shift and are in phase, while at the nonreciprocal port one wave has experienced twice this $\pi/2$ phase shift and the other one none, which yields a π phase difference. In a [3 × 3] coupler, each coupled wave experiences a $2\pi/3$ phase shift, and by connecting the coil on the two coupled output ports (Figure 10.1), both counter-propagating waves experience the $2\pi/3$ phase shift at the input, while, at the output, one is transmitted and the other is coupled with a second $2\pi/3$ phase shift. Therefore, they interfere, in absence of rotation, with a $2\pi/3$ phase difference at one

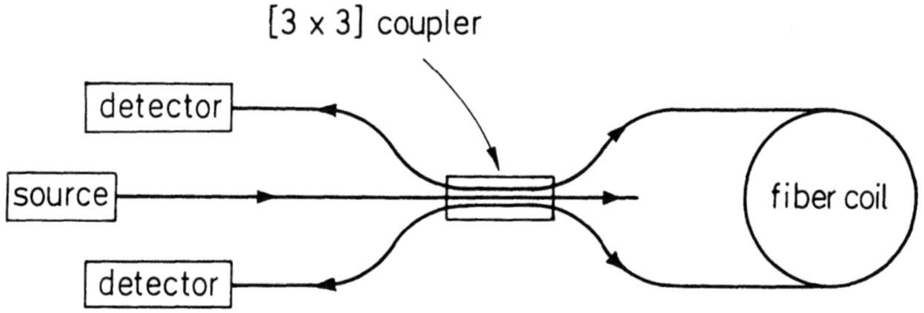

Figure 10.1 Fiber gyro using a [3 × 3] coupler.

output free port of the interferometer and with a $-2\pi/3$ phase difference at the other free port. By taking the power difference between these two ports, a sine biased signal is obtained, but with a very simple electronic circuit.

It is possible to easily check that the phase shift has to be $2\pi/3$. Let us consider a symmetrical [3 × 3] coupler where all the ports are equivalent (Figure 10.2(a)). They can be made with a fused-tapered technique [6] similar to the one used in a [2 × 2] coupler (see Appendix 2). Assuming that three waves are entering the coupler in phase and with the same input power, because of symmetry, the power must be equal in the three ports at the output. At each port there is interference between three waves with the same modulus $\sqrt{A_{in}A^*_{in}}/\sqrt{3}$ of their amplitude (where A_{in} is the input amplitude at the three ports). The transmitted wave does not experience any phase shift, while both coupled waves have the same additional phase shift. Considering a phasor diagram (Figure 10.2(b)), the output amplitude

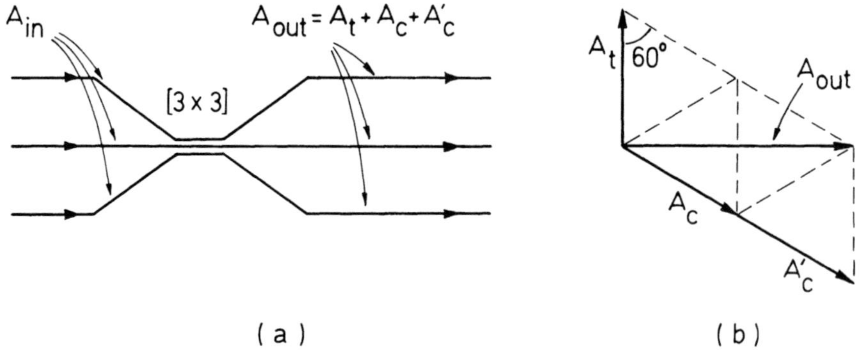

Figure 10.2 [3 × 3] coupler: (a) uniform excitation; (b) phasor diagram of the output amplitude.

A_{out} is the vectorial sum of a transmitted amplitude A_t and two coupled amplitudes A_c and A'_c. The moduli of these three amplitudes are equal. Simple geometrical rules of equilateral triangles show that if the phase shift of A_c and A'_c is $2\pi/3$, the modulus $\sqrt{A_{out}A^*_{out}}$ equals $\sqrt{3}\sqrt{A_tA^*_t}$ (i.e., the modulus $\sqrt{A_{in}A^*_{in}}$ of each input wave).

This gyro configuration is very attractive in terms of simplicity, but it faces the same problem as the use of the nonreciprocal free port of a [2 × 2] coupler (see Section 3.3.1): the coupler loss induces a spurious phase difference and there is a problem of polarization nonreciprocity, since a polarizer cannot be used at the common input-output port. This last situation is improved with the use of an unpolarized source (see Section 3.4.4) [5], but this approach remains limited to low-performance applications.

10.2 ALTERNATIVE SIGNAL PROCESSING SCHEMES

10.2.1 Open-Loop Scheme With Use of Multiple Harmonics

As we have already seen (Chapter 8), closed-loop schemes using phase ramp feedback provide by far the best performances of scale factor, but they require the use of wideband integrated optic phase modulators. The all-fiber approach (see Section 3.3.3) yields very good sensitivity, but wideband phase modulation has not been demonstrated in a practical all-fiber form.

Numerous signal processing schemes compatible with an all-fiber configuration avoid the use of integrated optics, even if they are not preferred anymore. A first solution uses the usual modulation-demodulation technique with a sine wave (see Section 3.2.2), but also considers the various harmonic components of the detector signal [7,8]. The first and other odd harmonics provide a biased sine signal of the rotation-induced phase difference $\Delta\phi_R$, while the even harmonics provide a cosine signal, and through calculation it is possible to recover the value of $\Delta\phi_R$. Such an open-loop approach, however, is limited in practice to a typical scale factor accuracy of 0.1% to 1% because of the unperfect stability of the gain of the demodulation chains.

10.2.2 Second Harmonic Feedback

An all-fiber gyro is not bound to an open-loop scheme. In particular, the first proposed closed-loop scheme compatible with an all-fiber piezoelectric phase modulator was the use of the second-harmonic feedback [9,10]. As seen in Section 3.2.3 and Figure 3.10, an additional second-harmonic phase modulation yields an unbalanced biasing modulation, which may be used to compensate for the rotation-induced phase difference. Such an approach, however, yields a nonlinear response, and performances depend on the stability of the phase modulator, which is difficult to control accurately.

10.2.3 Gated Phase Modulation Feedback

The gated phase modulation approach is conceptually derived from the harmonic feedback scheme. We have already seen that any phase modulation $\phi_m(t)$ yields a modulation $\Delta\phi_m(t)$ of the phase difference, with $\Delta\phi_m(t) = \phi_m(t) - \phi_m(t - \Delta\tau_g)$ (see Sections 3.2.2 and 8.2.2). Since the mean value of a difference is the difference of the mean values, the mean modulation $<\Delta\phi_m>$ of the phase difference is

$$<\Delta\phi_m(t)> = <\phi_m(t) - \phi_m(t - \Delta\tau_g)> = <\phi_m(t)> - <\phi_m(t - \tau)> \quad (10.1)$$

and both mean values $<\phi_m(t)>$ and $<\phi_m(t - \Delta\tau g)>$ are perfectly equal because of reciprocity; therefore,

$$<\Delta\phi_m> = <\phi_m> - <\phi_m> = 0 \quad (10.2)$$

This particularly applies to the case of the phase ramp (Section 8.2.2); however, because of the reset, the amplitude of $\Delta\phi_{PR}$ is then much larger than the linear part of the sine response of the interferometer, and the average of the detector signal is no longer equal to the null value of the average modulation of the phase difference $<\Delta\phi_{PR}>$. However, when the amplitude of $\Delta\phi_m$ remains in this linear range, its mean feedback effect should be zero as the mean value of the phase modulation. Harmonic feedback is working because the demodulation at the fundamental frequency takes into account only one half-period of the second harmonic component. The opposite half-period is in quadrature with respect to the demodulation at the fundamental frequency and can be considered as "gated out" (Figure 10.3). Therefore, the mean effect of the "gated" second harmonic modulation is not zero, even if its mean value does equal zero.

For a lower frequency of feedback modulation, an equivalent effect is obtained with an actual electronic gating of the detector signal over one half-period of feedback modulation [11]. This yields a nonlinear response similar to the harmonic feedback case and with the same problem of control of modulator stability.

It is possible to linearize the response with a combined phase modulation feedback [12] using the sum of a low frequency sine modulation and a specific amount of its second harmonic component, which "flattens" the feedback modulation during the time when the detector is turned on (Figure 10.4).

Another principle of gated modulation feedback may also be implemented with a high-frequency square-wave modulation when an integrated-optic modulator is used [13]. This has the advantage of a linear response without an accurate control of harmonic ratio, in contrast to the previous case.

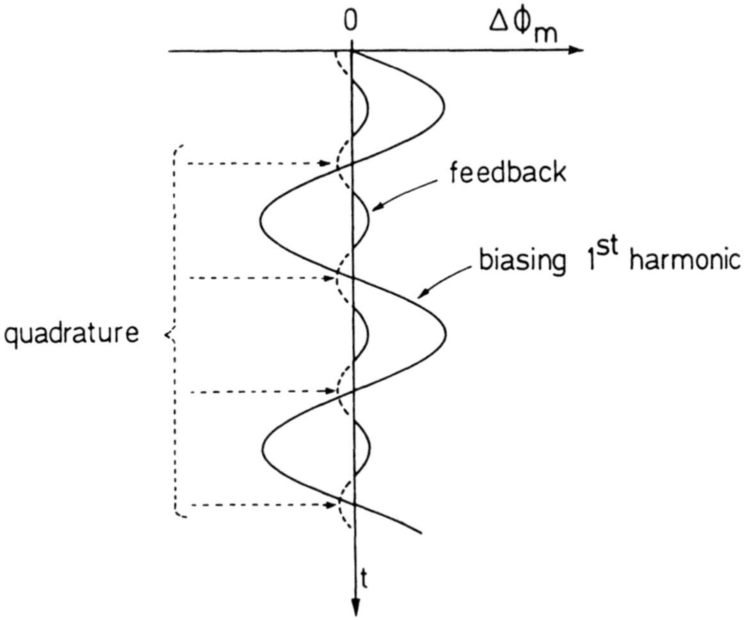

Figure 10.3 Actual gating of one half-period of the second harmonic with demodulation at the fundamental frequency.

All these schemes suffer, however, from the instability of the phase modulator response. They do not have the equivalent of the 2π-reset of the phase ramp (see Sections 8.2.2 and 8.2.3) to control it. Like the situations discussed in Section 8.3.2, these methods are, in principle, wavelength-independent, but the modulator response drift does not allow this fact to be taken advantage of in practice.

10.2.4 Heterodyne and Pseudo-Heterodyne Schemes

Heterodyne techniques are commonly used in interferometry to avoid the problem of the basic nonlinearity of the cosine response. A frequency shifter is placed on one arm of the interferometer, which yields a beating of the output signal, since two waves interfere with different frequencies. The output signal becomes (see Appendix 1)

$$I = I_1 + I_2 + 2\sqrt{I_1 I_2}\cos(2\pi\Delta f t + \Delta\phi) \qquad (10.3)$$

where Δf is the frequency difference and $\Delta\phi$ is the optical phase difference induced by the path imbalance of the interferometer. This value $\Delta\phi$ is then measured

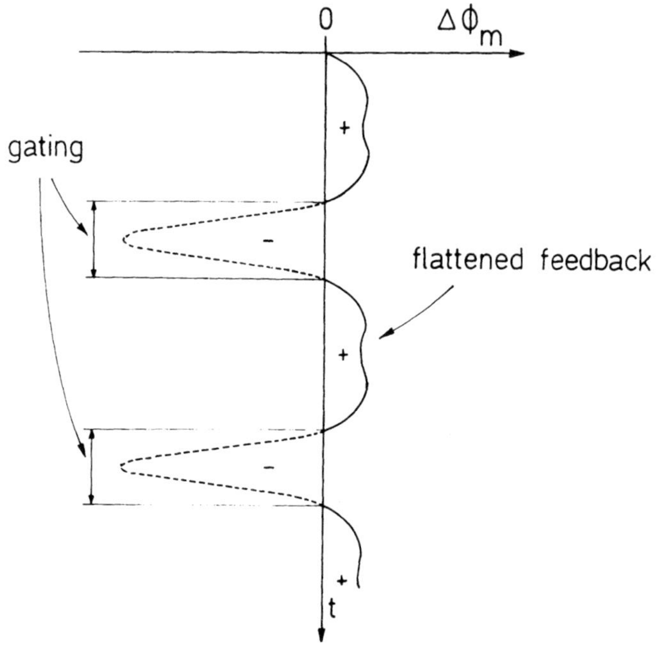

Figure 10.4 Combined phase modulation for gated feedback.

linearly with an electronic phase meter, which compares the phase of the interference beating and the reference phase of the modulation voltage of the frequency shifter.

Several schemes have been proposed to apply this technique to a ring interferometer [2,14,15]; but if a frequency shifter is used, both counterpropagating paths must be separated in order to apply the shifting on only one wave. This destroys reciprocity, and even if some common-mode rejection may be used [14,15], it does not provide a bias stability comparable to the one obtained with a reciprocal configuration. Note that the frequency shift feedback described in Section 8.2.1 is not a heterodyne technique, despite the use of an acousto-optic frequency modulator: both counterpropagating waves experience the frequency shift, and they interfere with the same frequency without beating.

To avoid this problem of nonreciprocity induced by separating the counterpropagating waves, the implementation of a heterodyne scheme with an integrated optic phase modulator in a typical configuration (i.e., placed at one end of the fiber coil) was proposed [16]. A sawtooth electrical generator is sent into an electrical integrator and the output voltage is then applied to the phase modulator.

As seen in Section 8.2.2 a sawtooth modulation is equivalent to a frequency shift, and, as seen in Section 6.1, the ring interferometer responds to phase modulation as a differentiator. Therefore, the combination of frequency shifting, integration, and differentiation results in a constant frequency shift between both counterpropagating waves, and the optical phase in the interferometer may be measured with an electronic phase meter, as in any heterodyne scheme. However, the value of this frequency shift depends on the amplitude of the sawtooth and on the gain of the electrical integrator, which modifies the scale factor. This method requires "hardware" comparable to that needed in phase ramp schemes (see Sections 8.2.2 and 8.2.3) without obtaining the same performances.

On the other hand, "pseudo-heterodyne" techniques have also been proposed [17,18], which use an all-fiber piezoelectric phase modulator and are therefore compatible with an all-fiber gyroscope. It is based on the use of a large amplitude of phase modulation that scans several fringes. Contrary to the usual demodulation technique (see Section 3.2.2), which compares the value of the detector signal at the peaks of the phase modulation, the "pseudo-heterodyne" technique analyzes the phase of the modulated detector signal with a gating when the optical phase modulation has a high slope. This yields the equivalent of a heterodyning frequency upshift for the positive slope and a downshift for the negative slope, since the frequency is the derivative of the phase. When the interferometer is balanced, both beatings are symmetrical; but when there is an additional optical phase difference, these two beatings are shifted in opposite directions (Figure 10.5), and by comparing their phases, the value of the optical phase difference can be retrieved. This technique, though, requires careful control of the amplitude of the phase modulation, which limits its practicability.

10.2.5 Beat Detection With Phase Ramp Feedback

An alternative approach to the simple phase ramp feedback (see Section 8.2.2) was proposed [19] that uses additional couplers to superimpose a Mach-Zehnder interferometer on the Sagnac interferometer. The phase ramp used to null out the phase difference in the gyro generates an interference beating in the Mach-Zehnder interferometer, and the counting of the beats yields an incremental angular measurement of the rotation.

If it were possible to generate infinite ramps, this approach would be interesting because the beats yield a direct calibration of the ramp slope. However, the ramp modulation has to be reset because of the limited range of driving voltage, and it requires control of the 2π-reset, as in the case of the simple serrodyne scheme. Therefore, this approach has no real advantage, particularly because it requires a much more complex integrated optic circuit (Figure 10.6) [20].

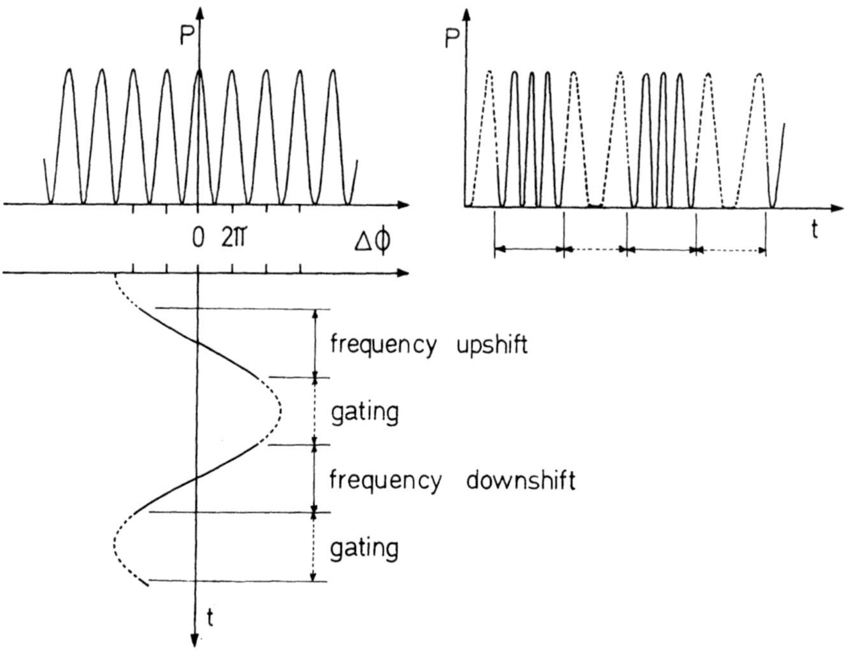

Figure 10.5 Principle of the "pseudo-heterodyne" scheme.

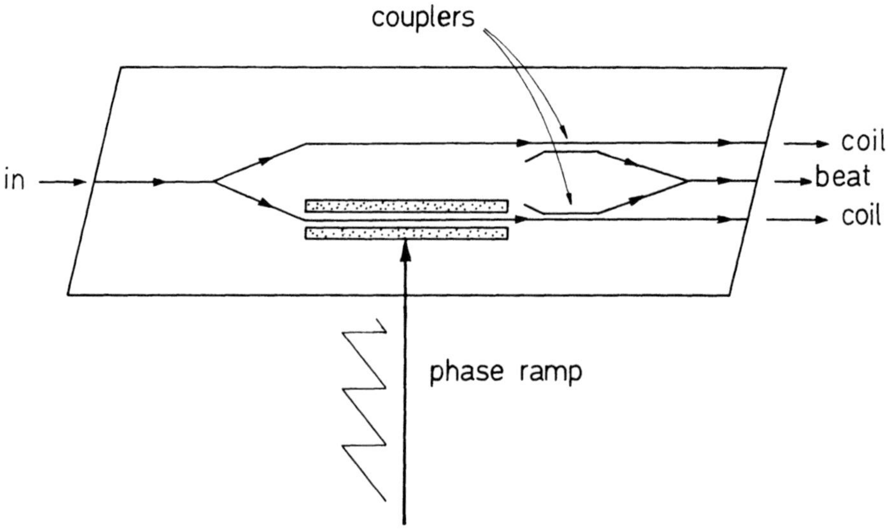

Figure 10.6 Integrated optic circuit for beat detection.

10.2.6 Dual Phase Ramp Feedback

Phase ramp feedback techniques require the use of integrated optics to get the high modulator bandwidth required by the resets. However, an alternative dual phase ramp feedback, using a triangular waveform instead of a sawtooth, has been proposed [21]. This approach, which has no fast reset, is compatible with the all-fiber piezoelectric modulator.

At rest, the positive slope of the triangular wave form $\phi_{PR}(t)$ induces a π rad phase difference $\Delta\phi_{PR}$ while the negative slope induces a $-\pi$ rad phase difference (Figure 10.7(a)). In rotation, the feedback loop keeps the system locked on $\pm\pi$, and it yields a difference of duration between the positive and negative slopes (Figure 10.7(b)), which is proportional to the rotation rate. This dual ramp technique may be also implemented in a digital form.

This approach requires a larger phase modulation amplitude than the more "conventional" phase ramp technique, but has some interest, particularly since it is compatible with the all-fiber configuration of the fiber gyroscope.

10.3 EXTENDED DYNAMIC RANGE WITH MULTIPLE WAVELENGTH SOURCE

Most fiber gyros work over an unambiguous dynamic range of $\pm\Omega_\pi$, which corresponds to a Sagnac phase difference of $\pm\pi$ rad (see Section 2.3.1). However, it is possible to work over several fringes, even with a broadband source, since the contrast is preserved over the coherence length that corresponds to several wavelengths; but there is an ambiguity. For applications where the gyro is turned on while being in the unambiguous range, it is possible to count the fringes that are passed and to keep a valid measurement over an extended dynamic range.

Now, if several wavelengths are used, the phase measurement varies with the wavelength, and it is possible to recognize the fringe order which increases the true unambiguous dynamic range [22].

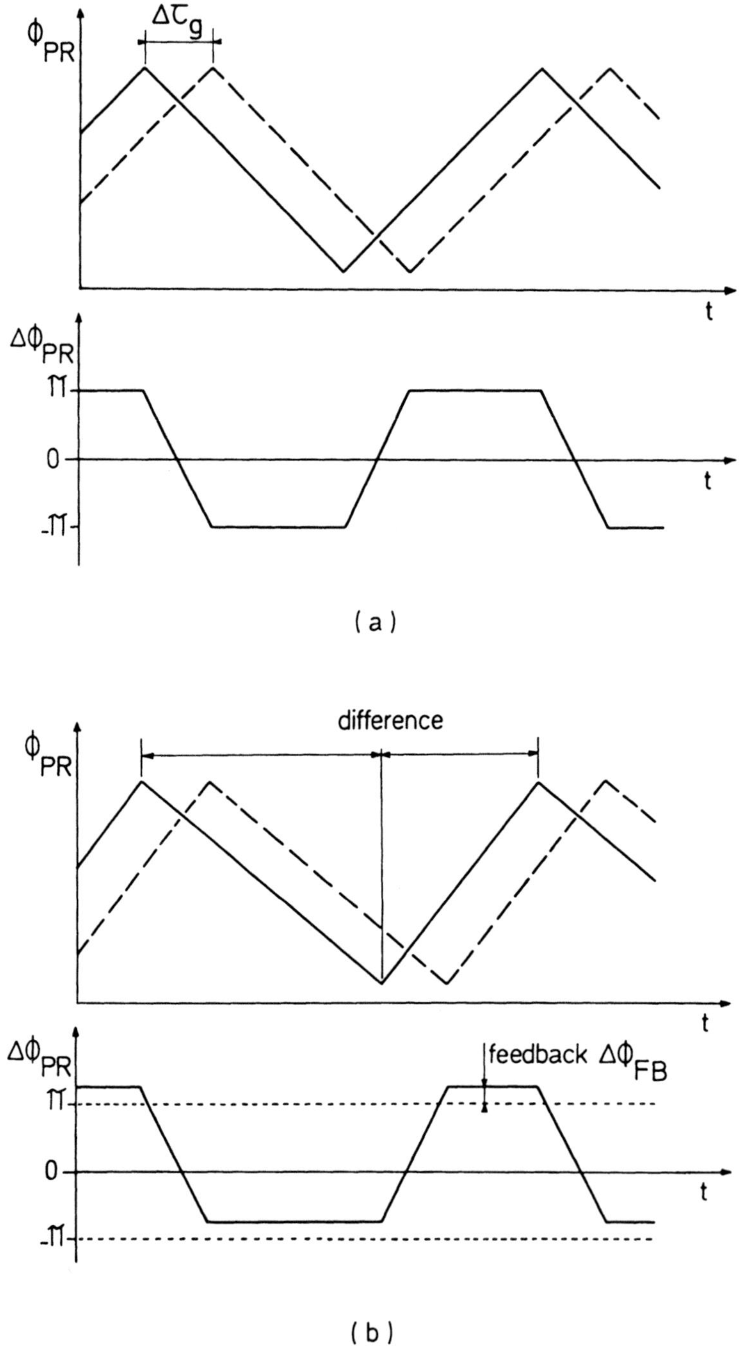

Figure 10.7 Dual phase ramp: (a) at rest; (b) in rotation.

REFERENCES

[1] Udd, E., and R. F. Cahill, "Compact Fiber-Optic Gyro," Springer Series in Optical Sciences, Vol. 32, 1982, pp. 189–194 (SPIE MS8, pp. 189–194).

[2] Hotate, K., N. Okuma, M. Higashiguchi, and N. Niwa, "Rotation Detection by Optical Heterodyne Fiber Gyro With Frequency Output," Optics Letters, Vol. 7, 1982, pp. 331–333 (SPIE MS8, pp. 414–416).

[3] Jackson, D. A., A. D. Kersey, and A. C. Lewin, "Fibre Gyroscope With Passive Quadrature Detection," Electronics Letters, Vol. 20, 1984, pp. 399–401 (SPIE MS8, 417–418).

[4] Sheem, S. K., "Fiber-Optic Gyroscope With [3 × 3] Directional Coupler," Applied Physics Letters, Vol. 37, 1980, pp. 869–871 (SPIE MS8, pp. 167–169).

[5] Burns, W. K., R. P. Moeller, and C. A. Villaruel, "Observation of Low Noise in a Passive Fibre Gyroscope," Electronics Letters, Vol. 18, 1982, pp. 648–650 (SPIE MS8, 167–169).

[6] Poisel, H., and G. F. Trommer, "Low Cost Fiber Optic Gyroscope," SPIE Proceedings, Vol. 1169, 1989, pp. 361–372.

[7] Böhm, K., P. Marten, E. Weidel, and K. Petermann, "Direct Rotation-Rate Detection With a Fibre-Optic Gyro by Using Digital Data Processing," Electronics Letters, Vol. 19, 1983, pp. 997–999 (SPIE MS8, pp. 378–379).

[8] Frigo, N. J., "A Constant Accuracy, High Dynamic Range Fiber Optic Gyroscope," SPIE Proceedings, Vol. 719, 1986, pp. 155–159 (SPIE MS8, pp. 437–441).

[9] Kim, B. Y., H. C. Lefèvre, R. A. Bergh, and H. J. Shaw, "Harmonic Feedback Approach to Fiber-Gyro Scale Factor Stabilization," Proceedings of OFS 1/'83, IEE, London, Vol. 221, 1983, pp. 136–137 (SPIE MS8, pp. 429–431).

[10] Kim, B. Y., H. C. Lefèvre, R. A. Bergh, and H. J. Shaw, "Response of Fiber Gyros to Signals Introduced at the Second Harmonic of the Bias Modulation Frequency," SPIE Proceedings, Vol. 425, 1983, pp. 86-89 (SPIE MS8, pp. 380–383).

[11] Kim, B. Y., and H. J. Shaw, "Gated Phase-Modulation Feedback Approach to Fiber-Optic Gyroscope," Optics Letters, Vol. 9, 1984, pp. 263–265 (SPIE MS8, pp. 387–389).

[12] Kim, B. Y., and H. J. Shaw, "Gated Phase-Modulation Approach to Fiber-Optic Gyroscope With Linearized Scale Factor," Optics Letters, Vol. 9, 1984, pp. 375–377 (SPIE MS8, pp. 393–395).

[13] Page, J. L., "Multiplexed Approach for the Fiber Optic Gyro Inertial Measurement Unit," SPIE Proceedings, Vol. 1367, 1990, pp. 93–102.

[14] Culshaw, B., and I. P. Giles, "Frequency Modulated Heterodyne Optical Fiber Sagnac Interferometer," Journal of Quantum Electronics, Vol. QE-18, 1982, pp. 690–693 (SPIE MS8, pp. 410–413).

[15] Ohtsuka, Y., "Optical Heterodyne Detection Schemes for Fiber-Optic Gyroscopes," SPIE Proceedings, Vol. 954, 1988, pp. 617–624 (SPIE MS8, pp. 421–428).

[16] Eberhard, D., and E. Voges, "Fiber Gyroscope With Phase-Modulated Single-Sideband Detection," Optics Letters, Vol. 9, 1984, pp. 22–24 (SPIE MS8, pp. 384–386).

[17] Kersey, A. D., A. C. Lewin, and D. A. Jackson, "Pseudo-Heterodyne Detection Scheme for the Fibre Gyroscope," Electronic Letters, Vol. 20, 1984, pp. 368–370 (SPIE MS8, pp. 419–420).

[18] Kim, B. Y., and H. J. Shaw, "Phase-Reading, All-Fiber-Optic Gyroscope," Optics Letters, Vol. 9, 1984, pp. 378–380 (SPIE MS8, pp. 390–392).

[19] Goss, W. C., "Fiber Optic Gyro Development at the Jet Propulsion Laboratory," SPIE Proceedings, Vol. 719, 1986, pp. 113–121.

[20] Minford, W. J., F. T. Stone, B. R. Youmans, and R. K. Bartman, "Fiber Optic Gyroscope Using an Eight-Component $LiNbO_3$ Integrated Optic Circuit," SPIE Proceedings, Vol. 1169, 1989, pp. 304–322.

[21] Bergh, R. A., "Dual-Ramp Closed-Loop Fiber-Optic Gyroscope," SPIE Proceedings, Vol. 1169, 1989, pp. 429–439.

[22] Kersey, A. D., A. Dandrige, and W. K. Burns, "Two-Wavelength Fibre Gyroscope With Wide Dynamic Range," Electronics Letters, Vol. 22, 1986, pp. 935–937 (SPIE MS8, pp. 442–443).

Chapter 11

Resonant Fiber-Optic Gyroscope (R-FOG)

11.1 PRINCIPLE OF OPERATION OF AN ALL-FIBER RING CAVITY

As described in Section 2.2.2, the resonant fiber-optic gyroscope, or R-FOG, uses a recirculating ring resonant cavity [1] to enhance the rotation-induced Sagnac effect. Theoretically, the sensitivity of a R-FOG is comparable to the sensitivity of an I-FOG that has a fiber length $\mathfrak{F}/2$ times larger (where $\mathfrak{F}$ is the finesse of the passive ring cavity). This possibility of using a shorter fiber length has looked attractive, but the "serendipity" of the two-wave interferometer does not apply to the resonant approach, and the R-FOG faces much more difficult technical challenges, particularly because it requires the use of a very coherent light source, and the various parasitic effects cannot be reduced as simply nor as efficiently as in the case of the I-FOG, where a broadband low-coherence source is a very good solution to these problems.

The principle of a ring cavity is very similar to that of a Fabry-Perot cavity (see Appendix 1). There is multiple interference between the recirculating waves instead of the reflected waves. An all-single-mode-fiber configuration uses low-loss fiber couplers instead of mirrors [1,2]. Assuming a lossless propagation and two similar couplers (Figure 11.1), the finesse $\mathfrak{F}$ of the cavity is

$$\mathfrak{F} = \frac{3}{C} \tag{11.1}$$

where C is the identical low coupling ratio of the two couplers, which replaces, in the formula, the low mirror transmissivity $\mathfrak{T}$ of ordinary Fabry-Perot interfero-

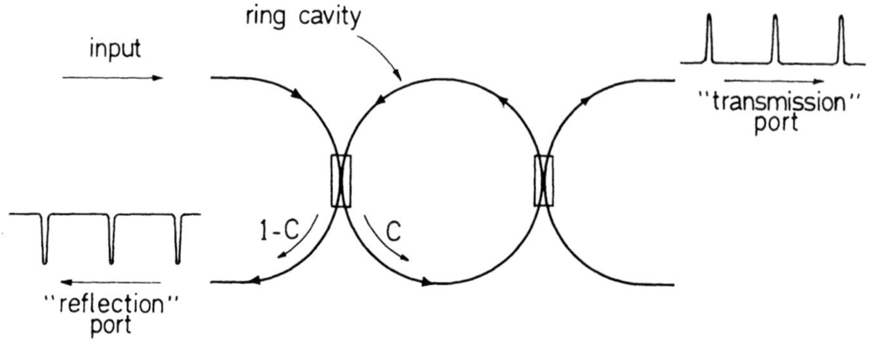

Figure 11.1 All-fiber ring cavity.

meters. There is a periodic spectral response (in spatial frequency σ; i.e., inverse of wavelength λ). In particular, the "transmission" response is

$$\mathfrak{T}_r(\sigma) = \frac{1}{1 + F\sin^2(\pi n L_r \sigma)} \tag{11.2}$$

where nL_r is the optical length of one circulation and the coefficient F is equal to $4(1 - C)/C^2$. There is a resonant effect, and the light is transmitted when the optical length nL_r of one round trip along the ring cavity is equal to an integral number of wavelengths. Like the Fabry-Perot cavities, the periodicity of the response is called the free spectral range $\Delta\sigma_{\text{free}} = 1/nL_r$. It can be expressed in terms of temporal frequency $\Delta f_{\text{free}} = c \cdot \Delta\sigma_{\text{free}} = c/nL_r$ or in terms of angular frequency $\Delta\omega_{\text{free}} = 2\pi\Delta f_{\text{free}} = 2\pi c/nL_r$. The full width at half maximum of the narrow transmission peaks is simply related to the free spectral range, with

$$\Delta\sigma_{FWHM} = \Delta\sigma_{\text{free}}/\mathfrak{F} \tag{11.3}$$

Assuming that the source frequency $\sigma_0 = 1/\lambda_0$ is matched on a resonance peak of the cavity, the transmitted power varies as a function of the rotation rate as

$$P(\Omega) = P_{\max}\frac{1}{1 + F\sin^2(\pi\Delta L_R'(\Omega)\sigma_0)} \tag{11.4}$$

where $\Delta L_R'(\Omega) = L_r D\Omega/2c$ is the length change (in one direction) of the cavity due to rotation. It is possible to define a free range $\Delta\Omega_{\text{free}}$ (Figure 11.2(a)) in terms of rate with

$$\Delta L_R'(\Delta\Omega_{\text{free}}) = \lambda_0 \tag{11.5}$$

that is,

$$\Delta\Omega_{\text{free}} = \frac{2\lambda_0 c}{L_r D} \tag{11.6}$$

Therefore, the full width at half maximum $\Delta\Omega_{\text{FWHM}}$ of the rate response is

$$\Delta\Omega_{\text{FWHM}} = \frac{\Delta\Omega_{\text{free}}}{\mathfrak{F}} = \frac{2\lambda_0 c}{\mathfrak{F} L_r D} \tag{11.7}$$

The equivalent of $\Delta\Omega_{\text{FWHM}}$ for the two-wave interferometer is the rate that corresponds to $\pm\pi/2$ rad (Figure 11.2(b)); that is, Ω_π, defined in Section 2.3.1 as

$$\Omega_\pi = \frac{\lambda_0 c}{2LD} \tag{11.8}$$

It can be seen that the resonance response curve (for one direction) is equivalent to the response of a two-wave interferometer, which would have a coil length L; that is, $\mathfrak{F}/4$ times larger than the length L_r of the ring cavity. As will be seen later,

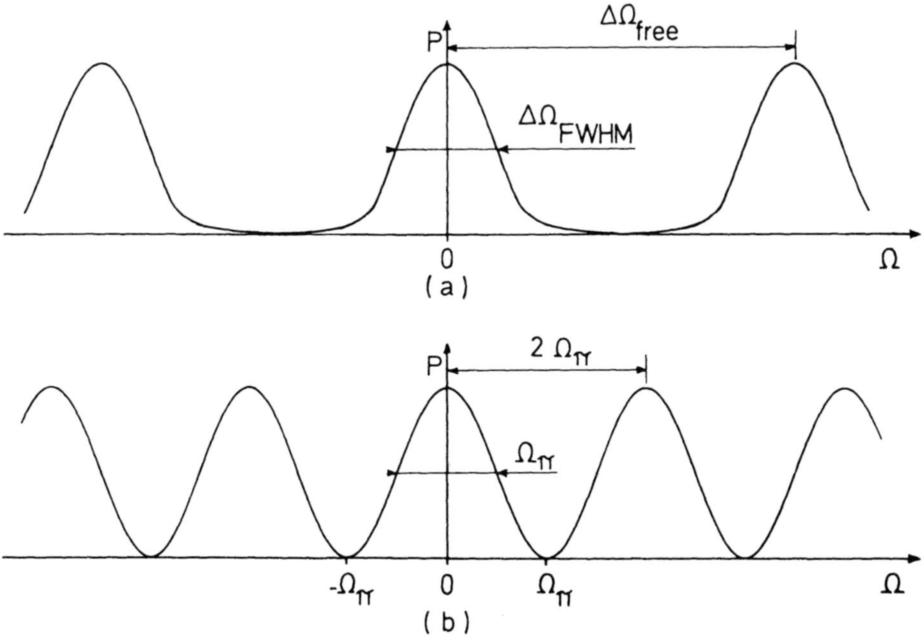

Figure 11.2 Comparison between (a) resonator response (transmission port); (b) interferometer response (reciprocal port).

the sensitivity is actually multiplied by two by comparing the resonances between the opposite directions, which divides the dynamic range by two.

It is also possible to use the "reflection" port (Figure 11.1), where the notch response is complementary to the transmission response composed of peaks. It is important to remember that a resonant cavity is perfectly contrasted if both mirrors (or couplers) have the same transmissivity (or the same coupling) and if the cavity has no loss. In practice, there is always a residual attenuation, but it is still possible to get a perfect contrast for high finesse by using a single-coupler ring with a coupler coupling ratio equal to the loss (Figure 11.3). This loss is thus equivalent to a loss resulting from a coupling in a second coupler, and the notch response is then perfectly contrasted. Note that such a single-coupler cavity may be realized without any fiber splice along the resonant path with the use of a high coupling ratio and low transmission in the coupler [3] (Figure 11.4).

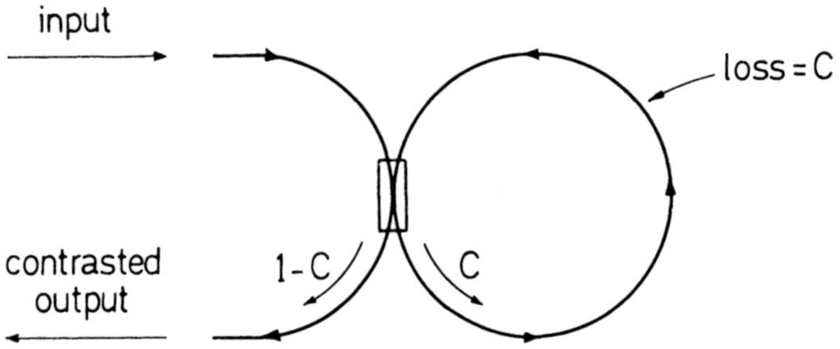

Figure 11.3 Single-coupler fiber ring cavity.

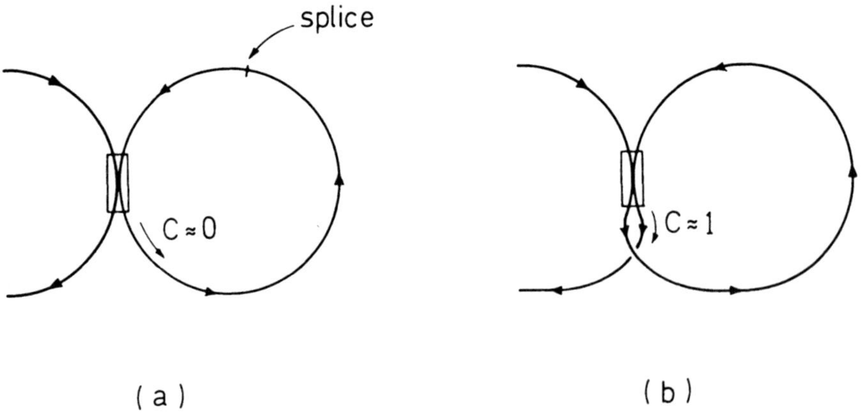

Figure 11.4 (a) Spliced and (b) unspliced single-coupler cavity.

To analyze the resonance of the ring cavity, the source spectrum width has to be narrower than the width of the response peaks. This corresponds to a coherence length longer than the coil length multiplied by the finesse. Therefore, the R-FOG requires a very long source coherence of a few kilometers, since the coil length is typically a few tens of meters and the finesse is on the order of 100. In terms of temporal frequency, this corresponds to a very narrow line width, on the order of 10 to 100 kHz.

11.2 SIGNAL PROCESSING METHOD

The principle of the signal processing method of the R-FOG has some similarities to the case of the I-FOG. In a first step, a modulation-demodulation scheme is used to get an open-loop biased signal that is a derivative of the unmodulated response (Figure 11.5), and in a second step this signal is used as the error signal of a closed-loop processing unit that linearizes and stabilizes the scale factor [4,5].

However, in contrast to the I-FOG, the response is not automaticaly centered on an extremum for zero rotation rate. The source frequency and the cavity length have first to be matched in resonance in one direction, and the rotation is detected in the opposite direction, where the sensitivity is then doubled. The "dithering" modulation required to get a biased demodulated signal is performed by modulating the cavity length with a piezoelectric modulator placed inside the cavity [4,5], or by modulating, outside the cavity, the input light frequency [6], or directly the

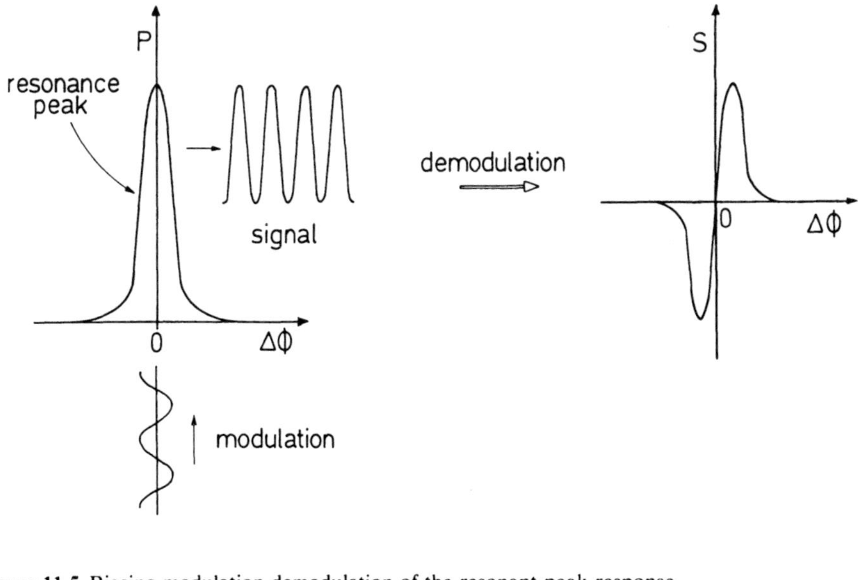

Figure 11.5 Biasing modulation-demodulation of the resonant peak response.

source frequency, particularly with the driving current of a semiconductor laser [7]. Note that a sine modulation of the frequency may of course be obtained with a sine modulation of a frequency shifter, but also with a sine modulation of a phase modulator. As a matter of fact, we have already seen that a frequency is the derivative of a phase, and a sine-modulated phase yields by differentiation a (co)sine modulation of the frequency.

This biasing modulation-demodulation is not defect-free, in contrast to that of the I-FOG, because it does not use the rejections of nonlinearities and spurious intensity modulation brought by the use of the coil as a delay line filter at the proper frequency (see Section 3.2.3). Furthermore, the interferometer response is perfectly symmetrical because it is an autocorrelation function, while the resonance peaks may carry some dissymetry, especially because of coupler loss [8]. Therefore, it is important to make the processing systems of the two opposite directions symmetrical to get a good common mode rejection of their defects.

The complete system is composed of a modulation of the cavity length or of the light frequency. The two counterrotating signals are demodulated, and one is used as an error signal to keep the system on the resonance peak, while the opposite path is used as an error signal, which applies an additional frequency shift through a closed-loop processing circuit (Figure 11.6). The value of this frequency shift is used as the rotation rate signal. It corresponds to the difference Δf_R of resonance

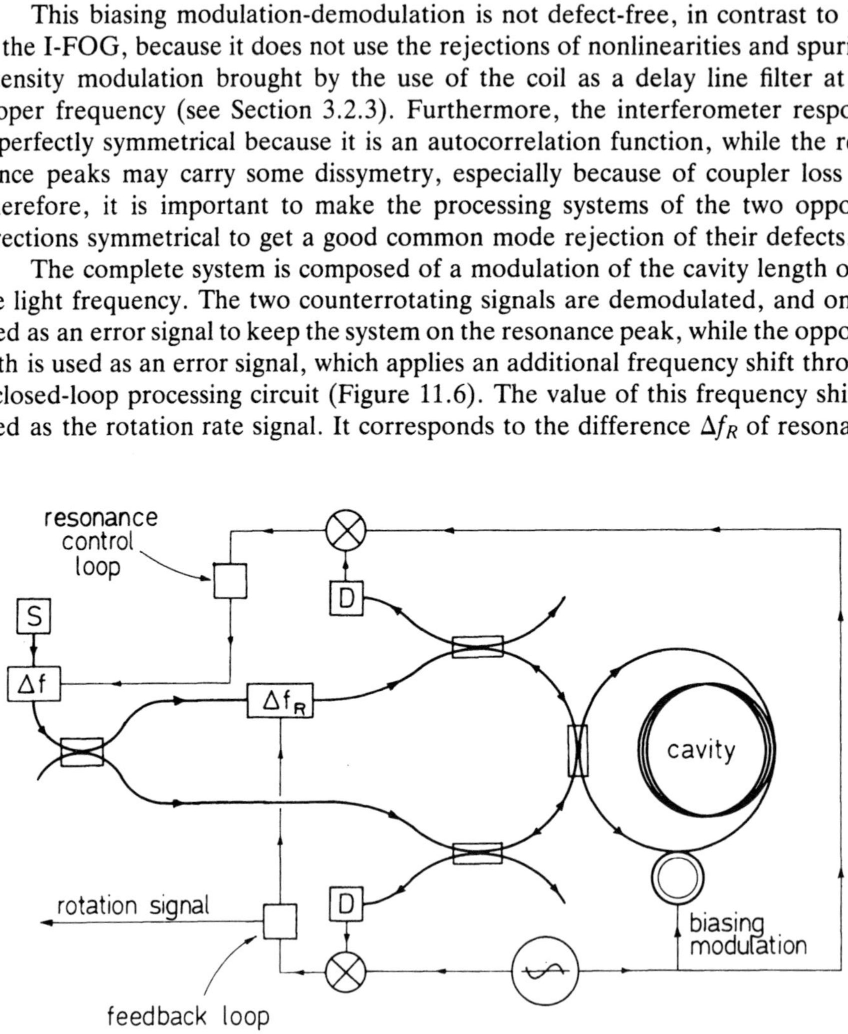

Figure 11.6 Architecture of a resonant fiber-optic gyroscope.

frequency, which is induced by the rotation rate Ω between both counterrotating paths (see Section 2.2.2):

$$\Delta f_R = \frac{ND}{n\lambda} \cdot \Omega \tag{11.9}$$

where D is the coil diameter, N the number of turns, n the index of the fiber, and λ the source wavelength. The frequency shifting required to close the two processing loops was originally performed with bulk acousto-optic Bragg cells [4,5], but it can also be done with sawtooth serrodyne modulation (see Section 8.2.2) applied on integrated-optic phase modulators [9], which preserves the ruggedness of an all-guided approach. It can be seen that if the R-FOG is using a shorter fiber coil than the I-FOG, this advantage is counterbalanced by a higher complexity, since the number of components (couplers and modulators) is nearly doubled in comparison to the minimum configuration of the two-wave ring interferometer.

However, note that, for the same sensitivity, the unambiguous dynamic range of the R-FOG is larger than that of the I-FOG. Their sensitivities are equivalent when $\Delta\Omega_{FWHM}$ is equal to Ω_π, but the range of the resonator is $\pm\Delta\Omega_{free}/2$, which is $\mathfrak{F}/2$ wider than the interferometer range of $\pm\Omega_\pi$.

11.3 RECIPROCITY OF A RING FIBER CAVITY

Reciprocity has been seen to be a fundamental feature of the I-FOG (see Chapter 3), and it is also possible to define a reciprocal configuration of an R-FOG. It should use only the "reflection" ports of the cavity, with an in-line polarizer on the two single-mode fiber leads [10] (Figure 11.7). With such a configuration, each

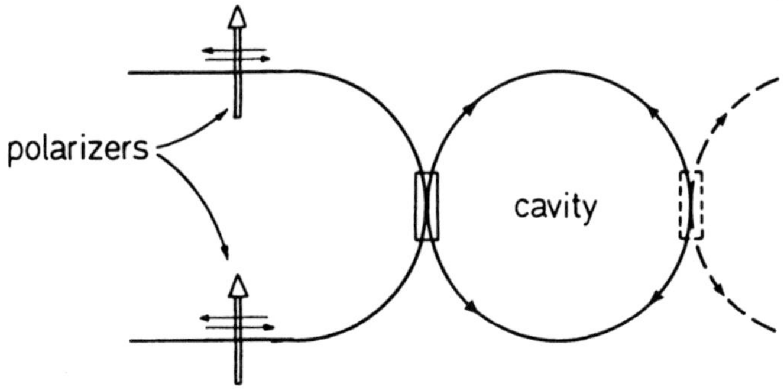

Figure 11.7 Reciprocity of a ring cavity.

couple of counterpropagating waves that recirculates the same number of times along the cavity follows exactly the same path, including the leads, in opposite directions: the two counterpropagating waves accumulate exactly the same phase, and they have exactly the same attenuation because of reciprocity. Since this is valid for all the recirculating couples, the resonance responses of a cavity using such a reciprocal configuration are perfectly identical in both of the opposite directions.

However, the use of a strictly reciprocal configuration is not as fundamental for the R-FOG as it is for the I-FOG, and similar performances are also obtained at the transmission ports of an all-fiber ring resonator [11]. It is possible to consider that the second coupler is just tapping off a small amount of the intensity of the light which resonates inside the cavity and that parasitic phase shifts in the tapping coupler do not affect the measurement of this intensity.

The main problem of reciprocity in the ring resonator comes from polarization and birefringence. As already seen, a single-mode fiber actually transmits two polarization modes with slightly different velocities because of birefringence (see Appendix 2). Therefore, a single-mode-fiber ring resonator has two resonances that correspond to these two polarization modes [3,11–13]. Along one round trip, from one point in the cavity back to the same point, there is a total amount of birefringence that can be represented by a rotation vector on the Poincaré sphere (see Appendix 2). The two orthogonal states of polarization that correspond to the intersection of the sphere with this vector are the two eigenstates of polarization of the resonator at the considered point. As the light propagates along the resonator fiber, these two states may change as a function of the birefringence, but they remain orthogonal and are back into their original state after one round trip. This condition is necessary to get resonance by multiple interference of the various recirculating waves, since they must be in the same state of polarization to be combined constructively.

Then the cavity has two sets of resonance corresponding to its two polarization modes that propagate over one round trip along different optical lengths, $n_1 L_r$ and $n_2 L_r$, because of birefringence. By reciprocity, these two sets are identical in both of the opposite directions (when the system does not rotate!). It seemed at first that an R-FOG could work with ordinary single-mode fiber by using only the resonance of one polarization mode [6]. However, a precise measurement of the central resonance frequency cannot tolerate a small dissymmetry of the peak or notch response. In practice, there is always some light coupled into the crossed polarization state because of parasitic misalignment of the polarization of the input wave, and the effect of the resonance response of this parasitic crossed mode degrades the symmetry of the main resonance signal. The reciprocal configuration described earlier does not avoid this dissymmetry, but ensures that it is identical in both of the opposite directions [10]. However, polarization fading may occur and polarizer rejection is limited, which yields parasitic amplitude-type effects because of the high coherence of the source. It is not possible to take advantage of depolarization effects, as in the case of I-FOG that works with a broadband source (see Chapter 5).

To avoid polarization fading and to reduce the influence of the crossed-mode resonance, it would be ideal to make the resonator with single-polarization fiber [13]; but such a fiber is delicate to use and most present systems employ, instead, polarization-preserving fibers [10,14].

With polarization conservation, most of the optical power remains in one mode of the resonator, but there is still some light in the crossed mode because parasitic polarization cross-coupling occurs, particularly in the coupler. There is still a small resonance dip in addition to the main signal. Furthermore, the position of this parasitic dip shifts as a function of temperature with respect to the resonance response of the main mode [14]. This yields an unstable dissymmetry of the resonance peak (Figure 11.8). As a matter of fact, the resonances of the two polarization modes have a periodicity equal to the free spectral ranges n_sL_r and n_fL_r, where n_s and n_f are, respectively, the indexes of the slow mode and of the fast mode. With stress-induced high-birefringence fibers, the index difference $n_s - n_f$ has a typical variation of 10^{-3}/°C, and when a temperature change induces a change of $(n_s - n_f)L_r$ equal to the source wavelength λ, the small parasitic dip shifts over the free spectral range of the main resonance signal. With a resonator length of 10m and a birefringence beat length of 1 mm, this is obtained with a temperature change of only 0.1°C. Note, however, that polarization-preserving fibers using an elliptical core have a much lower temperature dependence (about 10^{-5}/°C instead of 10^{-3}/

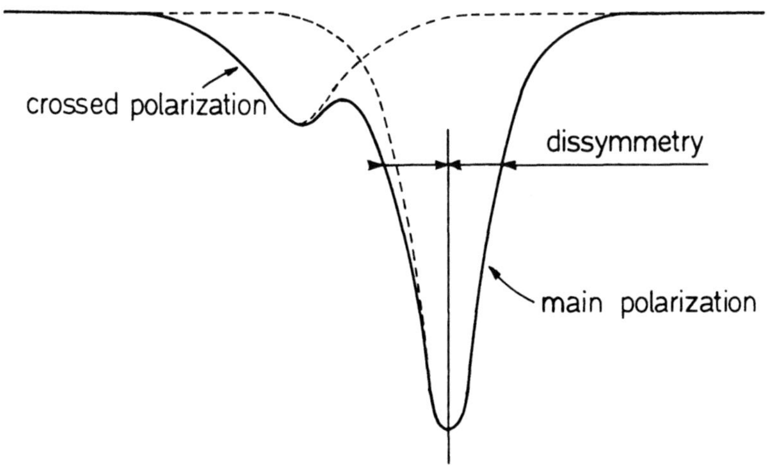

Figure 11.8 Dissymmetry of the response induced by the crossed-polarization response.

°C); but, in practice, they are not often used because of a lower polarization conservation.

To avoid this problem of resonance of the crossed polarization mode, placing a polarizing element inside the cavity has been proposed [13]; however, more recently, a very promising (and subtle) solution has been described [15,16] and demonstrated [11]. The resonator loop is still made of polarization-preserving fiber, but a splice with a 90-deg rotation of the principal axes of the fiber is added inside the cavity (Figure 11.9). With this configuration, despite the two polarization modes of the fiber, the cavity becomes single-mode in terms of polarization. A complete cavity round trip is now composed of a propagation along the cavity in the fast mode and a second propagation in the slow mode. The optical length of the cavity becomes $(n_f + n_s)L_r$ and the free spectral range $1/(n_f + n_s)L_r$. This idea is similar to the Moebius ring, which has only one "side," while a tape has two sides [17]!

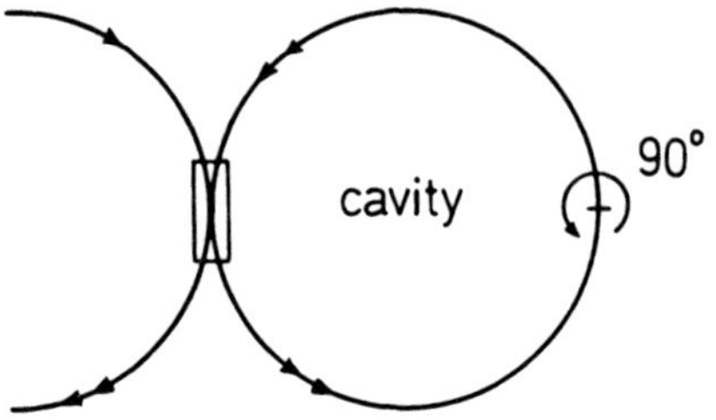

Figure 11.9 Polarization-preserving cavity with a 90-deg rotation of the fiber principal axes.

11.4 PARASITIC EFFECTS IN THE R-FOG

In addition to these problems of reciprocity, the R-FOG faces various parasitic effects similar to the ones encountered in the I-FOG. Solutions to these problems are usually derived from what was proposed for the I-FOG, except, as we have already seen, the fundamental drawback that a broadband source cannot be used because a R-FOG requires a source with a very long coherence length.

This makes Rayleigh backscattering and the Kerr effect very severe limitations to high performance. The effect of Rayleigh backscattering may be reduced by various techniques of phase or frequency modulation [4,5,18,19] to avoid, in the detection band, spurious interference signals between the primary waves and the backscattered waves; but this increases significantly the complexity of the system. The problem of the nonlinear Kerr effect is even worse, since the optical power stored in the cavity is much higher than the input power, and for the same relative power imbalance between both counterpropagating waves, an R-FOG has a Kerr sensitivity enhanced by about one-third the finesse compared to an equivalent I-FOG using a coherent source. Square-wave intensity modulation techniques proposed for the I-FOG to reduce the Kerr sensitivity (see Section 7.2) can also be used for the R-FOG [20], but the precision of modulation required is very difficult to get and the Kerr effect is still regarded as the main limitation to low drift for the R-FOG [21].

The other parasitic effects that are not related to coherence are not significantly different between the R-FOG and the I-FOG. Transient related effects (see Chapter 6) are reduced in the R-FOG [2], since the fiber coil is shorter, but this problem is easily solved in the I-FOG with quadrupolar winding and adequate potting of the fiber coil. Faraday effect (see Section 7.1) also induces a rate error in the R-FOG if the polarization-preserving fiber has a twist variation matched to the perimeter of one turn [22]. The R-FOG also has the same problem of wavelength dependence of the scale factor (see Section 8.3).

Despite similar theoretical performance, the R-FOG still faces very difficult technical problems, and it does not appear to be a very strong challenger of the I-FOG anymore.

Note that, based on this principle of passive ring cavity, an even more ambitious approach was pursued with an integrated-optic single-loop ring cavity, which could have made possible the fabrication of a fully integrated optical gyro by planar mass-duplication techniques [23–29]; but the technological challenges to be faced are very difficult, and this approach is not presently believed to be competitive.

More recently, there has been a proposal [30] to make an active fiber resonator similar to a ring laser gyro, but in a fiber form, using amplification by stimulated Brillouin scattering. Backscattering also induces lock-in around zero, but it will be interesting to see how this novel approach will progress, even if, at this stage, the subject requires more R&D to precisely evaluate its advantages.

REFERENCES

[1] Ezekiel, S., and S. R. Balsamo, "Passive Ring Resonator Laser Gyroscope," Applied Physics Letters, Vol. 30, 1977, pp. 478–480 (SPIE MS8, pp. 457–459).

[2] Schupe, D. M., "Fiber Resonator Gyroscope: Sensitivity and Thermal Non-Reciprocity," Applied Optics, Vol. 20, 1981, pp. 286–289 (SPIE MS8, pp. 460–463).

[3] Stokes, L. F., M. Chodorow, and H. J. Shaw, "All Single-Mode Fiber Resonator," Optics Letters, Vol. 7, 1982, pp. 288–290 (SPIE MS8, pp. 464–466).

[4] Sanders, G. A., M. G. Prentiss, and S. Ezekiel, "Passive Ring Resonator Method for Sensitive Inertial Rotation Measurements in Geophysics and Relativity," Optics Letters, Vol. 6, 1981, pp. 569–571.

[5] Meyer, R. E., S. Ezekiel, D. W. Stowe, and V. J. Tekippe, "Passive Fiber-Optic Ring Resonator for Rotation Sensing," Optics Letters, Vol. 8, 1983, pp. 644–646 (SPIE MS8, pp. 467–469).

[6] Carroll, R., C. D. Coccoli, D. Cardarelli, and G. T. Coate, "The Passive Resonator Fiber Optic Gyro and Comparison to the Interferometer Fiber Gyro," SPIE Proceedings, Vol. 719, 1986, pp. 169–177 (SPIE MS8, pp. 486–494).

[7] Ohtsu, M., and S. Araki, "Using a 1.5 μm DFB InGaAsP Laser in a Passive Ring Cavity-Type Fiber Gyroscope," Applied Optics, Vol. 26, 1987, pp. 464–470 (SPIE MS8, pp. 515–521).

[8] Youngquist, R. C., L. F. Stokes, and H. J. Shaw, "Effect of Normal Mode Loss in Dielectric Waveguide Directional Couplers and Interferometers," Journal of Quantum Electronics, Vol. QE-19, 1983, pp. 1888–1896 (SPIE MS8, pp. 352–360).

[9] Sanders, G. A., G. F. Rouse, L. K. Strandjord, N. A. Demma, K. A. Miesel, and Q. Y. Chen, "Resonator Fiber-Optic Gyro Using $LiNbO_3$ Integrated Optics at 1.5 μm," SPIE Proceedings, Vo! 985, 1988, pp. 202–210 (SPIE MS8, pp. 526–534).

[10] Schröder, W., P. Gröllmann, J. Herth, M. Kemmler, K. Kempf, G. Neumann, and S. Oster, "Progress in Fiber Gyro Development for Attitude and Heading Reference Systems," SPIE Proceedings, Vol. 719, 1986, pp. 162–168 (SPIE MS8, pp. 574–580).

[11] Strandjord, L. K., and G. A. Sanders, "Resonator Fiber Optic Gyro Employing a Polarization-Rotating Resonator," SPIE Proceedings, Vol. 1585, 1991, pp. 163–172.

[12] Lamouroux, B. F., B. S. Prade, and A. G. Orszag, "Polarization Effects in Optical-Fiber Ring Resonators," Optics Letters, Vol. 7, 1982, pp. 391–393.

[13] Iwatsuki, K., K. Hotate, and M. Higashiguchi, "Eigenstate of Polarization in a Fiber Ring Resonator and Its Effect in an Optical Passive Ring Resonator Gyro," Applied Optics, Vol. 25, 1986, pp. 2606–2612 (SPIE MS8, pp. 495–501).

[14] Sanders, G. A., N. Demma, G. F. Rouse, and R. B. Smith, "Evaluation of Polarization Maintaining Fiber Resonator for Rotation Sensing Applications," Proceedings of OFS 5, New Orleans, 1988, pp. 409–412 (SPIE MS8, pp. 522–525).

[15] Sanders, G. A., R. B. Smith, and G. F. Rouse, "Novel Polarization-Rotating Fiber Resonator for Rotation Sensing Applications," SPIE Proceedings, Vol. 1169, 1989, pp. 373–381.

[16] Mouroulis, P., "Polarization Fading Effects in Polarization-Preserving Fiber Ring Resonators," SPIE Proceedings, Vol. 1169, 1989, pp. 400–412.

[17] Schröder, W., Fachhochschule Offenburg, Germany, private conversation.

[18] Iwatsuki, K., K. Hotate, and M. Higashiguchi, "Effect of Rayleigh Backscattering in an Optical Passive Ring Resonator," Applied Optics, Vol. 23, 1984, pp. 3916–3924 (SPIE MS8, pp. 470–478).

[19] Hotate, K., K. Takiguchi, and A. Hirose, "Adjustment-Free Method to Eliminate the Noise Induced by the Backscattering in an Optical Ring-Resonator Gyro," IEEE Photonics Technology Letters, Vol. 2, 1990, pp. 75–77.

[20] Iwatsuki, K., K. Hotate, and M. Higashiguchi, "Kerr Effect in an Optical Passive Ring-Resonator Gyro," Journal of Lightwave Technology, Vol. LT4, 1986, pp. 645–651 (SPIE MS8, pp. 508–514).

[21] Hotate, K., and K. Takiguchi, "Drift Reduction in an Optical Passive Ring-Resonator Gyro," SPIE Proceedings, Vol. 1585, 1991, pp. 116–127.

[22] Hotate, K., and M. Murakami, "Drift of an Optical Passive Ring-Resonator Gyro Caused by the Faraday Effect," Proceedings of OFS 5, New Orleans, 1988, pp. 405–408.

[23] Haavisto, J., and G. A. Pajer, "Resonance Effects in Low-Loss Ring Waveguides," Optics Letters, Vol. 5, 1980, pp. 510–512.

[24] Walker, R. G., and C. D. W. Wilkinson, "Integrated Optical Ring Resonators Made by Silver Ion-Exchange in Glass," Applied Optics, Vol. 22, 1983, pp. 1029–1035.

[25] Honda, K., E. M. Garmire, and K. E. Wilson, "Characteristics of an Integrated Optics Ring Resonator Fabricated in Glass," Journal of Lightwave Technology, Vol. 2, 1984, pp. 714–719.

[26] Naumaan, A., and J. T. Boyd, "Ring Resonator Fabricated in Phosphosilicate Glass Film Deposited by Chemical Vaport Deposition," Journal of Lightwave Technology, Vol. 4, 1986, pp. 1294–1303.

[27] Connors, J. M., and A. Mahapatra, "High Finesse Ring Resonators Made by Silver Ion Exchange in Glass," Journal of Lightwave Technology, Vol. 5, 1987, pp. 1686–1689.

[28] Bismuth, J., P. Gidon, F. Revol, and S. Valette, "Low-Loss Ring Resonators Fabricated From Silicon Based Integrated Optics Technologies," Electronics Letters, Vol. 27, 1991, pp. 722–723.

[29] Adar, R., Y. Shani, C. H. Henry, R. C. Kistler, G. E. Blonder, and N. A. Olsson, "Measurement of Very Low-Loss Silica on Silicon Waveguides With a Ring Resonator," Applied Physics Letters, Vol. 58, 1991, pp. 444–445.
[30] Smith, S. P., F. Zarinetchi, and S. Ezekiel, "Fiber Laser Gyros Based on Stimulated Brillouin Scattering," SPIE Proceedings, Vol. 1585, 1991, pp. 302–308.

Chapter 12
Applications and Trends

12.1 PRESENT STATE OF DEVELOPMENT

If a theoretical analysis always remains relevant, a description of the present state of the art will become, by principle, obsolete. However, the potential of FOG technology is now ascertained, and present results provide a good prospect for future applications. The 15th Anniversary Conference on Fiber Optic Gyros held in September 1991 [1] gave a very complete update of the activity in the various companies, universities, and research centers working on the subject around the world.

In terms of architecture, some general trends may be outlined:

- Most companies are now using the Y-coupler configuration with a closed-loop processing scheme based on phase modulation feedback (usually phase ramp) applied on a multifunction integrated-optic circuit: Honeywell [2], Litton [3], Smith Industries [4] in the U.S., JAE [5], Mitsubishi [6] in Japan, British Aerospace [7], Litef [8], and Photonetics [9] in Europe. Alcatel-SEL [10] in Germany developed early on a product based on the Y-coupler configuration, but it is still using an open-loop approach.
- The all-fiber open-loop approach is pursued by Honeywell for its first products [11], and also by Hitachi [12] in Japan.
- Instead of using separate single-axis gyros, there is a tendency to make three-axis measurement units with the sharing of a single source (Litton [3] and Litef [8]) or with the multiplexing of this source (Smith Industries [4]). These inertial measurement units (IMU) are using, in addition, micro-machined silicon accelerometers.

- Most companies are using polarization-preserving fiber, except Smith Industries [4], which prefers the depolarized approach with an ordinary fiber coil.
- The [3 × 3] coupler approach is pursued by MBB Deutsche Aerospace [13] in Germany for very-low-cost low-performance applications.
- The only companies that are still studying the resonant fiber gyro are Honeywell [14] and British Aerospace [7].

The I-FOG technology is now mature, and a company like Alcatel-SEL [10] has delivered a few hundred gyro products of medium accuracy (10 deg/h range) over the last few years, and other companies like Hitachi [12], Honeywell [14], and Litton [3] are starting production. I-FOG technology is going to be used first for Attitude and Heading Reference Systems (AHRS) for aircraft, which requires a gyro accuracy of 10 to 0.1 deg/h. Honeywell will provide a fiber gyro-based AHRS for the new regional airliner Dornier 328 and has also been selected with this technology for the new Boeing 777 to be part of the Secondary Attitude Air Data Reference Unit (SAARU) [15]. This will be used as a redundant backup system of the primary flight controls using laser gyros. Fiber gyro-based AHRSs are also going to be used for military applications. Smith Industries won a C/AHRS (Magnetic Compass and AHRS) contract from the U.S. Air Force for the development of a fiber gyro system that will retrofit many previous mechanical reference systems [16]. The advantages outlined for these contracts are lower cost and lower maintenance due to the solid-state configuration of the I-FOG.

I-FOGs for such tactical grade applications (i.e., 10 to 0.1 deg/h) typically use 100m to 200m of fiber wound around a 30- to 60-mm diameter. Best bias performances [3,5,6,9,14] are below 1 deg/h (i.e., a few tenths of a microradian in terms of measured optical phase difference), with a dynamic range of 500 to 1500 deg/s and a scale factor accuracy better than 100 ppm. Higher performance is also pursued with larger fiber coils (0.5 to 2 km around 8- to 10-cm diameter). Bias stability in the inertial range (0.01 deg/h) has been demonstrated [2,3,5,6]. Present prototypes are now ruggedized to withstand difficult environments, particularly the temperature range of −55°C up to 90°C, and vibration of more than 10G rms.

12.2 TRENDS FOR THE FUTURE AND CONCLUDING REMARKS

After more than 15 years of research and development, the fiber gyro is now recognized as a crucial technology for many applications of inertial guidance and navigation. Its low-mass solid-state configuration brings unique technical advantages: reliability and lifetime, ability to withstand shocks and vibration, high dynamic range, large bandwidth, nearly instantaneous startup, and low power consumption. Its principle yield is a very useful design versatility to optimize the performances of a specific application by changing the length or the diameter of the sensing coil while keeping the same opto-electronic components and assembly techniques.

The main applications will be in the medium accuracy range (0.1 to 10 deg/h) with compact units (30 to 50 mm in diameter): AHRS for airplanes or helicopters, tactical guidance for missiles or smart ammunitions, but also bore-hole survey, robotics, and even guidance systems for automobiles. It is also a very good candidate for navigation systems aided by GPS (global positioning system by satellites). As the technology progresses, it will also become a significant competitor in the high-performance navigation-grade (0.01 deg/h) domain, particularly in space applications, where lifetime expectancy, low power consumption, and low disturbance of the surrounding structure are fundamental characteristics.

As a new technology using components that can be mass-produced, the fiber gyro brings great expectations of very significant cost reduction, which will further extend the field of inertial guidance techniques.

For the physicist and the signal processing specialist, the fiber gyro is a fascinating subject. The serendipity of the device, which has brought simple solutions to problems apparently complex, will continue to amaze: was it reasonable to expect to be able to measure a length difference of 10^{-5} nm (λ over 10^8) after several hundreds of meters of propagation? (This is the same order of magnitude as the classical definition of the diameter of electron (i.e., 0.6 10^{-5} nm)!)

REFERENCES

[1] Ezekiel, S., and E. Udd, eds., "Fiber Optic Gyros: 15th Anniversary Conference," SPIE Proceedings, Vol.1585, 1991.

[2] Liu, R. Y., T. F. El-Wailly, and R. C. Dankwort, "Test Results of Honeywell's First Generation High-Performance Interferometric Fiber-Optic Gyroscope," SPIE Proceedings, Vol. 1585, 1991, pp. 262–275.

[3] Pavlath, G. A., "Production of Fiber Gyros at Litton Guidance and Control Systems," SPIE Proceedings, Vol. 1585, 1991, pp. 2–6.

[4] Page, J. L., "Multiplexed Approach for the Fiber Optic Gyro Inertial Measurement Unit," SPIE Proceedings, Vol. 1367, 1990, pp. 93–102.

[5] Sakuma, K., "Fiber-Optic Gyro Production at JAE," SPIE Proceedings, Vol. 1585, 1991, pp. 8–16.

[6] Hayakawa, Y., and A. Kurokawa, "Fiber-Optic Gyro Production at Mitsubishi Precision Co.," SPIE Proceedings, Vol. 1585, 1991, pp. 30–39.

[7] Malvern, A. R., "Progress Towards Fibre Optic Gyro Production" SPIE Proceedings, Vol. 1585, 1991, pp. 48–64.

[8] Böschelberger, H. J., and M. Kemmler, "Closed-Loop Fiber Optic Gyro Triad," SPIE Proceedings, Vol. 1585, 1991, pp. 89–97.

[9] Lefèvre, H. C., P. Martin, J. Morisse, P. Simonpiétri, P. Vivenot, and H. J. Arditty, "Fiber Optic Gyro Productization at Photonetics," SPIE Proceedings, Vol. 1585, 1991, pp. 42–47.

[10] Auch, W., M. Oswald, and R. Regener, "Fiber Optic Gyro Production at Alcatel-SEL," SPIE Proceedings, Vol. 1585, 1991, pp. 65–79.

[11] Blake, J., J. Feth, J. Cox,, and R. Goettsche, "Design and Test of a Production Open Loop All-Fiber Gyroscope," SPIE Proceedings, Vol. 1169, 1989, pp. 337–346.
[12] Kajoka, H., T. Kumagai, H. Nakai, H. Tizuka, M. Nakamura, and K. Yamada, "Fiber Optic Gyro Productization at Hitachi," SPIE Proceedings, Vol. 1585, 1991, pp. 17–29.
[13] Hartl, E., G. Trommer, R. Möller, and H. Poisel, "Low Cost Passive Fiber Optic Gyroscope," SPIE Proceedings, Vol. 1585, 1991, pp. 405–416.
[14] Weed, G., G. A. Sanders, G. W. Adams, and A. P. Johnson, "Fiber Optic Gyro Productization at Honeywell, Inc.," Fiber Optic Gyro 15th Anniversary Conference, Paper 1585-01, SPIE, 1991, oral presentation only.
[15] Fiber Optics News, 16 December 1991, pp. 3,4.
[16] Military and Commercial Fiber Business, 10 January 1992, p. 3.

Appendix 1
Basics of Optics

A1.1 OPTICAL WAVE IN A VACUUM

An optical wave in a vacuum is actually an electromagnetic wave that is a solution of the differential propagation equation derived from Maxwell's equations in a vacuum:

$$(M_1)\ \nabla \cdot \mathbf{B} = \text{div}\, \mathbf{B} = 0 \qquad \text{(no free "magnetic charge")} \qquad (A1.1)$$

$$(M_2)\ \nabla \times \mathbf{E} = \mathbf{curl}\, \mathbf{E} = -\frac{\partial \mathbf{B}}{\partial t} \qquad \text{(Maxwell-Faraday law)} \qquad (A1.2)$$

$$(M_3)\ \nabla \cdot \mathbf{E} = \text{div}\, \mathbf{E} = \frac{\rho}{\epsilon_0} \qquad \text{(Maxwell-Gauss law)} \qquad (A1.3)$$

$$(M_4)\ \nabla \times \mathbf{B} = \mathbf{curl}\, \mathbf{B} = \mu_0\left[\mathbf{j} + \epsilon_0 \frac{\partial \mathbf{E}}{\partial t}\right] \qquad \text{(Maxwell-Ampère law)} \qquad (A1.4)$$

where the **E** vector is the electric field, the **B** vector is the magnetic field, the **j** vector is the electric current density, ρ is the electric charge density, ϵ_0 is the dielectric permittivity of a vacuum ($\epsilon_0 = 8.854 \times 10^{-12}\ F \cdot \text{m}^{-1}$), and μ_0 is the magnetic permeability of a vacuum ($\mu_0 = 4\pi \times 10^{-7}\ H \cdot \text{m}^{-1}$).

Note that the equation of conservation of charge, $\nabla \cdot \mathbf{j} + \partial\rho/\partial t = \text{div}\, \mathbf{j} + \partial\rho/\partial t = 0$, is included in Maxwell's equations, since it can be deduced by applying $\nabla \cdot \nabla\times = \text{div} \cdot \mathbf{curl} = 0$ to Maxwell-Ampère law M_4. Furthermore, equations

M_1 and M_2 express that the **E** and **B** fields are derived from an electromagnetic scalar potential V and an electromagnetic vector potential **A**, with:

$$\mathbf{E} = -\nabla \cdot V - \frac{\partial \mathbf{A}}{\partial t} = -\mathbf{grad}\ V - \frac{\partial \mathbf{A}}{\partial t} \quad \text{(A1.5)}$$

$$\mathbf{B} = \nabla \times \mathbf{A} = \mathbf{curl\ A} \quad \text{(A1.6)}$$

These equations use the divergence (div) and the curl (**curl**) operators on a vector function **U** of the spatial coordinates, as well as the gradient operator (**grad**) on a scalar function U. The notations can be unified with the vector differential operator ∇, the divergence being the scalar product $\nabla \cdot \mathbf{U}$, which gives a scalar function, the curl being the vector product $\nabla \times \mathbf{U}$, which gives a vector function, and the gradient being the product of the vector operator ∇ with a scalar U, which gives a vector function.

Another operator, the Laplacian operator, is of importance in vector analysis: it is a scalar differential operator equal to the scalar square of the vector operator ∇ (denoted by ∇^2) applied to a vector function **U**. It gives a vector function $\nabla^2\mathbf{U}$. In cartesian coordinates,

$$\nabla = \begin{bmatrix} \frac{\partial}{\partial x} \\ \frac{\partial}{\partial y} \\ \frac{\partial}{\partial z} \end{bmatrix} \qquad \nabla^2 = \frac{\partial^2}{\partial x^2} + \frac{\partial^2}{\partial y^2} + \frac{\partial^2}{\partial z^2} \quad \text{(A1.7)}$$

The propagation equation is deduced from $\nabla \times (\nabla \times \mathbf{U}) = \nabla(\nabla \cdot \mathbf{U}) - \nabla^2\mathbf{U}$ or $\mathbf{curl} \cdot \mathbf{curl\ U} = \mathbf{grad}\ \text{div}\ \mathbf{U} - \nabla^2\mathbf{U}$ in a vacuum without charges ($j = 0$, $\rho = 0$):

$$\begin{aligned} \nabla^2\,\mathbf{E} - \epsilon_0\,\mu_0 \frac{\partial^2\mathbf{E}}{\partial t^2} &= 0 \\ \nabla^2\,\mathbf{B} - \epsilon_0\,\mu_0 \frac{\partial^2\mathbf{B}}{\partial t^2} &= 0 \end{aligned} \quad \text{(A1.8)}$$

A simple solution to this propagation equation is the monochromatic (i.e., single-frequency) sinusoidal plane wave that can be expressed in cartesian coordinates (x, y, z and time t) (Figure A1.1). For easing the calculations, complex exponential notations are used, but it must be kept in mind that the true field is only the real part.

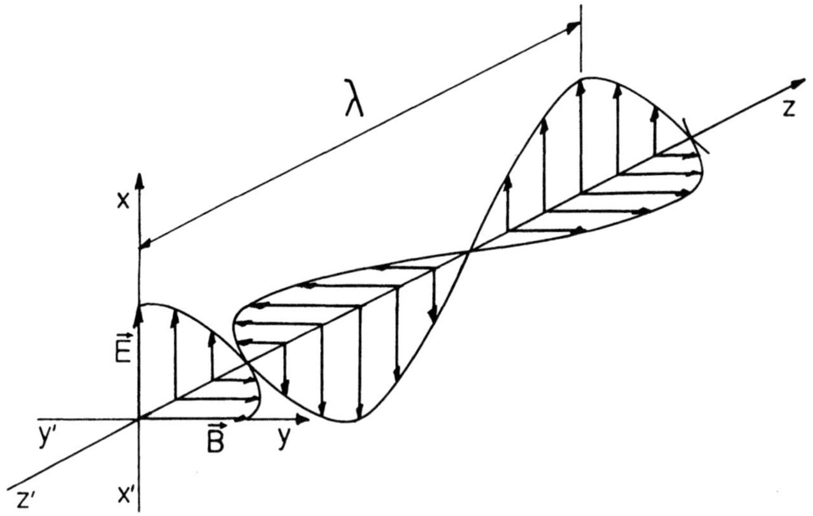

Figure A1.1 Monochromatic electromagnetic plane wave.

$$\mathbf{E} = \mathbf{E}_{0x}\, e^{i(\omega t - k_0 z)}$$
$$\mathbf{B} = \mathbf{B}_{0y}\, e^{i(\omega t - k_0 z)} \qquad \text{(A1.9)}$$

where ω is called the angular frequency and k_0 is the wave number in a vacuum. At a given position z, the electromagnetic field oscillates sinusoidally with a temporal frequency f; with $\omega = 2\pi f$. At a given time t, the electromagnetic field has a sinusoidal spatial distribution along the direction of propagation z, with a spatial period called the wavelength in a vacuum λ; with $k_0 = 2\pi/\lambda$. The inverse of the wavelength λ is called the spatial frequency and is denoted by σ.

The wave propagates at a velocity c in a vacuum, since it may be written

$$\mathbf{E}(z,t) = \mathbf{E}\left(t - \frac{z}{c}\right)$$
$$\mathbf{B}(z,t) = \mathbf{B}\left(t - \frac{z}{c}\right) \qquad \text{(A1.10)}$$

with $c = \omega/k_0 = (\epsilon_0\mu_0)^{-1/2}$. Its measured value is $c = 2.998 \times 10^8\ \text{m} \cdot \text{s}^{-1}$ and $c = \lambda \cdot f = f/\sigma$.

An electromagnetic plane wave in a vacuum is a transverse wave where the **E** and **B** fields are orthogonal to the direction of propagation. The magnitudes E and B of these two fields are proportional, and their ratio is equal to the light velocity:

$$\frac{E}{B} = c \tag{A1.11}$$

They oscillate in phase and are localized in two orthogonal planes. The electromagnetic field has the same value at any point in a plane that is perpendicular to the direction of propagation. Such planes are called the planar wavefronts of the plane wave.

This constant ratio $E/B = c$ implies that the effect of the electromagnetic field of an electromagnetic wave on matter is due mostly to the electric component. The electromagnetic force $\mathbf{f}_{EM}$ applied on a particle of charge q is the sum of an electric force $\mathbf{f}_E$ and a magnetic force $\mathbf{f}_M$:

$$\begin{aligned} \mathbf{f}_{EM} &= \mathbf{f}_E + \mathbf{f}_B \\ \mathbf{f}_E &= q\,\mathbf{E} \\ \mathbf{f}_M &= q\,(\mathbf{v}_p \times \mathbf{B}) \end{aligned} \tag{A1.12}$$

where $\mathbf{v}_p$ is the speed of the particle. Since, in matter, $\mathbf{v}_p << c$, this implies that $f_M << f_E$. Therefore, most problems of electromagnetic waves are treated by considering only the electric field.

The electromagnetic spectrum extends from radiowaves to gamma radiations: optical waves correspond to the visible range (i.e., a vacuum wavelength between 400 and 750 nm) and to the near-ultraviolet and near-infrared ranges. An important characteristic of these optical waves is their very high frequency: for a vacuum wavelength λ of 1 μm, the wave frequency f is 3×10^{14} Hz or 300.000 GHz! No electronic system can follow such a fast vibration, and optical detectors can only measure the mean power of the wave. It is not possible to measure directly the **E** or **B** field of the wave.

The power flow across unit area is called the intensity I of the wave and is proportional to the mean value of the scalar square of the **E** field:

$$I = c \cdot \epsilon_0 \langle \mathbf{E} \cdot \mathbf{E}^* \rangle \tag{A1.13}$$

where $\mathbf{E}^*$ is the complex conjuguate of **E**, and the brackets $\langle\ \rangle$ denote temporal averaging. Optical detectors are often called square-law or quadratic detectors.

When the exact physical or vectorial nature of the optical wave is not needed, the analysis of an optical scheme can be carried out with a scalar quantity, the complex wave amplitude A:

$$A = A_0 e^{i(\omega t - k_0 z)} \tag{A1.14}$$

The wave intensity is the mean value of the scalar square of the complex amplitude:

$$I = \langle A \cdot A^* \rangle = A_0^2 \tag{A1.15}$$

A1.2 POLARIZATION OF AN OPTICAL WAVE

An optical wave in a vacuum is a wave denoted by TEM (for transverse electromagnetic), since the **E** and **B** fields are both in a plane 0xy orthogonal to the direction of propagation 0z. At a given position 0 in space, a possible solution of the electric field (Figure A1.2) is

$$\mathbf{E} = \mathbf{E}_{0x} e^{i\omega t} \tag{A1.16}$$

Such a sinusoidal oscillation is said to be linearly polarized along the 0x axis. The **E** vector always remains parallel to $\mathbf{E}_{0x}$. Another possible solution is

$$\mathbf{E} = \mathbf{E}_{0y} e^{i\omega t} \tag{A1.17}$$

The wave is then linearly polarized along the perpendicular 0y axis.

The set of all the possible solutions is a two-dimensional linear space with complex coefficients. The general solution is a linear combination with complex coefficients:

$$\mathbf{E} = (\mathbf{E}_{0x} + \mathbf{E}_{0y} e^{i\Delta\phi})\, e^{i\omega t} \tag{A1.18}$$

where $\Delta\phi$ is the phase difference between both orthogonal components $\mathbf{E}_{0x}$ and $\mathbf{E}_{0y}$.

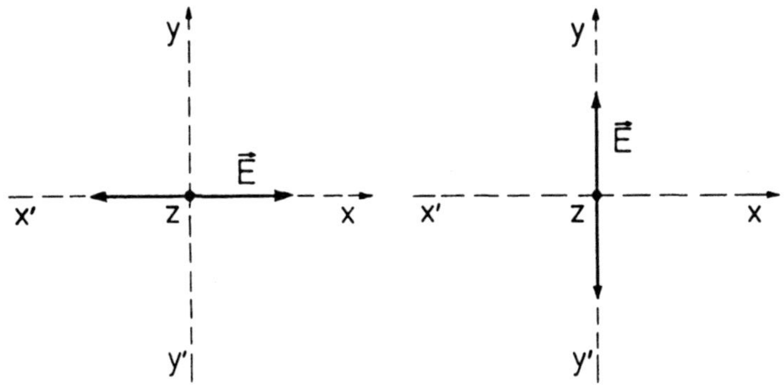

Figure A1.2 Linear polarization aligned along the x-axis or y-axis.

When $\Delta\phi$ is equal to 0 or π rad, the wave is linearly polarized along an intermediate direction (Figure A1.3), but generally the wave is elliptically polarized. The extremity of the **E** vector follows an ellipse with the frequency $f = \omega/2\pi$ (Figure A1.4). When two linear polarizations with the same amplitude are combined in quadrature (i.e., with a phase shift of $\pm\pi/2$ rad), a circular polarization results (Figure A1.5):

$$\mathbf{E} = [\mathbf{E}_{0x} + \mathbf{E}_{0y}\, e^{\pm i\pi/2}]\, e^{i\omega t} \tag{A1.19}$$

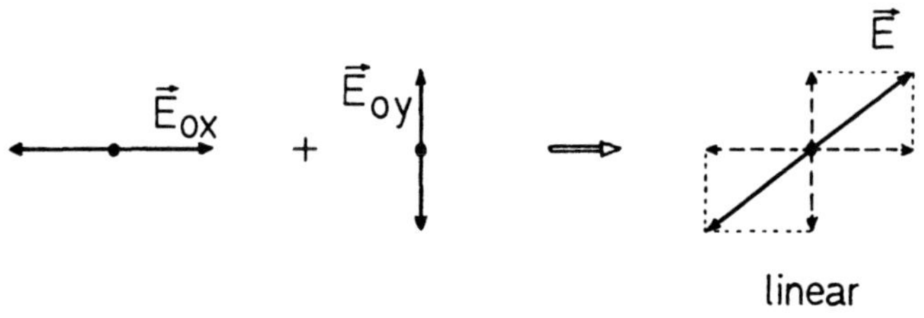

Figure A1.3 Linear polarization aligned along any axis.

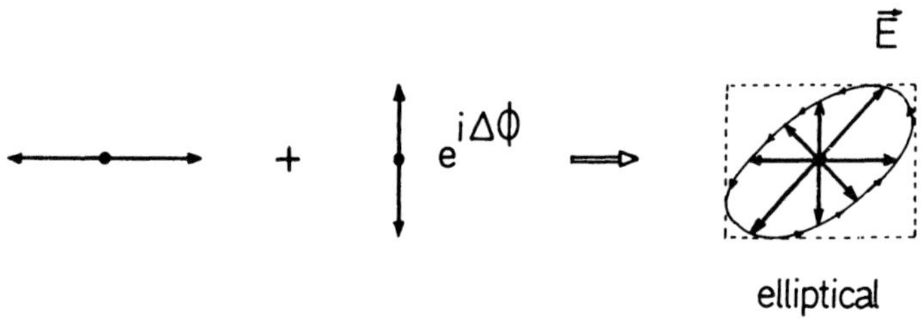

Figure A1.4 Elliptical polarization considered as the sum of two linear polarizations.

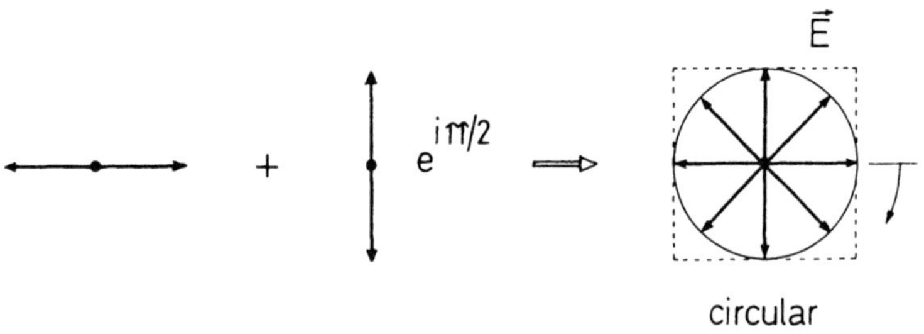

Figure A1.5 Circular polarization.

where the magnitudes of the $\mathbf{E}_{0x}$ and $\mathbf{E}_{0y}$ vectors are equal:

$$E_{0x} = E_{0y}$$

The tip of the **E** vector follows a circle. There are two states of circular polarization which turn in opposite directions depending on the sign of the $\pm\pi/2$ phase shift. They are called the right-handed and left-handed states of polarization.

Since the set of polarization states is a two-dimensional linear space, it can be analyzed with any orthonormal basis of eigenvectors. Two perpendicular linear polarizations are a possible orthonormal basis, but the two circular polarizations are also an orthonormal basis. An elliptical polarization can be expressed with linear polarizations (Figure A1.6(a)):

$$\mathbf{E} = (\mathbf{E}_{0L} + \mathbf{E}_{0s}\, e^{i\pi/2})\, e^{i\omega t} \tag{A1.20}$$

where $\mathbf{E}_{0L}$ is the peak value of the field along the large axis, and $\mathbf{E}_{0s}$ is the peak value of the field along the small axis. However, it can also be decomposed with circular polarizations (Figure A1.6(b)):

$$\mathbf{E} = [(\mathbf{E}_{0x} + \mathbf{E}_{0y}\, e^{i\pi/2}) + (\mathbf{E}'_{0x} + \mathbf{E}'_{0y} e^{-i\pi/2})] e^{i\omega t} \tag{A1.21}$$

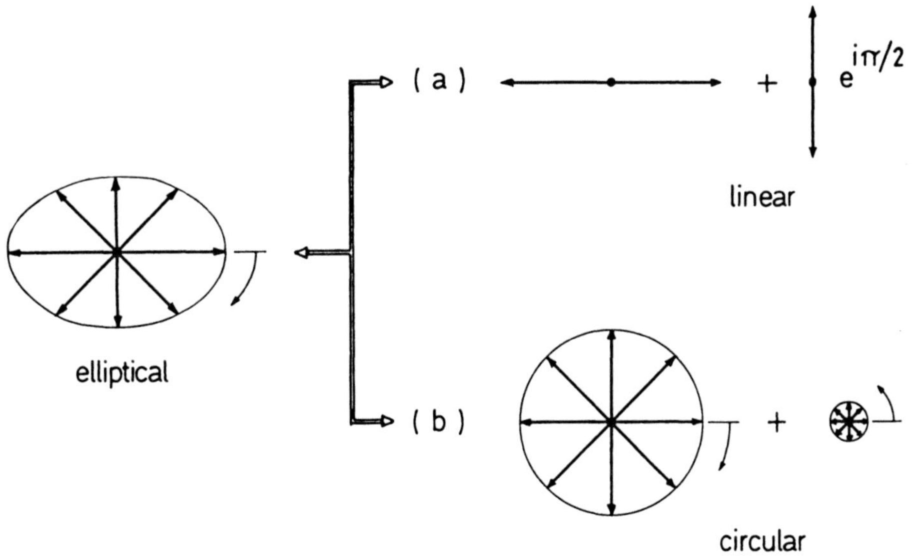

Figure A1.6 Decomposition of an elliptical polarization: (a) with two linear polarizations in quadrature; (b) with two circular polarizations.

where the magnitudes E of the various $\mathbf{E}$ vectors are given by

$$E_{0x} = E_{0y} = \frac{E_{0L} + E_{0s}}{2} = E_{0m} \tag{A1.22}$$

$$E'_{0x} = E'_{0y} = \frac{E_{0L} - E_{0s}}{2} = E_{0c} \tag{A1.23}$$

with $E_{0m} > E_{0c}$, and where E_{0m} is the magnitude of the rotating $\mathbf{E}_m$ field of the main circular component and E_{0c} is the magnitude of the rotating $\mathbf{E}_c$ field of the other crossed circular component. Also,

$$E_{0L} = E_{0m} + E_{0c} \tag{A1.24}$$

$$E_{0s} = E_{0m} - E_{0c} \tag{A1.25}$$

Note: When the quality of a given state of polarization has to be measured, the intensity ratio has to be measured between the component in this given state of polarization and the component in the orthogonal state of polarization.

With a linear state of polarization, a polarizer is used to select the main polarization component, and it is rotated by 90 deg to select the crossed orthogonal polarization. The ratio between both measured powers directly gives the ratio between the intensity I_c of the crossed linear state and the intensity I_m of the main linear state.

With a circular state of polarization, a polarizer is used to check that the transmitted power remains equal to half the input power when the polarizer is rotated. When the circular state is not perfect, a maximum power P_{max} is measured along the major axis of the polarization that is not perfectly circular and a minimum power P_{min} is measured along the minor axis. P_{max} is actually the power of the linear E_{0L} component of the elliptical polarization and is proportional to $E_{0L}{}^2$, while P_{min} is the power of E_{0s} and is proportional to $E_{0s}{}^2$. The contrast of this variation is

$$\frac{P_{max} - P_{min}}{P_{max} + P_{min}} = \frac{E_{0L}^2 - E_{0s}^2}{E_{0L}^2 + E_{0s}^2} \tag{A1.26}$$

and since $E_{0L} = E_{0m} + E_{0c}$, $E_{0s} = E_{0m} - E_{0c}$, and $E_{0c}/E_{0m} << 1$,

$$\frac{P_{max} - P_{min}}{P_{max} + P_{min}} \approx 2\frac{E_{0c}}{E_{0m}} = 2\sqrt{\frac{I_c}{I_m}} \tag{A1.27}$$

where I_m is now the intensity of the main circular state and I_c is the intensity of the crossed circular state. Then the contrast of this variation is proportional to the ratio of the amplitudes (or fields) of both orthogonal circular components; that is, the square root of the ratio of the intensities (or powers).

If, for example, the ratio $I_c/I_m = 1\%$, the contrast is as high as 20%. This does not mean that this main circular state of polarization is not as well preserved as the main linear state of polarization in the previous case. In both cases, the intensity or power ratio is I_c/I_m and the amplitude or field ratio is $A_c/A_m = \sqrt{I_c/I_m}$; but in the linear case the experiment yields an intensity ratio measurement of the defect, while in the circular case the experiment yields an amplitude ratio measurement which is much more sensitive!

A1.3 PROPAGATION IN A DIELECTRIC MEDIUM

In a medium, Maxwell equations can be averaged spatially to eliminate the strong variations of the electromagnetic field between the discrete charged particles of the material, and to be able to consider a macroscopically continuous material. The first two Maxwell equations M_1 and M_2 that are independent of ρ and $\mathbf{j}$ are preserved in the averaging:

$$(M_1)\ \nabla \cdot \mathbf{B} = \text{div}\,\mathbf{B} = 0 \qquad \text{(A1.28)}$$

$$(M_2)\ \nabla \times \mathbf{E} = \mathbf{curl}\,\mathbf{E} = -\frac{\partial \mathbf{B}}{\partial t} \qquad \text{(A1.29)}$$

The macroscopic effect of the charges of the medium can be taken into account with an electric polarization vector **P** and a magnetic polarization vector **M** (note that "polarization" has, in this case, a completely different meaning from the "state of polarization"). A derived electric field **D** and a derived magnetic field **H** may be defined as follows:

$$\mathbf{D} = \epsilon_0 \mathbf{E} + \mathbf{P} \qquad \text{(A1.30)}$$

$$\mathbf{H} = \frac{\mathbf{B}}{\mu_0} - \mathbf{M} \qquad \text{(A1.31)}$$

The last two Maxwell equations become

$$(M_3)\ \nabla \cdot \mathbf{D} = \text{div}\,\mathbf{D} = \rho_{\text{free}} \qquad \text{(A1.32)}$$

$$(M_4)\ \nabla \times \mathbf{H} = \mathbf{curl}\,\mathbf{H} = \mathbf{j}_{\text{free}} + \frac{\partial \mathbf{D}}{\partial t} \qquad \text{(A1.33)}$$

where ρ_{free} is the free charge density and $\mathbf{j}_{\text{free}}$ is the free current density vector.

In an isotropic linear dielectric (i.e., nonconducting) medium, $\rho_{\text{free}} = 0$, $\mathbf{j}_{\text{free}} = 0$, **P** is proportional to **E**, and **M** is proportional to **B**:

$$\mathbf{P} = \chi_e \epsilon_0 \mathbf{E} \tag{A1.34}$$

$$\mathbf{M} = \chi_m \frac{\mathbf{B}}{\mu_0} \tag{A1.35}$$

where χ_e and χ_m are the dielectric and magnetic susceptibilities, respectively. The **D** and **H** derived fields are also respectively proportional to the **E** and **B** fields:

$$\mathbf{D} = \epsilon_r \epsilon_0 \mathbf{E} \tag{A1.36}$$

$$\mathbf{H} = \frac{1}{\mu_r \mu_0} \mathbf{B} \tag{A1.37}$$

where ϵ_r is the relative dielectric permittivity and μ_r is the relative magnetic permeability. We have

$$\epsilon_r = 1 + \chi_e \tag{A1.38}$$

$$\frac{1}{\mu_r} = 1 + \chi_m \tag{A1.39}$$

Note that χ_e, χ_m, ϵ_r and μ_r are dimensionless values.

The propagation equation becomes

$$\begin{aligned} \nabla^2 \mathbf{E} - \epsilon_r \mu_r \epsilon_0 \mu_0 \frac{\partial^2 \mathbf{E}}{\partial t^2} &= 0 \\ \nabla^2 \mathbf{B} - \epsilon_r \mu_r \epsilon_0 \mu_0 \frac{\partial^2 \mathbf{B}}{\partial t^2} &= 0 \end{aligned} \tag{A1.40}$$

The plane wave solution is

$$\begin{aligned} \mathbf{E} &= \mathbf{E}_{0x} e^{i(\omega t - k_m z)} \\ \mathbf{B} &= \mathbf{B}_{0y} e^{i(\omega t - k_m z)} \end{aligned} \tag{A1.41}$$

The wave velocity $v = \omega / k_m$ in a medium becomes

$$v = (\epsilon_r \mu_r \epsilon_0 \mu_0)^{-1/2} = c\,(\epsilon_r \mu_r)^{-1/2} \tag{A1.42}$$

The index of refraction n of the material is defined by the ratio between the velocity in a vacuum and the velocity in the medium:

$$n = \frac{c}{v} = (\epsilon_r \, \mu_r)^{1/2} \tag{A1.43}$$

Even if this may look obvious, it is important to recall that when a lightwave propagates in a vacuum and in various media, its temporal frequency f and its related angular frequency $\omega = 2\pi f$ remain unchanged; but the actual spatial frequency σ_m and its inverse, the actual wavelength λ_m, in the medium are not equal anymore to the values σ and λ that the same wave would have in a vacuum. We have

$$\sigma_m = n \cdot \sigma \tag{A1.44}$$

$$\lambda_m = \lambda / n \tag{A1.45}$$

The wave number k_m in the medium is

$$k_m = \frac{2\pi}{\lambda_m} = \frac{2\pi n}{\lambda} = n \cdot k_0 \tag{A1.46}$$

where k_0 is the wave number that the same wave would have in a vacuum.

As can be seen, the important characteristic of a light beam that remains invariant is its temporal frequency, even if it is the habit to give its wavelength in a vacuum because a length of 1 μm is a quantity easier to grasp and to measure than a frequency of 300,000 GHz or a time of 3 femtoseconds.

Note that the relative dielectric permittivity ϵ_r depends on the frequency of the wave and that, at optical wave frequencies, the relative magnetic permeability μ_r is equal to unity. Then the index of refraction and the wave velocity depend only on ϵ_r and are frequency- (or wavelength-) dependent:

$$\begin{aligned} v &= \frac{c}{n(f)} \\ n(f) &= \sqrt{\epsilon_r(f)} \ (\text{or } n(\lambda) = \sqrt{\epsilon_r(\lambda)}) \end{aligned} \tag{A1.47}$$

As a matter of fact, a material is composed of electric particles that are bound together with electromagnetic forces. It can be considered a complex set of elastic mechanical oscillators. When the light frequency corresponds to a resonance frequency of the material, mechanical resonances are induced by the synchronized electromagnetic field of the wave, which is then absorbed. In the transparent region, the light frequency does not correspond to any resonance, but the oscillators vibrate and the macroscopic effect described by ϵ_r depends on the respective value of the light frequency and of the resonance frequencies of the material, which yields a frequency dependence of ϵ_r and of the index n.

This effect is called chromatic dispersion, since its first well-known application is the spatial decomposition of white light in prismatic colors. White light is composed of all the visible wavelengths, and since the angular deviation by a prism is proportional to the index n, the spectrum becomes spread angularly as a function of $n(f)$ (or $n(\lambda)$).

Dispersion yields two other important phenomena. If the optical wave is modulated in amplitude, frequency, or phase, the modulated "signal" does not propagate at the velocity $v = \omega/k_m$, often called phase velocity, but at the so-called group velocity v_g, the difference being proportional to the first-order derivative $dn/d\lambda$.

$$v_g = \frac{d\omega}{dk_m}$$

$$v_g = v\left(1 - \frac{k_0}{n}\frac{dn}{dk_0}\right) \tag{A1.48}$$

$$v_g = v\left(1 + \frac{\lambda}{n}\frac{dn}{d\lambda}\right)$$

Any nonmonochromatic wave is the integral summation of all its frequency components. At a given spatial position, the wave amplitude $A(t)$ is

$$A(t) = \int_{-\infty}^{+\infty} a(f)\, e^{2i\pi ft} df \tag{A1.49}$$

Its frequency components $a(f)$ are defined inversely by Fourier transform:

$$a(f) = \int_{-\infty}^{+\infty} A(t)\, e^{-2i\pi ft} dt \tag{A1.50}$$

Using the angular frequency ω instead of the temporal frequency f gives

$$A(t) = \frac{1}{2\pi}\int_{-\infty}^{+\infty} a(\omega)\, e^{i\omega t} d\omega \tag{A1.51}$$

When this wave propagates in a dispersive medium, each frequency component propagates independently at its own velocity, and the total wave amplitude $A(t,z)$ propagates accordingly to

$$A(t,z) = \frac{1}{2\pi}\int_{-\infty}^{+\infty} a(\omega)\, e^{i[\omega t - k_m(\omega)z]} d\omega \tag{A1.52}$$

The propagation argument $[\omega t - k_m(\omega)z]$ may be decomposed in the sum of a mean term $[\overline{\omega}t - k_m(\overline{\omega})z]$ that depends on the mean frequency $\overline{\omega}$ and a variable term $[\Delta\omega t - \Delta k_m(\Delta\omega)z]$ where $\Delta\omega = \omega - \overline{\omega}$ and $\Delta k_m(\Delta\omega) = k_m(\omega) - k_m(\overline{\omega})$. The wave amplitude becomes

$$A(t,z) = M(t,z) e^{i[\overline{\omega}t - k_m(\overline{\omega})z]} \tag{A1.53}$$

The second term has the same propagation argument as the one of a monochromatic wave and represents the propagation of the carrier at a velocity equal to the mean phase velocity $v = \overline{\omega}/k_m(\overline{\omega})$. The first term $M(t,z)$ is actually the modulation that propagates on the monochromatic carrier, and

$$M(t,z) = \frac{1}{2\pi}\int_{-\infty}^{+\infty} a(\overline{\omega} + \Delta\omega)\, e^{i[\Delta\omega t - \Delta k_m(\Delta\omega)z]} d(\Delta\omega) \tag{A1.54}$$

The modulation term $M(t,z)$ is the integral summation of frequency components $\Delta\omega$ that propagate respectively at a velocity $\Delta\omega/\Delta k_m(\Delta\omega)$. If this velocity is constant over the spectrum, $\Delta\omega/\Delta k_m$ is equal to the first-order derivative $d\omega/dk_m$, and the second-order derivative $d^2\omega/d^2k_m$ is zero. The modulation term $M(t,z)$ propagates without any change at the same velocity as the one of all its frequency components, the so-called group velocity $v_g = d\omega/dk_m$, since it can be written

$$M(t,z) = M\left(t - \frac{z}{v_g}\right) = \frac{1}{2\pi}\int_{-\infty}^{+\infty} a(\overline{\omega} + \Delta\omega)\, e^{i\Delta\omega\left[t - \frac{z}{v_g}\right]} d(\Delta\omega) \tag{A1.55}$$

The value of v_g is independent of the spectrum width, and this result is very general. It applies for any kind of function M (i.e., amplitude, phase, or frequency modulation signals), but it also applies if the spectrum is broad because of the natural emission of the source.

If $\Delta\omega/\Delta k_m$ is not constant over the spectrum (i.e., $d^2\omega/d^2k_m \neq 0$), the modulation term $M(t,z)$ propagates at the mean group velocity $v_g(\overline{\omega}) = d\omega/dk_m(\overline{\omega})$, and there is an additional second-order effect of signal temporal spreading, which limits the transmission bandwidth. The limitation induced by this effect over the intrinsic spectrum broadening due to a signal modulation of a monochromatic source is in practice negligible, but it has to be taken into account when the natural emission spectrum of the source is not narrow enough.

Let us give orders of magnitude to get a better understanding: the mean light frequency is 300,000 GHz (for a 1-μm wavelength), and the spectrum broadening due to electronically controlled signal modulation is at maximum several gigahertz (i.e., at maximum, about 10^{-5} of the central frequency), while the natural emission spectrum width of the source may be as much as 10^{-3} to 5×10^{-2} of the central frequency (i.e., a width of 1 to 50 nm in wavelength). In this case, the modulation term $M(t,z)$ is a complex combination of the effect of a modulated signal $S(t,z)$ and a random term $M_r(t,z)$ that takes into account the effect of the broad emission spectrum of the source. One may consider that each emission frequency ω_e of the source carries the signal that propagates on this frequency at the group velocity $v_g(\omega_e)$. The total signal is actually decomposed in elementary signals that propagate at a different velocity, comprised in $\Delta v_g = v_g[\overline{\omega}_e + (\Delta\omega_e/2)] - v_g[\overline{\omega}_e - (\Delta\omega_e/2)]$, where $\overline{\omega}_e$ is the mean frequency of the emission spectrum and $\Delta\omega_e$ is the width of the emission spectrum. These velocity differences yield, over a propagation length L, a temporal spreading $\Delta\tau_s$ of the elementary signals:

$$\Delta\tau_s = -\frac{\Delta v_g}{v_g^2} \cdot L \tag{A1.56}$$

since the group propagation time is $\tau_g(\omega_e) = L/[v_g(\omega_e)]$. Expressed in terms of index n and mean emission wavelength $\overline{\lambda}_e$,

$$\Delta\tau_s \approx -\frac{L}{c} \cdot \overline{\lambda}_e \cdot \frac{d^2n}{d\lambda^2} \cdot \Delta\lambda_e \tag{A1.57}$$

where $\Delta\lambda_e$ is the width of the emission spectrum, in wavelength.

For example, for silica used in optical fibers, with $\overline{\lambda}_e = 850$ nm and $\Delta\lambda_e = 30$ nm,

$$\Delta\tau_s/L = 2.5 \text{ ns} \cdot \text{km}^{-1}$$

which limits the transmission bandwidth to less than 400 MHz for 1 km of propagation.

On the other hand, since $d^2n/d\lambda^2 = 0$ around 1300 nm in silica, this wavelength is the optimal choice for large-bandwidth telecommunications that require minimum signal spreading. Note that when $d^2n/d\lambda^2 = 0$, it is often said that there is no dispersion. More precisely, it should be said that there is no "propagation dispersion." The reason for this is that even with a second-order derivative $d^2n/d\lambda^2 = 0$, it may still be possible to get chromatic dispersion in a prism, because this depends on the first-order derivative $dn/d\lambda$.

Note: The basic electromagnetic field is the set of the electric **E** field and the magnetic **B** field. Relativity shows that **E** and **B** are connected through Lorentz transformation for moving frames of reference. The derived fields **D** and **H** are only auxiliary fields that allow the effect of a medium to be treated macroscopically.

This is particularly seen in the electromagnetic theory of the Sagnac effect detailed in Appendix 4.

However, for historical reasons, many optics and antenna text books that are mostly devoted to motionless systems are considering **H** as the basic magnetic field instead of **B.** The advantage of the (**E, H**) approach is that it offers a good analogy with other wave phenomena. As a matter of fact, it is usually possible in a wave to consider a first physical measurand that is a "potential" and a second physical measurand that is a "flow." A hand-waving argument to explain wave propagation is the consideration that a variation of the "potential" induces a variation of the "flow." Then this variation of the "flow" induces a variation of the "potential," and so on. The power P of the wave is the product "potential" by "flow," and the impedance Z of the propagation medium is the ratio "potential" over "flow."

In electric circuits, the "potential" is the voltage V and the "flow" is the intensity of the electric current i, and the power and the impedance are

$$\begin{aligned} P &= V \cdot i \\ Z &= V/i \end{aligned} \tag{A1.58}$$

For acoustic waves, the "potential" is the variation of pressure Δp, the "flow" is the actual matter flow f_{matter}, and

$$\begin{aligned} P &= \Delta p \cdot f \\ Z &= \frac{\Delta p}{f_{\text{matter}}} \end{aligned} \tag{A1.59}$$

With electromagnetic waves, since the dimension of the **E** field is voltage $\times$ length^{-1} and the dimension of the **H** field is current $\times$ length^{-1}. The impedance Z is defined by

$$Z = \frac{E}{H} \tag{A1.60}$$

and has the same unit as V/i (i.e., the unit is the ohm (Ω)). In particular, for waves in a vacuum, there is the so-called impedance of a vacuum:

$$Z_0 = \sqrt{\frac{\mu_0}{\epsilon_0}} \approx 376.7\Omega \tag{A1.61}$$

It can be shown that in a medium with an index n, the impedance becomes

$$Z = \frac{Z_0}{n} \tag{A1.62}$$

and partial reflection on an interface may be interpreted as an impedance mismatch. The **E** field would be the "potential" and the **H** field would be the "flow." Furthermore, the product "potential" by "flow" does have a power significance, since it corresponds to the so-called Poynting vector **Π**, which is equal to the power flow per unit area. In a vacuum,

$$\mathbf{\Pi} = \mu_0^{-1} (\mathbf{E} \times \mathbf{B}) \quad \text{(A1.63)}$$

then

$$\mathbf{\Pi} = \mathbf{E} \times \mathbf{H} \quad \text{(A1.64)}$$

As we can see, this analogy using **E** as a "potential" and **H** as a "flow" has some interest, but the general tendency nowadays in physics is to consider the (**E**,**B**) couple as the basic electromagnetic field. The fact that the E/B ratio of an electromagnetic wave in a vacuum is equal to the velocity of light is, after all, a more fundamental result than the fact that the E/H ratio has the same unit as the one of electrical impedance.

Furthermore, note that so far we have used four physical constants of electromagnetism that are not independent:

- ϵ_0, the dielectric permittivity of a vacuum;
- μ_0, the magnetic permeability of a vacuum;
- c, the light velocity in a vacuum;
- Z_0, the impedance of a vacuum.

If relativistic arguments are predominate, then there are actually only two basic physical constants:

- ϵ_0, because without motion of charges there is only an electric field;
- c, because it is the basic constant of relativity, which explains transformation by motion.

As we have already seen, the magnetic **B** field is deduced from the electric **E** field through the relativistic Lorentz transformation, which is using c. Then μ_0 should not be a basic constant: it should be deduced from $\mu_0 = 1/\epsilon_0 c^2$, even if, for historical and practical reasons, μ_0 is a basic constant in the international unit system because of the definition of the ampere. Similarly, Z_0 is also a secondary physical constant deduced from $Z_0 = 1/\epsilon_0 c$. Now, secondary does not mean it is not useful: historically, it was fundamental to check that the measured value of the light velocity c equaled $1/\sqrt{\epsilon_0 \mu_0}$, ϵ_0 and μ_0 being defined independently in electrostatic and magnetostatic experiments.

As an aside, it is interesting to note that if c is considered a fundamental constant of physics, using (ct) as a temporal coordinate with the dimension of a length and $(c\mathbf{B})$ as a magnetic field with the dimension of an electric field, Maxwell equations in a vacuum, and without charge or current, appear very symmetric:

$$(M_1)\ \nabla \cdot (c\mathbf{B}) = \text{div}\,(c\mathbf{B}) = 0 \tag{A1.65}$$

$$(M_2)\ \nabla \times \mathbf{E} = \mathbf{curl}\,\mathbf{E} = -\frac{\partial(c\mathbf{B})}{\partial(ct)} \tag{A1.66}$$

$$(M_3)\ \nabla \times \mathbf{E} = \text{div}\,\mathbf{E} = 0 \tag{A1.67}$$

$$(M_4)\ \nabla \times (c\mathbf{B}) = \mathbf{curl}\,(c\mathbf{B}) = \frac{\partial \mathbf{E}}{\partial(ct)} \tag{A1.68}$$

A1.4 GEOMETRICAL OPTICS

A lot of optical problems, particularly optical imaging, can be treated without considering the wavelength that is very small macroscopically. They use the laws of geometrical optics, which consider that the optical energy propagates along light rays. In a plane wave, these rays are parallel to the direction of propagation and perpendicular to the planar wavefronts.

Another important solution of the propagation equation is the spherical wave:

$$\mathbf{E} = \mathbf{E}_0 \frac{r_0}{r} e^{i(\omega t - kr)} \tag{A1.69}$$

where r is the radial coordinate and r_0 is a normalization radius. The phase fronts are concentric spheres. In geometrical optics, such a wave is described with rays that are radii of the spherical phase fronts. Note that the amplitude of the wave has to decrease in $1/r$, since the conservation of energy requires that the intensity (the square of the amplitude) decreases in $1/r^2$ as the phase front area increases in r^2.

The divergence of spherical waves may be modified with lenses, particularly convergent lenses. When the distance d between the source point and the lens is equal to the focal length f_l, the divergent spherical wave is transformed into a plane wave (Figure A1.7). When the distance d is larger, the wave becomes convergent and is refocused at a distance d' (Figure A1.8), with

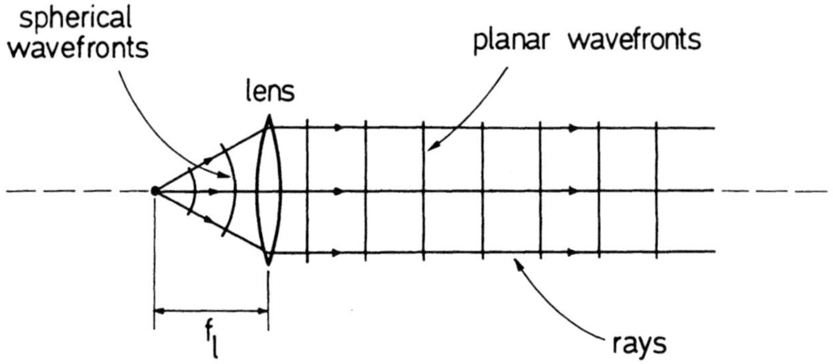

Figure A1.7 Transformation of a divergent spherical wave by a convergent lens into a plane wave.

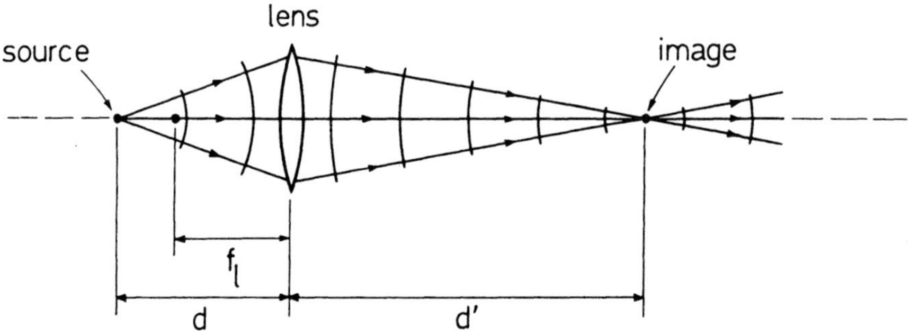

Figure A1.8 Transformation of a divergent spherical wave by a convergent lens into a converging wave.

$$\frac{1}{d} + \frac{1}{d'} = \frac{1}{f_l} \tag{A1.70}$$

Because of reciprocity, in the reverse operation the light rays follow the same path in opposite direction.

A1.5 DIELECTRIC INTERFACE: REFLECTION, REFRACTION, AND WAVEGUIDANCE

The basic laws of reflection and refraction at an interface between two dielectric media can be formulated very simply in geometrical optics (Figure A1.9):

- The reflected and refracted rays lie in the plane of incidence (i.e., the plane that is orthogonal to the interface and that contains the incident ray).

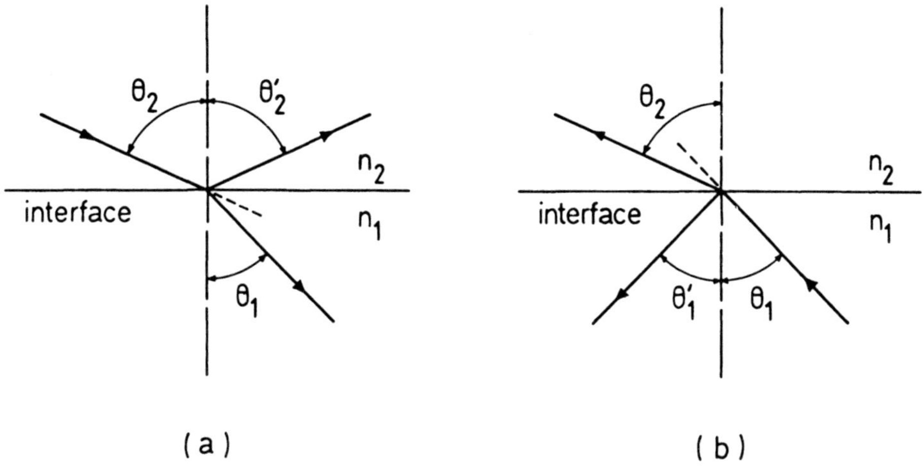

Figure A1.9 Reflected and refracted rays on a dielectric interface: (a) from low index n_2 to high index n_1; (b) from high index n_1 to low index n_2.

- The angle of reflection θ_2' is equal to the angle of incidence θ_2.
- The angle of refraction θ_1 is given by the Snell law:

$$n_1 \sin\theta_1 = n_2 \sin\theta_2 \tag{A1.71}$$

where n_1 and n_2 are the respective indexes of the dielectric media.

These laws give the angular deviation of the rays, but the calculation of the intensity splitting ratio at the interface, called Fresnel coefficients, require a more complicated analysis that uses the boundary conditions of the electromagnetic field at the interface.

For normal incidence (i.e., $\theta_1 = \theta'_1 = \theta_2 = 0$), the intensity ratio between the incident beam and the reflected beam, called Fresnel reflectivity $\mathfrak{R}_F$ is

$$\mathfrak{R}_F = \frac{(n_2 - n_1)^2}{(n_2 + n_1)^2} \tag{A1.72}$$

The intensity ratio between the refracted beam and the incident beam, called transmissivity $\mathfrak{T}_F$, is

$$\mathfrak{T}_F = \frac{4n_1n_2}{(n_1 + n_2)^2} \tag{A1.73}$$

It can be seen that

$$\lim_{n_1 \to n_2} \mathfrak{R}_F = 0 \qquad \lim_{n_1 \to n_2} \mathfrak{T}_F = 1$$

In particular, at the interface between silica fiber ($n_1 = 1.45$) and air ($n_2 = 1$),

$$\mathfrak{R}_F = 0.04 \text{ and } \mathfrak{T}_F = 0.96$$

Note that the transmissivity and the reflectivity have the same values in both directions (i.e., when the incident beam is coming from the low index medium or from the high index medium). This is a first case of the very basic principle of reciprocity of light propagation.

For other incidences, the result is more complicated and depends on the state of polarization, which may be parallel to the plane of incidence, denoted by a subscript // or p, or may be perpendicular to the plane of incidence, denoted by a subscript $\perp$ or s (from the German senkrecht). Several important results can be pointed out (Figure A1.10):

- The reflectivity and the transmissivity remains about constant in a cone of about 10 deg around normal incidence.
- The reflectivity for an s-polarized wave increases monotonically up to unity for 90-deg incidence in the low index medium.

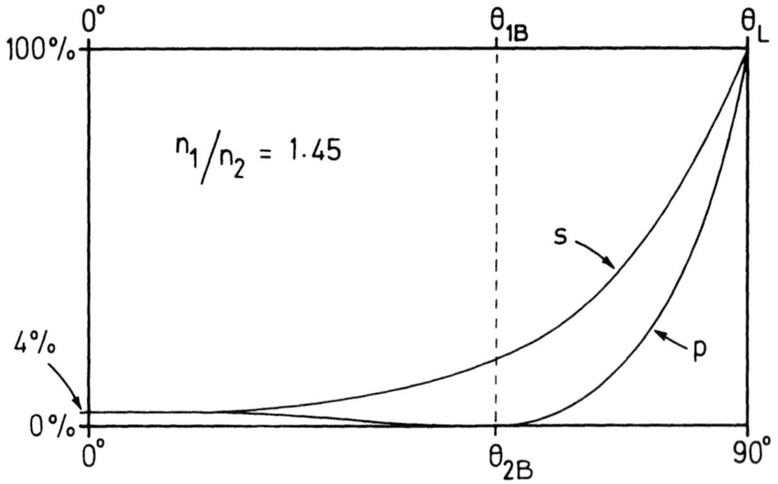

Figure A1.10 Fresnel reflection coefficients for an index of 1.45 for the s polarization and the p polarization.

- The reflectivity for a p-polarized wave decreases to zero for an incidence angle called the Brewster angle, and then increases up to unity. At Brewster incidence, the reflected beam is perpendicular to the refracted beam (Figure A1.11):

$$n_1 \sin\theta_{1B} = n_2 \sin\theta_{2B}$$
$$\theta_{1B} = \theta'_{1B} = 90 \text{ deg} - \theta_{2B} \tag{A1.74}$$

and since $\sin(90 \text{ deg} - \theta) = \cos\theta$,

$$tg\theta_{1B} = \frac{n_1}{n_2}$$
$$tg\theta_{2B} = \frac{n_2}{n_1} \tag{A1.75}$$

Between silica ($n_1 = 1.45$) and air ($n_2 = 1$), there is

$$\theta_{1B} = 34.6 \text{ deg} \quad \theta_{2B} = 55.4 \text{ deg}$$

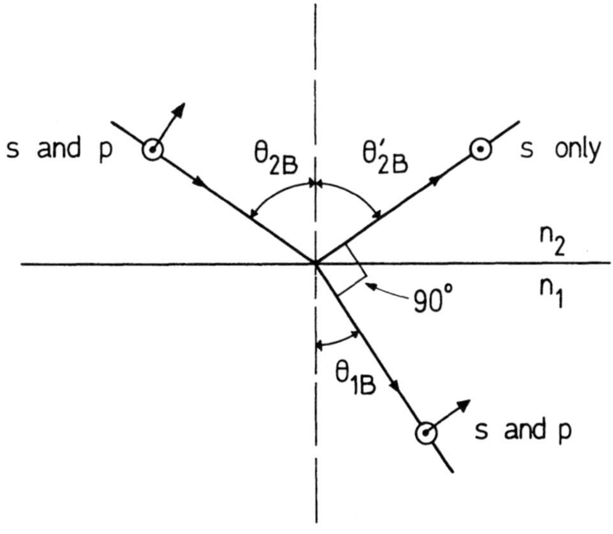

Figure A1.11 Brewster incidence.

There is a limit refraction angle θ_L for a 90-deg incidence angle in the low n_2 index medium, for which the reflectivity is 1 (Figure A1.12):

$$n_2 \sin 90 \text{ deg} = n_1 \sin\theta_L \tag{A1.76}$$

and

$$\sin\theta_L = \frac{n_2}{n_1} \tag{A1.77}$$

If a beam coming from the high index medium has an incidence angle larger than θ_L, the light cannot be refracted in the low index medium and it is entirely reflected. This phenomenon is known as total internal reflection, and it is the basis of dielectric waveguidance. A light beam being coupled through the edge of a dielectric plate, it will propagate inside the plate because its incidence on the interfaces yields multiple total internal reflections (Figure A1.13). The sine of the maximum input angle θ_{max} in a vacuum yielding internal reflection is called the numerical aperture (NA), defined by

$$\sin\theta_{max} = \text{NA} = \sqrt{n_1^2 - n_2^2} \tag{A1.78}$$

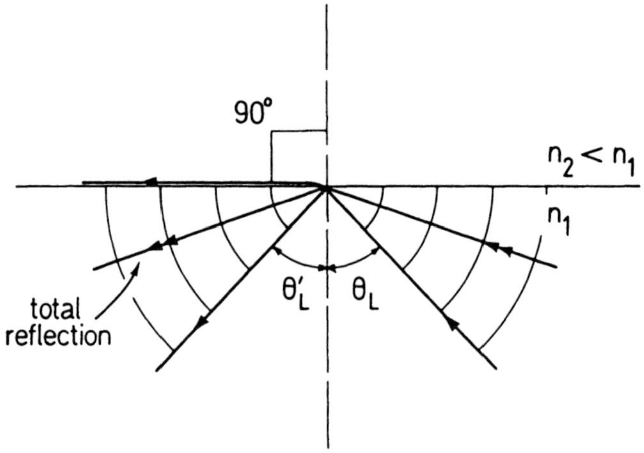

Figure A1.12 Total internal reflection on a dielectric interface.

A1.6 INTERFERENCES

As we have seen, the frequency of optical waves is so large that it is not possible to measure directly the modulated electromagnetic field, particularly its phase.

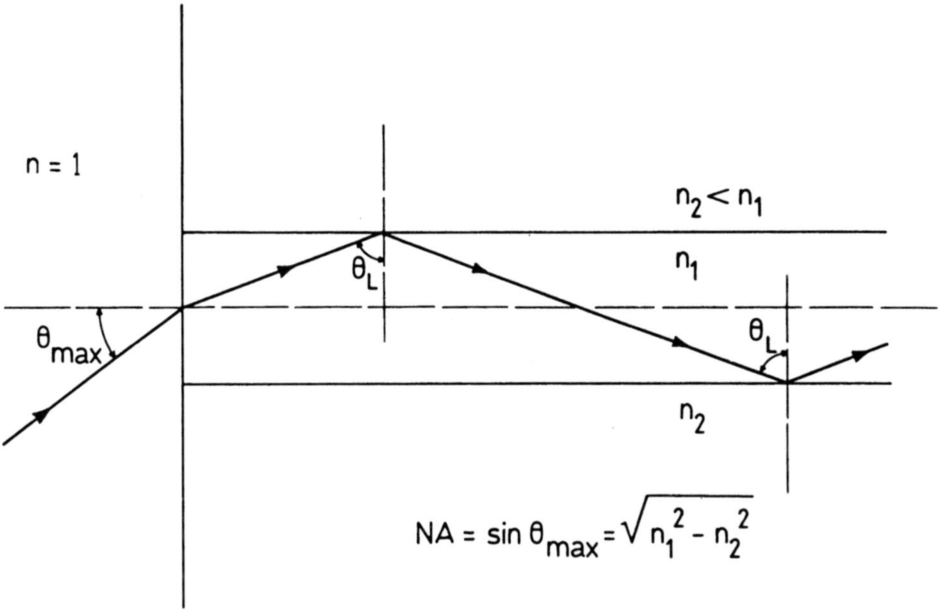

Figure A1.13 Dielectric waveguide with its numerical aperture NA.

However, an indirect intensity measurement is possible through the phenomenon of interferences.

In an interferometer, the input light wave is split, propagates along two different paths, and is recombined at the output. The total field of the interference wave is the vectorial sum of the two fields that have propagated along both paths:

$$\begin{aligned} \mathbf{E} &= \mathbf{E}_1 + \mathbf{E}_2 \\ \mathbf{E}_1 &= \mathbf{E}_{10}\, e^{i(\omega t + \phi_1)} \\ \mathbf{E}_2 &= \mathbf{E}_{20}\, e^{i(\omega t + \phi_2)} \end{aligned} \tag{A1.79}$$

where ϕ_1 and ϕ_2 are the respective phases of the fields $\mathbf{E}_1$ and $\mathbf{E}_2$.

The intensity of the interference wave is proportional to the temporal averaging of the scalar square of the field:

$$I = c\epsilon_0 \langle \mathbf{E} \cdot \mathbf{E}^* \rangle \tag{A1.80}$$

and since

$$\langle \mathbf{E}\,\mathbf{E}^* \rangle = \langle \mathbf{E}_1\,\mathbf{E}^*_1 \rangle + \langle \mathbf{E}_2\,\mathbf{E}^*_2 \rangle + \langle \mathbf{E}_1\,\mathbf{E}^*_2 \rangle + \langle \mathbf{E}^*_1\,\mathbf{E}_2 \rangle \tag{A1.81}$$

The general formula of interferences is

$$I = I_1 + I_2 + 2\sqrt{I_1 I_2}\cos\Delta\phi \tag{A1.82}$$

where I_1 and I_2 are the intensities of the interfering waves, and where $\Delta\phi = \phi_1 - \phi_2$ is their phase difference induced by the difference ΔL_{op} between both optical paths n_1L_1 and n_2L_2:

$$\begin{aligned} \Delta\phi &= 2\pi\frac{\Delta L_{op}}{\lambda} \\ \Delta L_{op} &= n_1 L_1 - n_2 L_2 \end{aligned} \tag{A1.83}$$

There is a maximum intensity I_{max} when both waves are in phase ($\Delta\phi = 2m\pi$, m being an integer); that is, the two field magnitudes are added:

$$E = E_1 + E_2 \tag{A1.84}$$

and

$$I_{max} = I_1 + I_2 + 2\sqrt{I_1 I_2} \tag{A1.85}$$

and there is a minimum intensity I_{min} when both waves are in phase opposition ($\Delta\phi = (2m + 1)\pi$); that is, the two field magnitudes are substracted:

$$E = E_1 - E_2 \tag{A1.86}$$

and

$$I_{min} = I_1 + I_2 - 2\sqrt{I_1 I_2} \tag{A1.87}$$

Note that these results assume that both fields have the same state of polarization. If they have orthogonal states of polarization, the interferences are suppressed, since the scalar product $\mathbf{E}_1 \cdot \mathbf{E}_2^* = \mathbf{E}_1^* \cdot \mathbf{E}_2 = 0$.

Interferometry is a very sensitive method for measuring various parameters, because the whole dynamic range of measurement between I_{max} and I_{min} is scanned for a change of π in the phase difference, which is induced by a change of $\lambda/2$ of the optical path difference (i.e., a change of less than a micrometer).

Two cases are of particular interest:

1. $I_1 = I_2$, then (Figure A1.14(a))

$$I(\Delta\phi) = 2I_1(1 + \cos\Delta\phi) \tag{A1.88}$$

The interference is "perfectly" contrasted, since

$$I_{\min} = 0$$

and the contrast or fringe visibility ϑ is defined by

$$\vartheta = \frac{I_{\max} - I_{\min}}{I_{\max} + I_{\min}} = 1 \qquad \text{(A1.89)}$$

2. $I_1 >> I_2$, then (Figure A1.14(b))

$$I(\Delta\phi) \approx I_1\left(1 + 2\sqrt{\frac{I_2}{I_1}}\cos\Delta\phi\right) \qquad \text{(A1.90)}$$

The fringe visibility is

$$\vartheta = 2\sqrt{\frac{I_2}{I_1}} \qquad \text{(A1.91)}$$

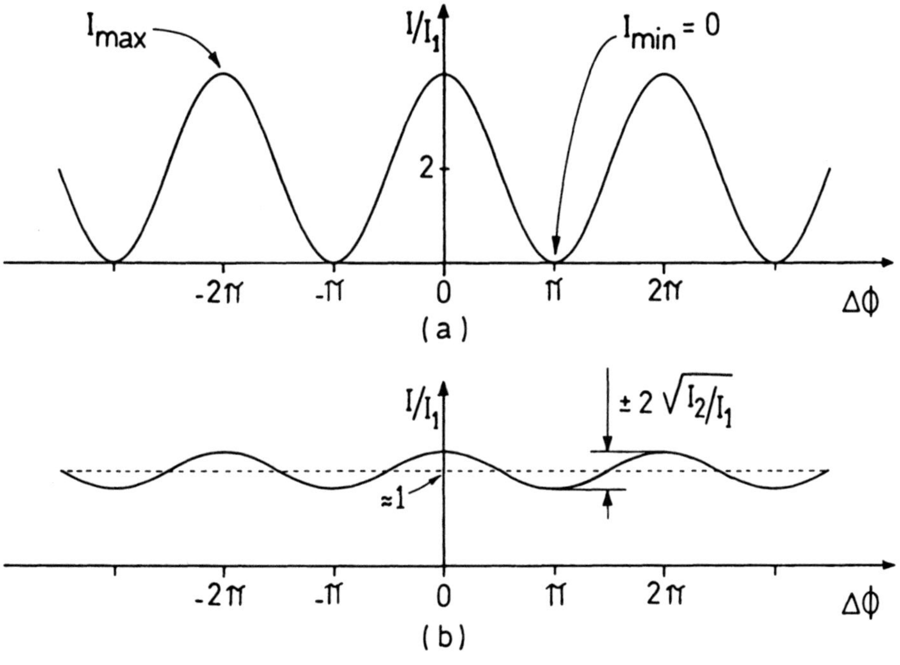

Figure A1.14 Response of an interferometer: (a) perfect contrast; (b) low contrast.

with $I(\Delta\phi) = I_1(1 + \vartheta \cos\Delta\phi)$; that is, the relative peak-to-peak variation of the interference signal is

$$\pm 2\sqrt{\frac{I_2}{I_1}}$$

It is proportional to the square root of the intensity ratio (i.e., the field or amplitude ratio) and not the power or intensity ratio. For example, an intensity ratio of 10^{-4} still yields a fringe visibility of 2×10^{-2}. This effect is used in the so-called coherent detection scheme, where a high-power local oscillator of intensity I_L interferes with a low-power signal of intensity I_S. Instead of measuring the low-intensity I_S directly, one measures the interference signal:

$$\pm 2I_L\sqrt{\frac{I_S}{I_L}} = \pm 2\sqrt{\frac{I_L}{I_S}} \cdot I_S \qquad \text{(A1.92)}$$

The signal I_S is "amplified" by a coefficient $\pm 2\sqrt{I_L/I_S}$.

Among the most well-known interferometers, there is the Michelson interferometer (Figure A1.15), in which the light is separated by a beamsplitting plate, back-reflected on two mirrors, and recombined on the beamsplitting plate. The scanning of one mirror along a distance d allows the optical path difference ΔL_{op}

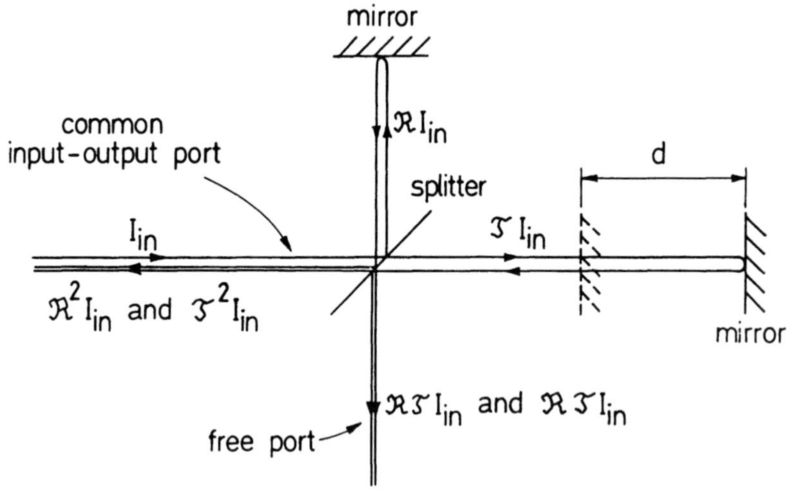

Figure A1.15 Michelson interferometer.

to be changed very simply to equal the geometrical path difference ΔL, since light propagates in a vacuum:

$$\Delta L_{op} = \Delta L = 2d$$

Note that the general formula for interference, in which the intensity I may be larger than the sum $I_1 + I_2$, does not violate the principle of conservation of energy. The input intensity I_{in} is split in two waves of respective intensities:

$$\mathfrak{R} I_{\text{in}} \quad \text{and} \quad \mathfrak{T} \cdot I_{\text{in}}$$

where $\mathfrak{R}$ is the splitter reflectivity and $\mathfrak{T}$ the splitter transmissivity, with $\mathfrak{R} + \mathfrak{T} = 1$. These two waves are split again when they come back on the beamsplitter. At the free output port, there is interference between

$$\mathfrak{R}\mathfrak{T} I_{\text{in}} \quad \text{and} \quad \mathfrak{R}\mathfrak{T}\, I_{\text{in}}$$

and

$$I_{\text{free}} = 2\mathfrak{R}\mathfrak{T} I_{\text{in}} + 2\mathfrak{R}\mathfrak{T} I_{\text{in}} \cos\Delta\phi \tag{A1.93}$$

At the common input-output port there is interference between

$$\mathfrak{R}^2 I_{\text{in}} \text{ and } \mathfrak{T}^2 I_{\text{in}}$$

and

$$I_{\text{common}} = \mathfrak{R}^2 I_{\text{in}} + \mathfrak{T}^2 I_{\text{in}} + 2\mathfrak{R}\mathfrak{T} I_{\text{in}} \cos\Delta\phi' \tag{A1.94}$$

Because of conservation of energy, the two output ports must be complementary:

$$\begin{aligned} I_{\text{free}} &= 2\mathfrak{R}\mathfrak{T} I_{\text{in}} + 2\mathfrak{R}\mathfrak{T} I_{\text{in}} \cos\Delta\phi \\ I_{\text{common}} &= \mathfrak{R}^2 I_{\text{in}} + \mathfrak{T}^2 I_{\text{in}} - 2\mathfrak{R}\mathfrak{T} I_{\text{in}} \cos\Delta\phi \end{aligned} \tag{A1.95}$$

and

$$I_{\text{free}} + I_{\text{common}} = I_{\text{in}} \tag{A1.96}$$

since

$$\mathfrak{R}^2 + \mathfrak{T}^2 + 2\mathfrak{R}\mathfrak{T} = (\mathfrak{R} + \mathfrak{T})^2 = 1 \tag{A1.97}$$

This implies that $\Delta\phi = \Delta\phi' + \pi$. This π difference is due to intrinsic phase shifts induced by reflections.

Note that the fringe visibility is always unity at the free port of a Michelson interferometer where both interfering waves have the same intensity, but that it depends on the balance of the splitter at the common port.

This result of complementary outputs respecting the principle of conservation of energy can be generalized to any lossless interferometer.

Another well-known interferometer is the Mach-Zehnder interferometer (Figure A1.16). Light is separated by a first beamsplitter, propagates along two different paths, and is recombined on a second beamsplitter.

So far, we have considered an interferometer as a system that uses a monochromatic source with a given wavelength λ_0 (and a given spatial frequency $\sigma_0 = 1/\lambda_0$) and that has a periodic response as a function of the optical path length difference ΔL_{op}:

$$I\,(\Delta\phi) = I\,(\Delta L_{op}) = I\left(\cos\left[2\pi\,\frac{\Delta L_{op}}{\lambda_0}\right]\right) = I\,(\cos(2\pi\sigma_0 \cdot \Delta L_{op})) \quad \text{(A1.98)}$$

Note that an interferometer can be alternatively considered as a wavelength filter. As a matter of fact, for a given path difference ΔL_{op0}, the response as a function of the wavelength λ (or the spatial frequency σ) is

$$\begin{aligned} I\,(\Delta\phi) = I\,(1/\lambda) = I\,(\sigma) &= I\left(\cos\left(2\pi\,\Delta L_{op0} \cdot \frac{1}{\lambda}\right)\right) \\ &= I\,(\cos(2\pi\Delta L_{op0} \cdot \sigma)) \end{aligned} \quad \text{(A1.99)}$$

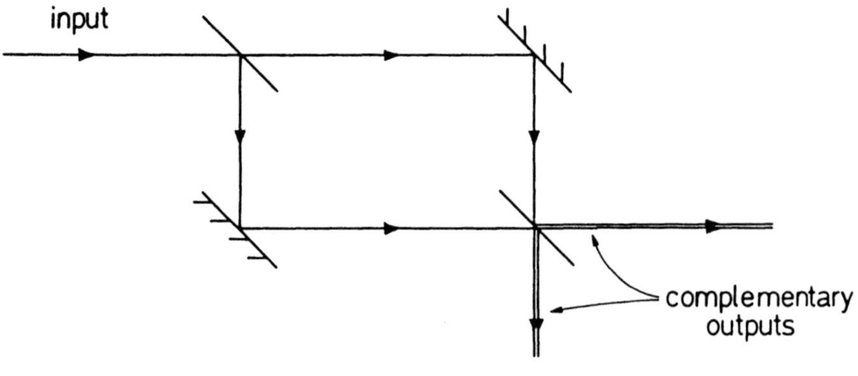

Figure A1.16 Mach-Zehnder interferometer.

A1.7 MULTIPLE-WAVE INTERFERENCES

The Michelson and Mach-Zehnder interferometers are two-wave interferometers with a cosine response. Some other interferometers are multiple-wave interferometers, and their response is also periodic, but not (co)sinusoidal anymore. They behave like resonators with narrow response peaks. Such interferometric resonators are often called optical cavities. Their behavior is not straightforward, and it is an opportunity to detail the fundamental phenomenon of resonance, which has numerous applications in optics.

A good example of such a multiple-wave interferometer is the Fabry-Perot interferometer, which is composed of two parallel high-reflectivity mirrors (Figure A1.17). At first, it seems that the added reflectivities of both mirrors should completely reflect the incident light, but there is a resonance effect when the thickness d of the interferometer is equal to a multiple number of half wavelengths, and for specific resonance wavelengths λ_r the light is fully transmitted.

This can be explained with hand-waving arguments. Let us assume that the reflectivity of each mirror is 99%. When an input intensity I_{in} is send in the Fabry-Perot interferometer, 1% of I_{in} is transmitted inside the cavity. On the second mirror, 0.01% of I_{in} is transmitted and 0.99% remains trapped inside the cavity. This trapped light oscillates back and forth, losing 1% of its own power at each reflection. At the output, there is a series of transmitted waves with a respective intensity i_{t_i}, which exponentially decays from 0.01%, the value of the first transmitted wave. These multiple waves interfere together, and for a resonance wavelength λ_r they are all in phase. Thus, the amplitude A_t of the transmitted wave is the sum of the interfering amplitude a_{t_i} of all the transmitted waves, and this is therefore not the intensity that is the sum of the interfering intensities.

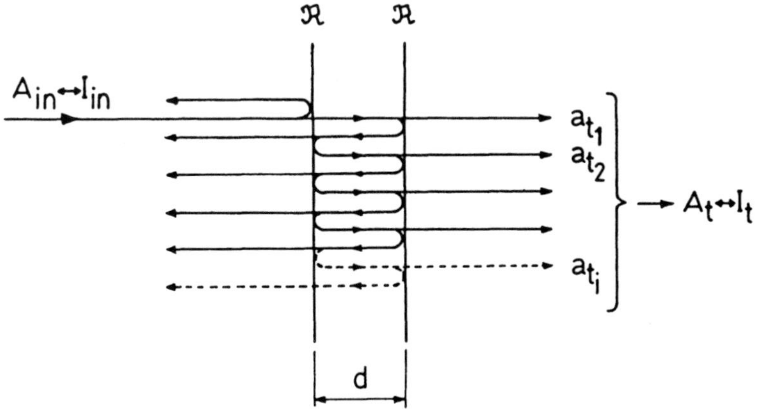

Figure A1.17 Fabry-Perot interferometer.

With mirrors with 1% of transmissivity, we can consider that at the output there are 100 waves of intensity $i_t = 10^{-4} I_{\text{in}}$, but of amplitude $a_t = 10^{-2} A_{\text{in}}$, since the amplitude is the square root of the intensity. The constructive interferences of 100 waves having an amplitude equal to $10^{-2} A_{\text{in}}$ give a total transmitted amplitude A_t equal to the input amplitude A_{in}: the specific resonance wavelengths are fully transmitted despite the two high-reflectivity mirrors, since on resonance the amplitudes and not the intensities of the interfering waves are added.

Calculations show precisely that the transmitted intensity of a Fabry-Perot interferometer is (Figure A1.18)

$$I_t = I_{\text{in}} \frac{1}{1 + F \sin^2(\Delta\phi/2)} \tag{A1.100}$$

where $\Delta\phi = 2\pi\Delta L_{op}/\lambda$ is the phase shift induced by a complete path $\Delta L_{op} = 2d$ inside the cavity. The parameter F is defined by

$$F = \frac{4\mathfrak{R}}{(1 - \mathfrak{R})^2} \tag{A1.101}$$

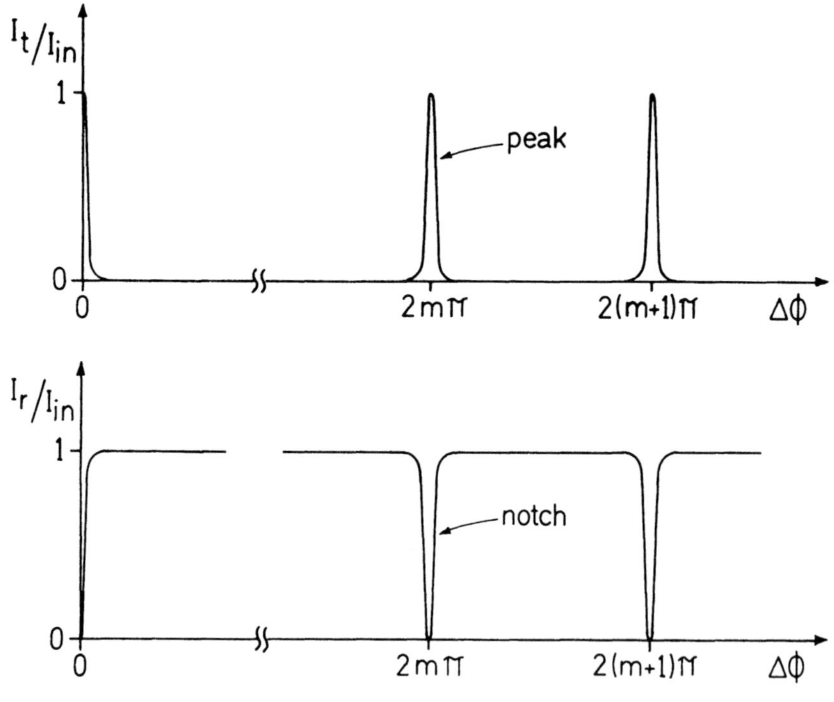

Figure A1.18 Transmission and reflection responses of a Fabry-Perot interferometer as a function of the phase difference.

where $\mathfrak{R}$ is the reflectivity of the mirrors (the transmissivity $\mathfrak{T} = 1 - \mathfrak{R}$). Because of conservation of energy, the reflected intensity I_r and the transmitted intensity I_t are complementary:

$$I_r = I_{in} - I_t = I_{in} \frac{F \sin^2(\Delta\phi/2)}{1 + F \sin^2(\Delta\phi/2)} \tag{A1.102}$$

As we have already seen, a two-wave interferometer may be considered a wavelength filter. For a multiple-wave Fabry-Perot interferometer, this filtering property also exists and is, in fact, one of its main applications. With a vacuum cavity and for a given thickness d_0, the transmission wavelength response is

$$\mathfrak{T}_{FP}(\lambda) = I_t(\lambda)/I_{in} = \frac{1}{1 + F \sin^2\left(2\pi d_0 \dfrac{1}{\lambda}\right)} \tag{A1.103}$$

or, with respect to the spatial frequency $\sigma = 1/\lambda$,

$$\mathfrak{T}_{FP}(\sigma) = \frac{1}{1 + F \sin^2(2\pi d_0 \sigma)} \tag{A1.104}$$

The filtering characteristic of a Fabry-Perot interferometer is given by the full width at half maximum (FWHM) $\Delta\sigma_{FWHM}$ of the filtering transmitted peaks, and by the free spectral range $\Delta\sigma_{free} = 1/(2d_0)$ between two response peaks (Figure A1.19). The ratio $\Delta\sigma_{free}/\Delta\sigma_{FWHM}$ depends only on the reflectivity of the mirror. It is called the finesse $\mathfrak{F}$ (finesse means "sharpness" in French) of the Fabry-Perot cavity:

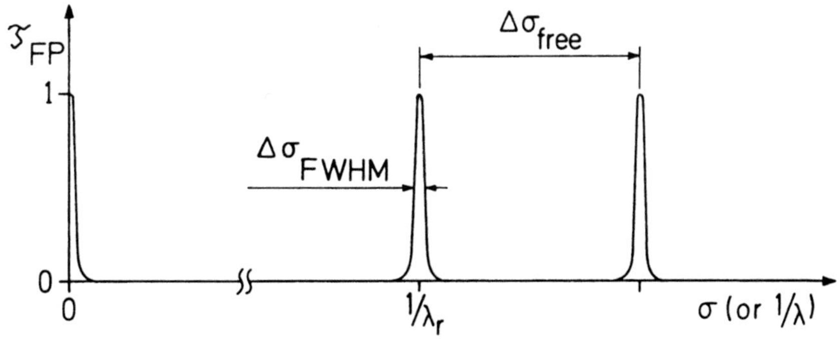

Figure A1.19 Filtering transmission of a Fabry-Perot interferometer as a function of spatial frequency σ (inverse of wavelength λ).

$$\mathfrak{F} = \frac{\pi\sqrt{\mathfrak{R}}}{1 - \mathfrak{R}} \approx \frac{3}{\mathfrak{T}} \qquad \text{(A1.105)}$$

For example, with $\mathfrak{R} = 0.99$ and $\mathfrak{T} = 0.01$, the finesse $\mathfrak{F}$ is 300.

Filtering width and free spectral range may alternatively be given in terms of wavelength, with

$$\frac{\Delta\lambda_{\text{free}}}{\lambda_r^2} \approx \Delta\sigma_{\text{free}} = \frac{1}{2d_0} \qquad \text{(A1.106)}$$

$$\Delta\lambda_{FWHM} = \frac{\Delta\lambda_{\text{free}}}{\mathfrak{F}} \qquad \text{(A1.107)}$$

but the wavelength response obtained by inversing the frequency is not perfectly periodic anymore.

Note that a very important application of the Fabry-Perot interferometer is its use as a resonant cavity to make a laser source with an amplifying medium. In this particular case, it is called an active resonator instead of a passive resonator for the basic device. The emission wavelengths λ_e of a laser are submultiples of the optical length $\Delta L_{op} = 2nd$ of the cavity (where n is the index):

$$\Delta L_{op} = m \cdot \lambda_e \qquad (m \text{ is an integer}) \qquad \text{(A1.108)}$$

A1.8 DIFFRACTION AND GAUSSIAN BEAM

Diffraction is a fundamental phenomenon related to the wave nature of light propagation and showing the limit of geometrical ray optics. It is not possible to isolate one ray in an optical wave. When light is selected by a small filtering hole, it can be observed diverging with a "mean" angle θ_D, which increases as the hole size a_h decreases (Figure A1.20):

$$\theta_D \approx \frac{\lambda}{a_h} \qquad \text{(A1.109)}$$

Diffraction can be explained by considering that, according to Huygens' principle, each point of the hole re-emits a spherical wave and that all these spherical waves interfere together. The amplitude of the diffracted wave at a given point is the integral summation of all the amplitudes of these waves that interfere.

Diffraction at a finite distance, known as Fresnel diffraction, is complicated mathematically, but diffraction at an infinite distance (or in the focal plane of a

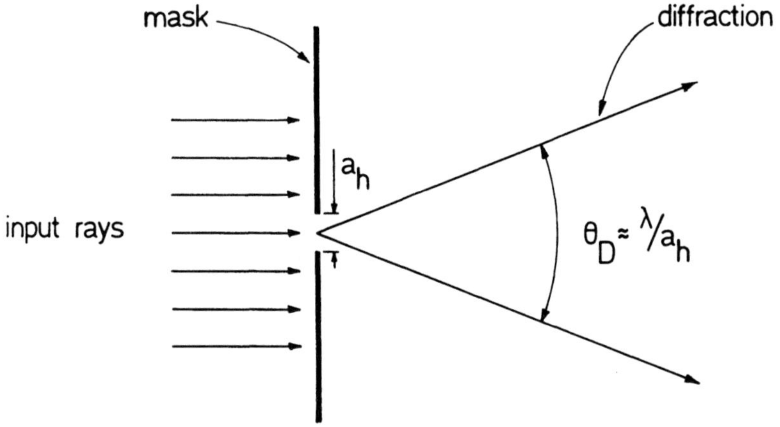

Figure A1.20 Diffraction of a wave.

lens that is making "an image of infinity"), known as Fraunhoffer diffraction, is simpler to solve. In a given direction θ, the waves interfere with a respective path length difference (Figure A1.21):

$$\Delta L_{op} = x \cdot \sin\theta \tag{A1.110}$$

The total diffracted amplitude A_d in a direction θ is then

$$A_d\,(\sin\theta) = \int_{-a_h/2}^{+a_h/2} A_f(x/\lambda)e^{2i\pi\cdot\sin\theta\cdot x/\lambda}dx \tag{A1.111}$$

where A_f is the amplitude of the spatially filtered wave. Hence, the diffracted amplitude $A_d(\sin\theta)$ is the Fourier transform of the spatially filtered amplitude $A_f(x/\lambda)$. The relation $\theta_D = \lambda/a_h$ corresponds to the fact that the product of the widths of a pair of Fourier transforms is constant.

This interferometric analysis of diffraction can be explained with hand-waving arguments: in a direction angle such as $\theta > \theta_D = \lambda/a_h$, each spherical wave emitted in the filtering hole has a corresponding wave with a path difference of $\lambda/2$, and then both waves interfere destructively. Note that diffraction is a phenomenon intrinsically related to the wavelength; therefore, in a medium, the same effect occurs, but as a function of the actual wavelength $\lambda_m = \lambda/n$ in the medium.

Since the far-field diffraction pattern is the Fourier transform of the spatially filtered amplitude distribution, a Gaussian distribution is of particular interest because the Gaussian function is invariant under Fourier transform. Such Gaussian beams are generated particularly in gas lasers. A Gaussian beam that propagates along the $0z$ axis has a waist with a radial amplitude distribution (Figure A1.22):

$$A(r) = A_0 e^{-r^2/w_0^2} \tag{A1.112}$$

and an intensity distribution that is the square of the amplitude distribution:

$$\begin{aligned} I(r) &= I_0 e^{-2r^2/w_0^2} \\ I_0 &= A_0^2 \end{aligned} \tag{A1.113}$$

The term w_0, called the waist, is the radius at $1/e$ in amplitude and at $1/e^2$ in intensity (and $2w_0$ the diameter at $1/e^2$), since

$$\begin{aligned} A(w_0)/A_0 &= 1/e \\ I(w_0)/I_0 &= 1/e^2 \end{aligned} \tag{A1.114}$$

From the beam waist, light is diffracted but keeps a Gaussian amplitude distribution:

$$A(r,z) = A_0(z) e^{-r^2/w(z)^2} \tag{A1.115}$$

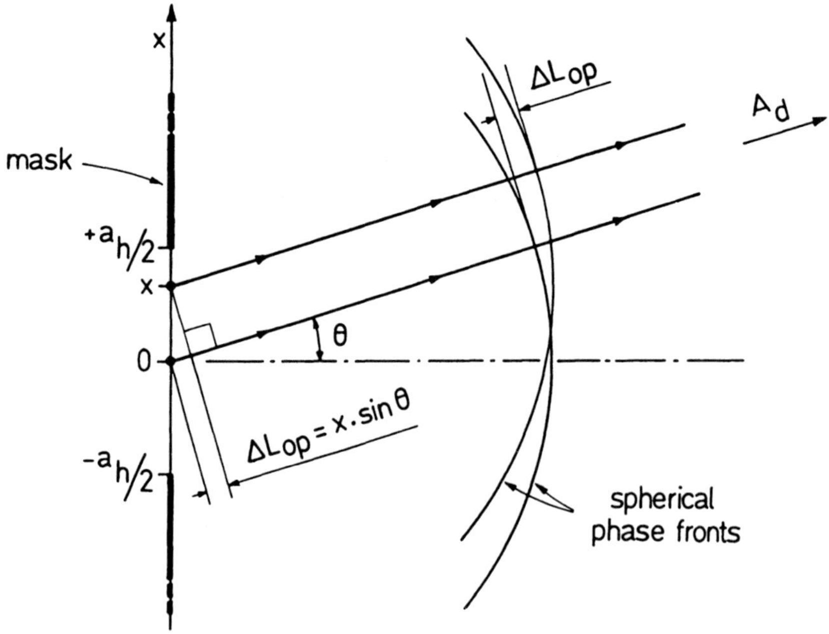

Figure A1.21 Calculation of Fraunhoffer diffraction.

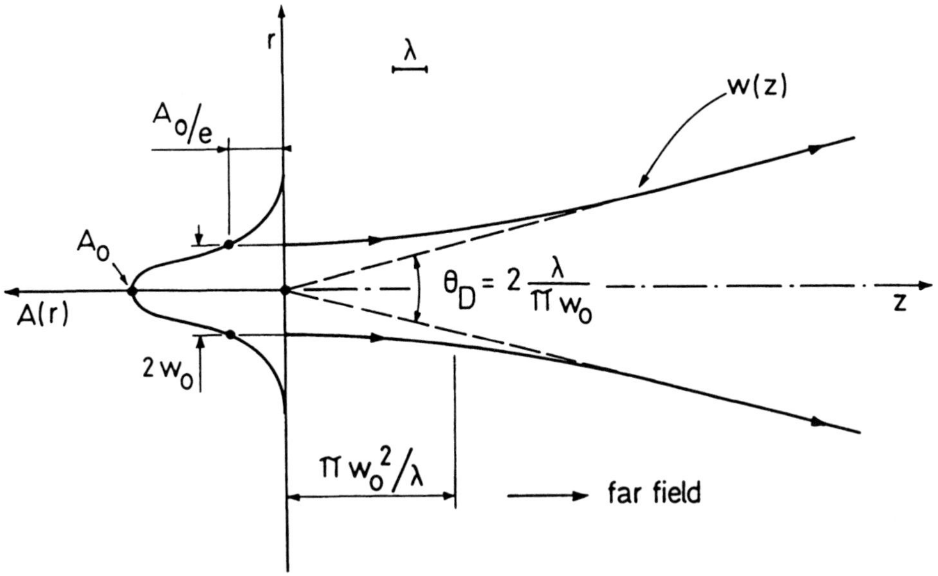

Figure A1.22 Gaussian beam.

with

$$w(z) = w_0 \left[1 + \left(\frac{\lambda z}{\pi w_0^2} \right)^2 \right]^{1/2} \quad \text{(A1.116)}$$

This formula may be simplified in two cases:

1. $\lambda z / \pi w_0^2 << 1$ (or $z << \pi w_0^2/\lambda$) and $w(z)$ is about constant:

$$w(z) \approx w_0$$

2. $\lambda z / \pi w_0^2 >> 1$ (or $z >> \pi w_0^2/\lambda$) and $w(z)$ is proportional to z:

$$w(z) \approx \frac{\lambda}{\pi w_0} \cdot z \quad \text{(A1.117)}$$

The Gaussian beam diverges with a full angle θ_D at $1/e^2$ in intensity:

$$\theta_D = \frac{2w(z)}{z} \quad \text{(A1.118)}$$

$$\theta_D = \frac{4}{\pi} \cdot \frac{\lambda}{(2w_0)} \quad \text{(A1.119)}$$

and it is found that

$$\theta_D \approx \frac{\lambda}{a_h} \tag{A1.120}$$

as stated before, if we consider that $a_h = 2w_0$ is about the "width" of the aperture.

In this last case, light is diffracted with a constant divergence angle, and when the condition $\lambda z/\pi w_0^2 \geq 1$ is fulfilled, it is possible that the result of diffraction may be approximated by Fraunhoffer diffraction, even if the distance remains finite. This condition may be generalized for any filtering hole, and when $z \geq a_h^2/\lambda$, it is possible that Fraunhoffer diffraction is observed. This diffraction pattern is often called the far field.

A1.9 COHERENCE

In the case of the ideal monochromatic plane wave, the phase of the wave at a given point can be deduced from the phase at any other point:

- The phase is the same at any point on the same phase front transverse to the direction of propagation. This perfect "transverse correlation" is called spatial coherence.
- The phase difference $\Delta\phi$ between two different phase fronts can be deduced from the distance d between both phase fronts, with $\Delta\phi = 2\pi nd/\lambda$. This perfect "longitudinal correlation" is called temporal coherence.

In practice, this ideal case is not possible, and a wave has a limited coherence. The complete theory of coherence is complicated and involves the mathematical apparatus of stochastic processes. However, we are going to analyze the problem of temporal coherence, which is very important for the interferometric fiber-optic gyroscope. On the other hand, the fiber gyro uses single-mode waveguide, and spatial coherence is automatically ensured because all the points of a mode have a perfect phase correlation in the transverse direction.

When a broad-spectrum source is sent into an interferometer, it is observed that there is good interference constrast about a null path difference, and as this path difference increases (in absolute value), the fringe visibility decreases and finally vanishes completely (Figure A1.23). When the path difference is longer than a given correlation length, called coherence length L_c of the source in optics, both interfering waves are not correlated anymore, and their phase difference varies as a function of time, which averages out the $\cos\Delta\phi$ term of the formula of interferences.

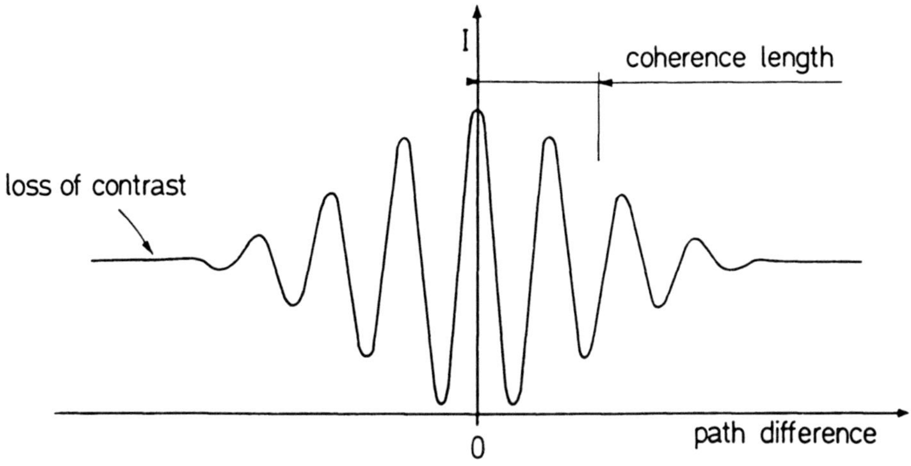

Figure A1.23 Interference signal with a broad spectrum source.

varies as a function of time, which averages out the cos$\Delta\phi$ term of the formula of interferences.

Another way to understand the phenomenon is to decompose the broad spectrum into separate wavelengths. Each wavelength λ_i creates its own interference pattern with a period λ_i. About zero path difference, all the interference patterns coincide; but as the path difference increases, they lose their coincidence because they have slighty different periods (Figure A1.24). Because some wave-

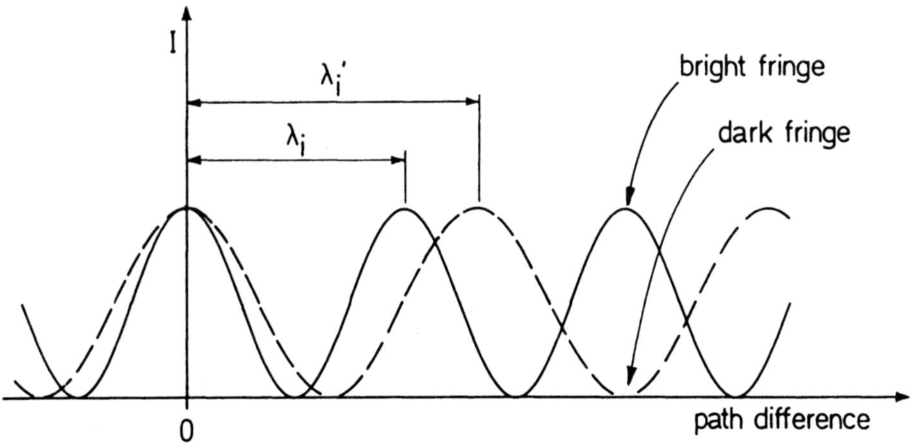

Figure A1.24 Interference signals with different wavelengths.

lengths are on a bright fringe and other ones are on a dark fringe, on average the total intensity becomes constant, and the contrast disappears. The coherence length L_c is inversely proportional to the spectrum width $\Delta\lambda$.

The exact result may be derived rigorously with Fourier transforms. Let us consider the wave amplitude $A(t)$, which varies as a function of time t at a given spatial position. As we have already seen, the frequency components $a(f)$ of this amplitude $A(t)$ may be defined by Fourier transform:

$$a(f) = \int_{-\infty}^{+\infty} A(t)e^{-2i\pi ft}dt \quad \text{(A1.121)}$$

and, inversely, the wave amplitude $A(t)$ is the integral summation of all its frequency components:

$$A(t) = \int_{-\infty}^{+\infty} a(f)e^{2i\pi ft}df \quad \text{(A1.122)}$$

In general, the frequency components $a(f)$ are complex and may be written with a positive real modulus $\alpha(f)$ and a phase term $e^{i\phi(f)}$:

$$a(f) = \alpha(f)e^{i\phi(f)} \quad \text{(A1.123)}$$

When this wave is sent in an interferometer with two 50-50 (in intensity) splitters, there are interferences between the wave $A(t)$ and itself, but they are shifted by a temporal delay τ, $A(t - \tau)$. Considering that light propagates in a vacuum along both paths, the temporal delay τ is related to the geometrical path difference ΔL by

$$\Delta L = c \cdot \tau \quad \text{(A1.124)}$$

To respect energy conservation, each 50-50 splitter reduces the intensity by 1/2 and the amplitude by $1/\sqrt{2}$. Therefore, after passing two splitters, there are interferences between $1/2(A(t))$ and $1/2(A(t - \tau))$. The intensity I of the interference wave is

$$I = \frac{1}{4} <|A(t) + A(t - \tau)|^2> \quad \text{(A1.125)}$$

$$I = \frac{1}{4} <[A(t) + A(t - \tau)] \cdot [A^*(t) + A^*(t - \tau)]> \quad \text{(A1.126)}$$

$$I = \frac{1}{4} [<A(t) \cdot A^*(t)> + <A(t - \tau) A^*(t - \tau)> \quad \text{(A1.127)}$$

$$+ <A(t) A^*(t - \tau)> + <A(t - \tau) A^*(t)>]$$

where the brackets < > denote temporal averaging. The term $<A(t) \cdot A^*(t - \tau)>$ is proportional to the autocorrelation function $\Gamma(\tau)$ of the function $A(t)$, defined in signal processing theory as

$$<A(t) \cdot A^*(t - \tau)> \propto \qquad \Gamma(\tau) = \int_{-\infty}^{+\infty} A(t) \cdot A^*(t - \tau) dt \tag{A1.128}$$

Then the intensity I of the interferences is proportional to

$$I \propto \frac{1}{4}\left[2\,\Gamma(0) + \Gamma(\tau) + \Gamma^*(\tau)\right] \tag{A1.129}$$

The basic autocorrelation theorem, also called the Wiener-Khinchin theorem, states: "if $A(t)$ has the Fourier transform $a(f)$, then its autocorrelation function $\Gamma(\tau) = \int_{-\infty}^{+\infty} A(t) \cdot A^*(t - \tau)dt$ has a real and positive Fourier transform equal to the power spectral density $|a(f)|^2 = \alpha^2(f)$." Therefore,

$$\alpha^2(f) = \int_{-\infty}^{+\infty} \Gamma(\tau) e^{-2i\pi f\tau} d\tau \tag{A1.130}$$

and, inversely,

$$\Gamma(\tau) = \int_{-\infty}^{+\infty} \alpha^2(f) e^{2i\pi f\tau} df \tag{A1.131}$$

In practice, the power spectral density $\alpha^2(f)$ is centered about a mean frequency $\bar{f}$, and a "centered" spectrum α_c^2 may be defined by (Figure A1.25)

$$\alpha_c^2(f) = \alpha^2(\bar{f} + f) \tag{A1.132}$$

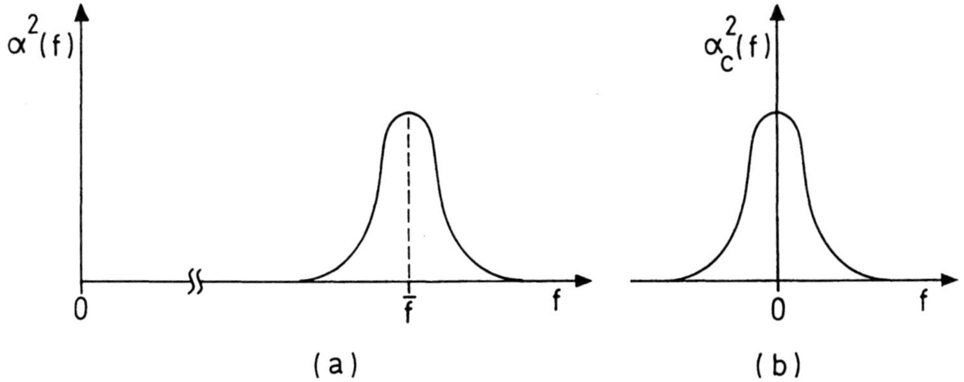

Figure A1.25 Intensity spectrum as a function of temporal frequency f: (a) actual spectrum $\alpha^2(f)$; (b) centered spectrum $\alpha_c^2(f)$.

Then

$$\Gamma(\tau) = \int_{-\infty}^{+\infty} \alpha_c^2(f) e^{2i\pi(\bar{f}+f)\tau} df \qquad \text{(A1.133)}$$

and

$$\begin{aligned} \Gamma(\tau) &= e^{2i\pi\bar{f}\tau} \cdot \Gamma_c(\tau) \\ \Gamma_c(\tau) &= \int_{-\infty}^{+\infty} \alpha_c^2(f) e^{2i\pi f\tau} df \end{aligned} \qquad \text{(A1.134)}$$

where $\Gamma_c(\tau)$ is the inverse Fourier transform of the centered intensity spectrum $\alpha_c^2(f)$.

If the spectrum $\alpha^2(f)$ is symmetrical about the mean frequency $\bar{f}$, the centered spectrum is an even real function and its inverse Fourier transform $\Gamma_c(\tau)$ is also an even real function (i.e., $\Gamma_c(\tau) = \Gamma_c(-\tau)$ and $\Gamma_c(\tau) = \Gamma_c^*(\tau)$). Since

$$\Gamma_c(0) = \Gamma(0) = \langle AA^* \rangle = \int_{-\infty}^{+\infty} |A(t)|^2 dt = \int_{-\infty}^{+\infty} \alpha^2(f) df \propto I_{in} \qquad \text{(A1.135)}$$

where I_{in} is the intensity of the input wave. The intensity I of the interference wave is

$$I = \frac{I_{in}}{4}\left[2 + e^{2i\pi\bar{f}\tau}\frac{\Gamma_c(\tau)}{\Gamma_c(0)} + e^{-2i\pi\bar{f}\tau}\frac{\Gamma_c(\tau)}{\Gamma_c(0)}\right] \qquad \text{(A1.136)}$$

It is possible to define a normalized centered autocorrelation function:

$$\gamma_c(\tau) = \frac{\Gamma_c(\tau)}{\Gamma_c(0)} \qquad \text{(A1.137)}$$

where $\gamma_c(0) = 1$ and $\lim_{\tau \to \infty} \gamma_c(t) = 0$. And finally,

$$I = \frac{I_{in}}{2}\left[1 + \gamma_c(\tau) \cdot \cos(2\pi \cdot \bar{f} \cdot \tau)\right] \qquad \text{(A1.138)}$$

In terms of spatial coordinates, there is

$$I = \frac{I_{in}}{2}\left[1 + \gamma_c(\Delta L) \cos\left(2\pi \frac{\Delta L}{\bar{\lambda}}\right)\right] \qquad \text{(A1.139)}$$

where the path length difference $\Delta L = c\tau$ and the mean wavelength $\bar{\lambda} = c/\bar{f}$.

Then, the effect of a broad spectrum is to yield a visibility decrease of the cosinusoidal modulation of the interference fringes, with a decrease of $\gamma_c(\tau)$ as $|\tau|$ increases (or a decrease of $\gamma_c(\Delta L)$ as $|\Delta L|$ increases). The fringe visibility actually becomes

$$\vartheta(\tau) = \gamma_c(\tau) \tag{A1.140}$$

where $\gamma_c(\tau)$, called the coherence function of the source, is the normalized inverse Fourier transform of the centered intensity spectrum α_c^2 (Figure A1.26).

The "half-width" of the $\gamma_c(\tau)$ function is called the coherence time τ_c of the input wave $A(t)$. This "half-width" is classicaly defined as the root-mean-square (rms) half-width of $\gamma_c(\tau)$, or half-width "at 1σ" (see Born and Wolf):

$$\tau_c^2 = \frac{\int_{-\infty}^{+\infty} \tau^2 \gamma_c^2(\tau) d\tau}{\int_{-\infty}^{+\infty} \gamma_c^2(\tau) d\tau} \tag{A1.141}$$

When the centered intensity spectrum $\alpha_c^2(f)$ and its normalized Fourier transform $\gamma_c(\tau)$ are Gaussian, calculations are simple and give

$$\gamma_c(\tau) = e^{-\tau^2/4\,\tau_c^2} \tag{A1.142}$$

since

$$\int_{-\infty}^{+\infty} e^{-x^2} dx = \sqrt{\pi} \tag{A1.143}$$

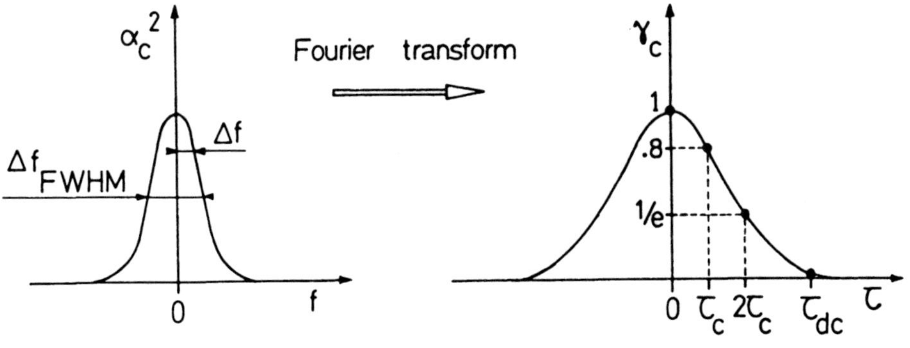

Figure A1.26 Centered spectrum $\alpha_c^2(f)$ and coherence function $\gamma_c(\tau)$ (assuming Gaussian functions).

and

$$\int_{-\infty}^{+\infty} x^2 e^{-x^2} dx = \sqrt{\pi}/2 \tag{A1.144}$$

Then the coherence time τ_c is the time that yields (Figure A1.26)

$$\gamma_c(\tau_c) = e^{-1/4} \approx 0.8$$

Note that the half-width at 2σ (i.e., $2\tau_c$) is the half-width at $1/e$, since

$$\gamma_c(2\tau_c) = e^{-1} \tag{A1.145}$$

It is also possible to define a rms half-width of the intensity spectral density $\alpha^2(f)$:

$$\Delta f^2 = \frac{\int_{-\infty}^{+\infty} (f - \bar{f})^2 \, [\alpha^2(f)]^2 df}{\int_{-\infty}^{+\infty} [\alpha^2(f)]^2 df} \tag{A1.146}$$

In the case where $\alpha^2(f)$ is Gaussian, there is

$$\alpha^2(f) = e^{-(f-\bar{f})^2/4\Delta f^2} \tag{A1.147}$$

and

$$\alpha^2(\bar{f} + \Delta f) = e^{-1/4} \approx 0.8 \tag{A1.148}$$

When $e^{-\pi x^2}$ and $e^{-\pi y^2}$ are a pair of Fourier transforms, the product of their rms halfwidths is 1. Therefore,

$$\tau_c \cdot \Delta f = \frac{1}{4\pi} \tag{A1.149}$$

This result has been derived exactly for the Gaussian spectrum, but it remains approximatively true for other spectrum shapes.

The coherence length L_c is defined by

$$L_c = c \cdot \tau_c = \frac{c}{4\pi \Delta f} \tag{A1.150}$$

and since $\Delta f = \Delta\lambda/\bar{\lambda}^2$, where $\Delta\lambda$ is the rms half-width of the spectrum in wavelength,

$$L_c = \frac{1}{4\pi} \cdot \frac{\bar{\lambda}^2}{\Delta\lambda} \quad \text{(A1.151)}$$

As we have seen, the coherence time τ_c and the coherence length L_c correspond to the maximum temporal or spatial delay that preserves a good fringe visibility:

$$\vartheta(\tau_c \text{ or } L_c) = \gamma_c\,(\tau_c \text{ or } L_c) \approx 0.8 \quad \text{(A1.152)}$$

As can be seen in coherent noise analysis in the fiber gyro, the time or length above which the wave loses its coherence is also of interest. A good definition of this "decoherence" time τ_{dc} is the inverse of the full width at half maximum Δf_{FWHM} of the spectrum (Figure A1.26):

$$\tau_{dc} = \frac{1}{\Delta f_{FWHM}}$$

The "decoherence" length L_{dc} is then

$$L_{dc} = c \cdot \tau_{dc} \approx \frac{\bar{\lambda}^2}{\Delta\lambda_{FWHM}} \quad \text{(A1.153)}$$

since $\Delta f_{FWHM} \approx \dfrac{c\Delta\lambda_{FWHM}}{\bar{\lambda}^2}$

Going back to the simple case of Gaussian functions, it is found that

$$\Delta f_{FWHM} = 4 \cdot \sqrt{\text{Ln } 2} \cdot \Delta f \approx 3.3\ \Delta f$$

$$\tau_{dc} = \frac{\pi}{\sqrt{\text{Ln } 2}}\,\tau_c \approx 3.8\ \tau_c \quad \text{(A1.154)}$$

and then

$$\gamma_c(\tau_{dc}) = 0.03$$

which means that for τ_{dc} (or L_{dc}) the contrast has been reduced to 3%.

The coherence time τ_c is the half-width at 1σ, and the "decoherence" time τ_{dc} defined by $1/\Delta f_{FWHM}$ is actually the half-width at 3σ to 4σ (the half-width at $1/e$ being the half-width at 2σ).

These results derived exactly for Gaussian functions may be approximately extended to any "bell-shaped" spectrum. However, to solve certain parasitic effects in the fiber gyro, it is necessary to know precisely the whole coherence function of the source and not simply its half-width.

The very important result of this analysis is the fact that the autocorrelation function and the fringe visibility depends on the Fourier transform of the power spectral density $|a(f)|^2 = \alpha^2(f)$ and not on the amplitude spectral density $a = \alpha e^{i\phi}$. As is usually stated, the process of autocorrelation loses the information about the phase of the spectral components of $A(t)$.

This means that waves, with the same modulus $\alpha(f)$ but a different phase $\phi(f)$ of their frequency components, yield the same fringe visibility. In particular, a broad-spectrum source has frequency components with a random phase, while a pulse with pure amplitude modulation has frequency components that have the same phase; but both yield the same fringe visibility if they have the same power spectrum $\alpha^2(f)$. Very often, the temporal coherence behavior of a broad-spectrum source is explained by considering wave trains of duration equal to the "decoherence" time τ_{dc} and a length equal to the "decoherence" length L_{dc}. They are actually image pulses with pure amplitude modulation; but since the phase is lost in the autocorrelation, they would yield the same interference contrast as the broad-spectrum source. This allows one to analyze simply the effect in the time domain, and, particularly when a wave train is sent in an unbalanced interferometer, there are two wave trains at the output. If the path imbalance ΔL_{op} is larger than the wave train length L_{dc}, the two output wave trains do not overlap and cannot interfere (Figure A1.27).

Note: These results are simple if the spectrum is symmetrical, which yields an even-centered spectrum. When the spectrum is not symmetrical, as is the case with the broadband sources used in the fiber gyro (superluminescent diode or rare-earth doped fiber source), the analysis is more complicated. The centered spectral density α_c^2 has to be decomposed in an even density $\alpha^2_{ce}(f)$ and a residual odd density $\alpha^2_{co}(f)$ (Figure A1.28):

$$\begin{aligned}
\alpha_c^2(f) &= \alpha^2(\bar{f} + f) \\
\alpha_c^2(f) &= \alpha_{ce}^2(f) + \alpha_{co}^2(f) \\
\alpha_{ce}^2(f) &= \alpha_{ce}^2(-f) = \frac{\alpha_c^2(f) + \alpha_c^2(-f)}{2} \\
\alpha_{co}^2(f) &= -\alpha_{co}^2(-f) = \frac{\alpha_c^2(f) - \alpha_c^2(-f)}{2}
\end{aligned} \tag{A1.155}$$

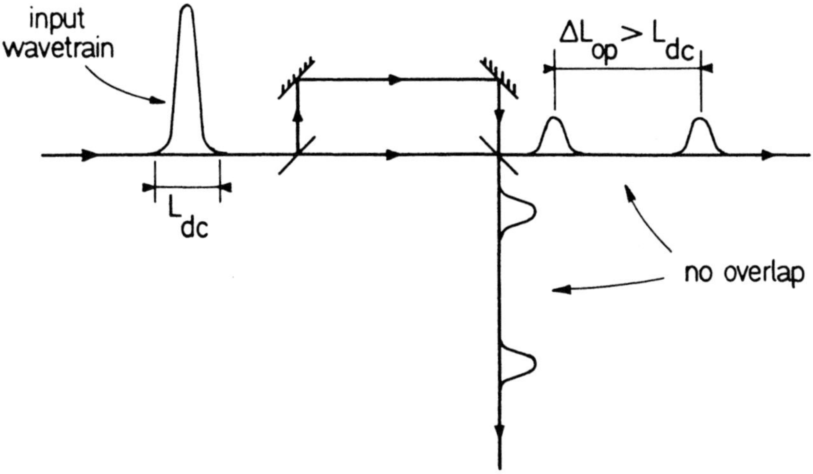

Figure A1.27 Propagation of a wave train in an unbalanced interferometer.

The mean frequency $\bar{f}$ is of course defined as the frequency that minimizes $\alpha_{co}(f)$. The even component $\alpha^2{}_{ce}(f)$ does yield through Fourier transform an even and real autocorrelation function $\Gamma_{ce}(\tau)$, but the residual odd component $\alpha^2{}_{co}(f)$ yields through Fourier transform an odd and purely imaginary autocorrelation function $i \cdot \Gamma_{co}(\tau)$. The interferences become

$$I = \frac{I_{in}}{2}\left[1 + \gamma_{ce}(\tau)\cos(2\pi\bar{f}\tau) + \gamma_{co}(\tau)\sin(2\pi\bar{f}\tau)\right] \tag{A1.156}$$

where $\gamma_{ce}(\tau)$ and $\gamma_{co}(\tau)$ have been accordingly normalized to

$$\gamma_{ce}(\tau_c) = \frac{\Gamma_{ce}(\tau)}{\Gamma(0)} \tag{A1.157}$$

$$\gamma_{co}(\tau) = \frac{\Gamma_{co}(\tau)}{\Gamma(0)} \tag{A1.158}$$

with

$$\Gamma(\tau) = \Gamma_{ce}(\tau) + i\Gamma_{co}(\tau) \tag{A1.159}$$

and

$$\Gamma(0) = \Gamma_{ce}(0)$$

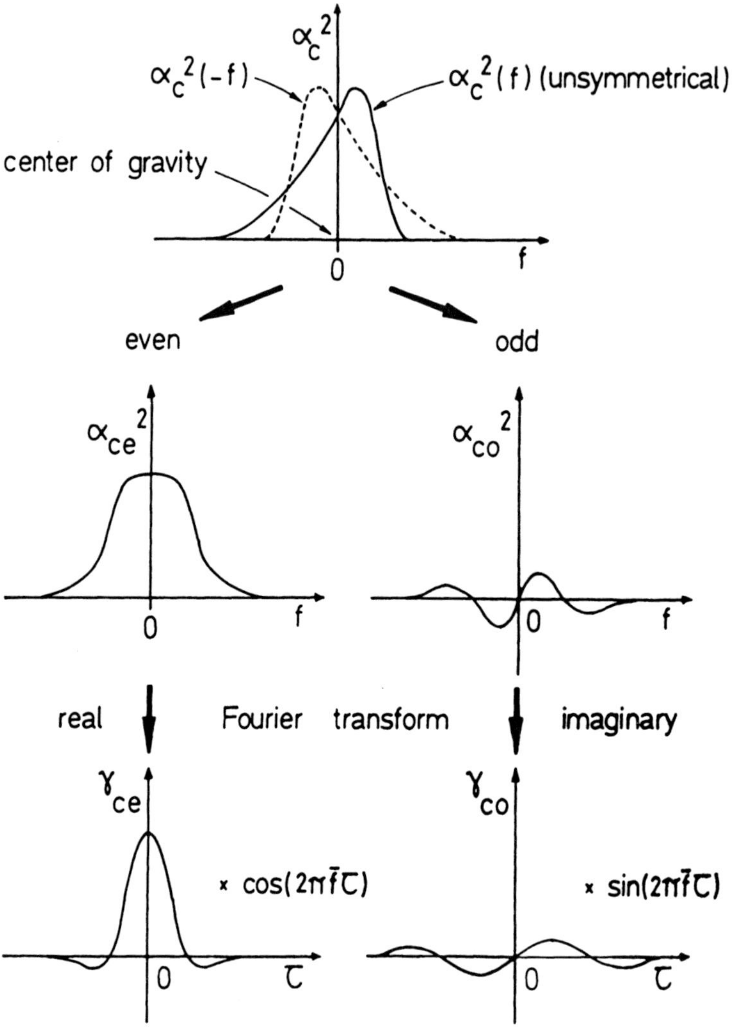

Figure A1.28 Decomposition of an asymmetrical spectrum and related coherence functions.

since, as an odd function,

$$\Gamma_{co}(0) = 0$$

Compared to the case of a symmetrical spectrum, there is an additional term $[\gamma_{co}(\tau) \cdot \sin(2\pi\bar{f}\tau)]$ that is an even function of τ, since it is the product of two odd functions: γ_{co} and sine. Even with an asymmetrical spectrum, the interference remains symmetrical with respect to zero. However, because of this additional term, the zero crossings of the fringe modulation have lost the periodicity of the cosine term for $2\pi\bar{f}\tau = \pi/2, 3\pi/2$, and so on. This effect is usually negligible in an interferometric fiber gyro that works on the central fringe about zero phase difference, but it may have to be taken into account with certain spectrum stabilization schemes.

Note that this analysis has been carried out with respect to the frequency spectrum, which has the same shape if we consider the temporal frequency or the spatial frequency $\sigma = f/c$. In practice, the spectrum is often given with respect to the wavelength λ which is inversely proportional to the frequencies, $\lambda = 1/\sigma = c/f$; and a symmetrical frequency spectrum does not strictly give a symmetrical wavelength spectrum (nor, inversely, a symmetrical wavelength spectrum does not strictly give a symmetrical frequency spectrum), particularly with the large relative width (a large percentage) of the broadband sources desired in fiber gyro applications (Figure A1.29).

Note: This analysis has yielded the definition of the coherence time τ_c (and decoherence time τ_{dc}) which is a characteristic of the source spectrum. To observe this effect, an interferometer has been used in a vacuum, where the temporal delay τ is related simply to the difference of geometrical length ΔL between both paths of this interferometer:

$$\Delta L = c \cdot \tau \tag{A1.160}$$

The spatial equivalent of the temporal quantities τ_c and τ_{dc} have been defined as the coherence length $L_c = c \cdot \tau_c$ and the decoherence length $L_{dc} = c\tau_{dc}$, which are also a characteristic of the source.

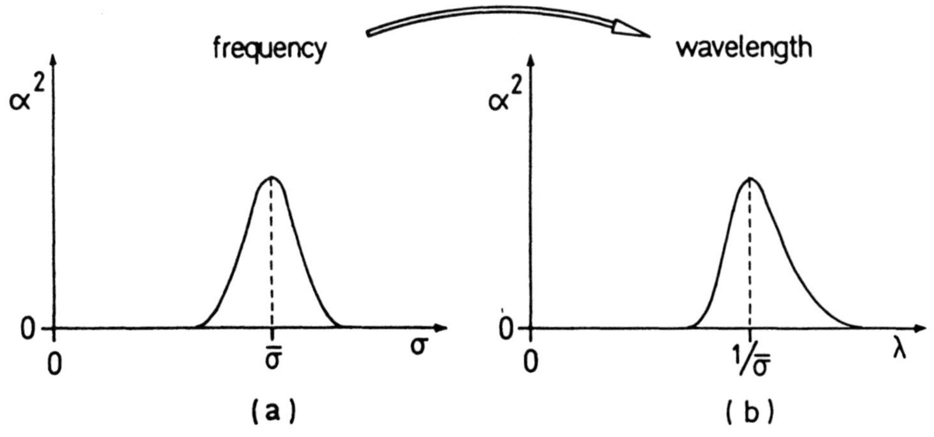

Figure A1.29 Problem of spectrum symmetry: (a) case of symmetry with respect to spatial frequency σ; (b) related asymmetry with respect to wavelength λ (inverse of σ).

Now when light propagates in a medium, the result is more complicated because of dispersion effects. As we saw in Section A1.3, the amplitude A may be decomposed as the product of:

- A monochromatic wave with a frequency equal to the mean frequency $\overline{\omega}$ and that propagates at the phase velocity $v = \overline{\omega}/k_m(\overline{\omega})$;
- A modulation term M that propagates at the group velocity $v_g = d\omega/dk_m$.

Temporally, there is

$$A(t) = M(t)\, e^{i\overline{\omega}t} \tag{A1.161}$$

and taking into account spatial propagation, there is

$$A(t,z) = M\left(t - \frac{z}{v_g}\right) e^{i\overline{\omega}\left(t - \frac{z}{v}\right)} \tag{A1.162}$$

It can be shown that the autocorrelation function of M is actually the inverse Fourier transform of the centered spectrum α_c:

$$<M(t)M^*(t - \tau)> = \Gamma_c(\tau) \tag{A1.163}$$

Going back to the concept of a wave train, it can be shown that a wave train is a sinusoidal wave modulated in amplitude with an envelop $M_{wt}(t)$ which has an autocorrelation function also equal to $\Gamma_c(t)$. The wave train amplitude is

$$A_{wt}(t) = M_{wt}(t)e^{i\overline{\omega}t} \tag{A1.164}$$

Now the following formula is still valid, since it has been derived only with temporal coordinates:

$$I = \frac{I_{\text{in}}}{2}\left[1 + \gamma_{ce}(\tau)\cos(2\pi\bar{f}\tau) + \gamma_{co}(\tau)\sin 2\pi\bar{f}\tau\right] \tag{A1.165}$$

However, when light propagates in a medium, the simple relation $\Delta L = c \cdot \tau$ is not valid anymore. The temporal delay τ does not correspond to the same spatial delay for the autocorrelation terms γ_{ce} and γ_{co} and for the "fringe" terms cosine and sine. Then, in an interferometer with one path in a medium 1 with a geometrical length L_1 and the other path in a medium 2 with a geometrical length L_2, the actual temporal delay is different in both cases. It is:

- $\dfrac{L_1}{v_{g1}} - \dfrac{L_2}{v_{g2}}$ for the autocorrelation terms γ_{ce} and γ_{co}, since they are derived from $M(t)$, which propagates at the group velocity v_g. This delay is called the group temporal delay, denoted by τ_g. It is also the propagation time of the envelop of a wave train.

- $\frac{L_1}{v_1} - \frac{L_2}{v_2}$ for the "fringe" terms, since they are derived from $e^{i\overline{\omega}t}$, which propagates at the phase velocity v. This delay is called the phase temporal delay, denoted by τ_ϕ, with $c \cdot \tau_\phi = n_1(\overline{\omega})L_1 - n_2(\overline{\omega})L_2$.

Therefore, in the most general case, the interference intensity is

$$I = \frac{I_{in}}{2}\left[1 + \gamma_{ce}(\tau_g)\cos(2\pi\bar{f}\tau_\phi) + \gamma_{co}(\tau_g)\sin(2\pi f\tau_\phi)\right] \qquad \text{(A1.166)}$$

In practice, material dispersion yields only a small difference between τ_g and τ_ϕ, but interferometric measurements are very sensitive, and this may yield spurious effects. For example, this problem is avoided in Michelson interferometers with an additional compensating plate, which cancels out the effect of dispersion of the material supporting the 50-50 splitting coating.

In a fiber, there are some cases where τ_g may be very different from τ_ϕ, as will be seen in Appendix 2.

A1.10 BIREFRINGENCE

As we have seen, in an isotropic linear dielectric medium, the derived electric field **D** is proportional to the electric field **E**:

$$\mathbf{D} = \epsilon_r\epsilon_0\,\mathbf{E} \qquad \text{(A1.167)}$$

Gas, liquids, and amorphous solids like glass or fused silica are isotropic because their structure is random and they do not have any predominate axis of orientation. Conversely, crystals have an ordered lattice with predominate axes, and the permittivity ϵ_r may depend on the orientation of the field. They have three principal axes x, y, and z that are orthogonal. The permittivity depends on the principal axes:

$$D_x = \epsilon_{rx}\epsilon_0 E_x$$

$$D_y = \epsilon_{ry}\epsilon_0 E_y \qquad \text{(A1.168)}$$

$$D_z = \epsilon_{rz}\,\epsilon_0 E_z$$

Each value ϵ_{ri} corresponds to an index value $n_i = \sqrt{\epsilon_{ri}}$, with $i = x, y$, or z.

Crystals may be classified into three groups:

- Crystals of the cubic system like diamond, for example, where $n_x = n_y = n_z$. They are optically isotropic and behave optically like amorphous glasses.
- Uniaxial crystals that have two ordinary axes with $n_x = n_y = n_o$, and one extraordinary axis with $n_z = n_e \neq n_o$. The birefringence index difference is $\Delta n_b = n_e - n_o$. The birefringence is said to be positive when $\Delta n_b > 0$ (i.e., $n_e > n_o$), and negative when $n_e < n_o$. The extraordinary z-axis is often called the C-axis of the uniaxial crystal.
- Biaxial crystals that have different indexes for each axis: $n_x \neq n_y \neq n_z \neq n_x$.

When a wave is linearly polarized along a principal axis, it propagates with an index of refraction $n_i = \sqrt{\epsilon_{ri}}$ (with $i = x, y$, or z) that corresponds to the axis of the electric field, since, as we have seen, the effect of the **B** field of the wave on matter is usually negligible. For example, a wave that is linearly polarized along the x-axis and that propagates along the y- or z-axis has a velocity c/n_x.

Propagation along an intermediate direction is complicated, but birefringence problems that we will have to analyze in fiber optics are simpler because they are concerned only with propagation along one principal axis. The main effect of birefringence is to modify the state of polarization during the propagation when the wave is not linearly polarized along a principal axis.

Let us consider a birefringent plate with the x-axis and z-axis parallel to the interface and the y-axis parallel to the direction of propagation (Figure A1.30). When an input state of polarization is sent through this plate, it will be modified at the output. The input state $\mathbf{E}_{in}$ has to be decomposed along the principal x-axis and z-axis of the plate:

$$\mathbf{E}_{in} = [\mathbf{E}_{0x} + \mathbf{E}_{0z} e^{i\Delta\phi_{in}}]\, e^{i\omega t} \tag{A1.169}$$

where $\Delta\phi_{in}$ is the phase difference between both components at the input. Each component $\mathbf{E}_{0x}$ and $\mathbf{E}_{0z}$ propagates without change of polarization. At the output, the state of polarization becomes

$$\mathbf{E}_{out} = [\mathbf{E}_{0x} e^{-2i\pi n_x d/\lambda} + \mathbf{E}_{0z} e^{-2i\pi n_z d/\lambda + i\Delta\phi_{in}}] e^{i\omega t} \tag{A1.170}$$

The phase difference between both components is at the output:

$$\Delta\phi_{out} = \Delta\phi_{in} - 2\pi \frac{\Delta n_b \cdot d}{\lambda} \tag{A1.171}$$

where d is the thickness of the plate and $\Delta n_b = n_z - n_x$.

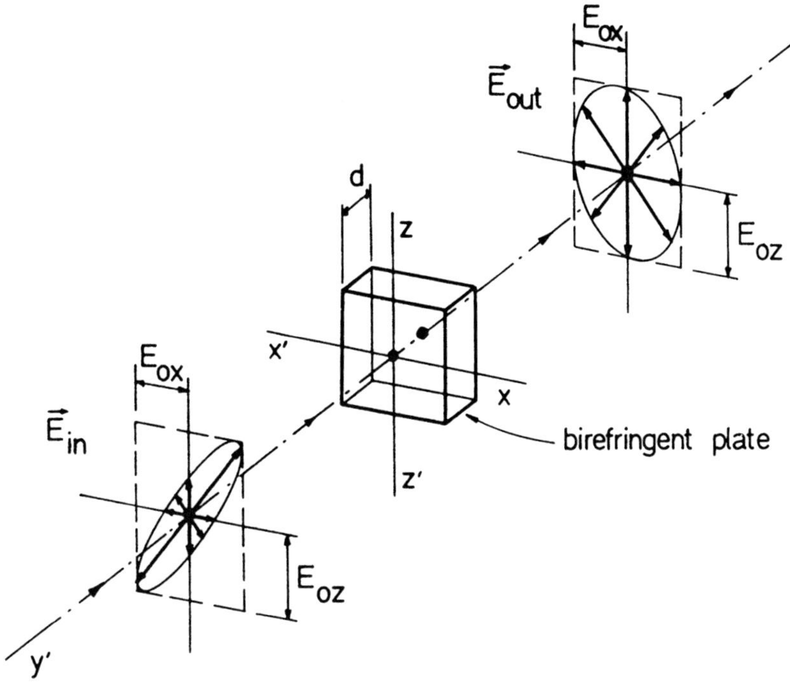

Figure A1.30 Change of polarization due to propagation in a birefringent medium.

The change of the state of polarization along propagation through the plate is periodic. The phase difference increases linearly with respect to the thickness d, and the input state of polarization is retrieved when the accumulated phase difference is 2π rad. The spatial period of the change is called the birefringence beat length Λ, with

$$\Lambda = \frac{\lambda}{\Delta n_b}$$

Two kinds of birefringent plates are particularly useful:

- Half-wave plates where $d = m\Lambda + \Lambda/2$ and $\phi_{out} - \phi_{in} = 2m\pi + \pi$ (m being integral);
- Quarter-wave plates where $d = m\Lambda + \Lambda/4$ and $\phi_{out} - \phi_{in} = 2m\pi + \pi/2$.

The effect of a half-wave plate is to generate an output state of polarization symmetrical to the input state with respect to the principal axes (Figure A1.31).

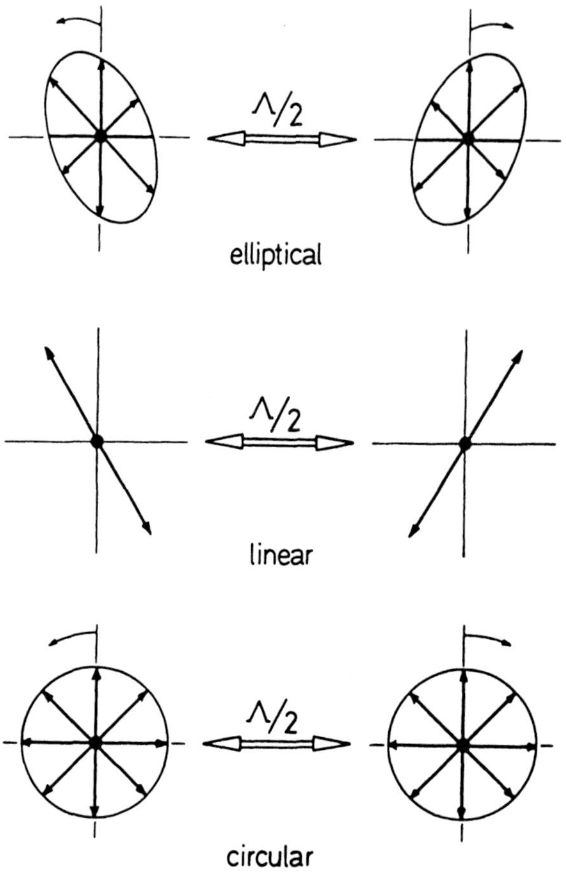

Figure A1.31 Change of polarization with a half-wave plate (Λ/2).

This symmetry reverses the direction of rotation of an elliptical or circular state of polarization. Note that, strictly speaking, a half-wave plate does not rotate a linear state of polarization, but symmetrizes it with respect to the principal axes. However, the rotation of a half-wave plate does rotate a linear state of polarization just as the rotation of a mirror rotates an image.

The effect of a quarter-wave plate is to change the ellipticity of the state of polarization. In particular, when the principal axes are parallel to the axes of the ellipse, it yields a linear polarization. Inversely, a linear polarization yields an elliptical polarization aligned along the principal axes (Figure A1.32). In the particular case of a linear polarization at 45 deg of the principal axes, this yields a circular polarization.

Mathematically, this linear transformation of the state of polarization is a product of matrixes, known as Jones formalism. The state of polarization can be

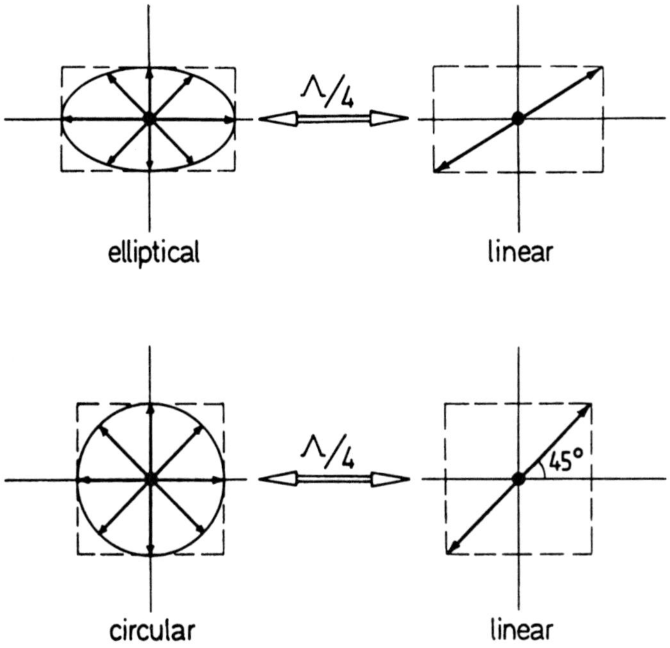

Figure A1.32 Change of polarization with a quarter-wave plate (Λ/4).

represented by a 1 × 2 column matrix, and the effect of the plate is a multiplication by a 2 × 2 square matrix. Decomposing the polarization along the principal axes is actually the usual mathematical decomposition along the eigen axes of the 2 × 2 matrix, which yields a diagonal matrix that becomes much easier to handle. The effect of the birefringent plate can be written

$$[E_{\text{out}}] = [M] \times [E_{\text{in}}] \qquad \text{(A1.172)}$$

where the column matrices $[E_{\text{out}}]$ and $[E_{\text{in}}]$ are called the Jones vectors:

$$\begin{bmatrix} E_{0x_{\text{out}}} \\ E_{0z_{\text{out}}} \end{bmatrix} = \begin{bmatrix} e^{-2i\pi n_x d/\lambda} & 0 \\ 0 & e^{-2i\pi n_z d/\lambda} \end{bmatrix} \times \begin{bmatrix} E_{0x_{\text{in}}} \\ E_{0z_{\text{in}}} \end{bmatrix} \qquad \text{(A1.173)}$$

The mathematical interest of this matrix formalism is due to the fact that the propagation through several plates can be calculated with a matrix product:

$$[E_{\text{out}}] = [M] \times [E_{\text{in}}] \qquad \text{(A1.174)}$$

with

$$[M] = [M_m] \times \ldots \times [M_2] \times [M_1] \quad \text{(A1.175)}$$

Note that the phase delay created by birefringence is inversely proportional to the wavelength, and therefore a change of polarization through a birefringent medium may be modified by wavelength variation, even with a stable medium.

So far, we have described birefringent effects related to orthogonal principal axes of a crystal, which is actually called linear birefringence. There is also a phenomenon of circular birefringence, sometimes called optical activity. With circular birefringence, the eigen states of polarization that propagates without change are the two circular states of polarization. Then an input state of polarization has to be decomposed with its two circular components. Each circular component propagates with its own index of refraction, and their recombination yields the output state of polarization. The effect of a circular birefringence with an index difference Δn_c is to rotate any input state of polarization with an angle θ (Figure A1.33) equal to half the phase difference change between both circular components:

$$\theta = \frac{\Delta\phi_{out} - \Delta\phi_{in}}{2} \quad \text{(A1.176)}$$

$$\theta = \pi\frac{\Delta n_c \cdot d}{\lambda} \quad \text{(A1.177)}$$

Note that the ellipticity of the state of polarization remains unchanged by circular birefringence.

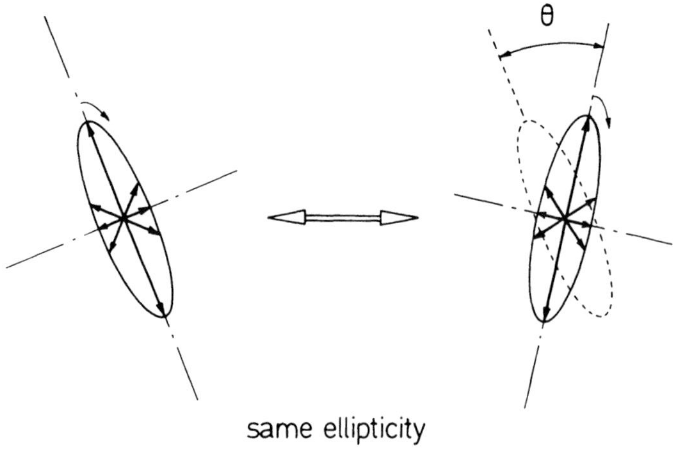

Figure A1.33 Change of polarization with circular birefringence.

BIBLIOGRAPHY

About optics:
- Born, M., and E. Wolf, *Principles of Optics,* Pergamon Press, 1975.

About electrodynamics:
- Feynman, R., *Lectures on Physics,* Addison-Wesley, 1965.

About Gaussian beam:
- Kogelnik, H., and T. Li, "Laser Beams and Resonators," *Proceedings of the IEEE,* Vol. 54, 1966, pp. 1312–1329.

About Fourier transform:
- Bracewell, R., *The Fourier Transform and Its Applications,* McGraw-Hill, 1965.

Appendix 2
Basics of Single-Mode Fiber Optics

A2.1 DISCRETE MODAL GUIDANCE IN A STEP INDEX FIBER

A step index fiber is composed of a cylindrical core with an index of refraction n_1 and a radius a, and of a surrounding cladding with a lower index $n_2 < n_1$ (Figure A2.1). Because of the boundary conditions at the core-cladding interface, there is a discrete number of eigen solutions of the general propagation equation which are guided in the fiber. These eigen solutions are called the modes of the fiber, which is said multimode. They can be written:

$$\begin{aligned} \mathbf{E}_{m_i}(x,y,z,t) &= \mathbf{E}_{0m_i}(x,y)\ e^{i(\omega t - \beta_i z)} \\ \mathbf{B}_{m_i}(x,y,z,t) &= \mathbf{B}_{0m_i}(x,y)\ e^{i(\omega t - \beta_i z)} \end{aligned} \tag{A2.1}$$

where x and y are the transverse spatial coordinates, and z is the longitudinal spatial coordinate corresponding to the direction of propagation.

Contrary to the case of the plane wave, each mode has specific transverse distributions of the fields $\mathbf{E}_{0m_i}(x,y)$ and $\mathbf{B}_{0m_i}(x,y)$ that tend to zero far from the core. The propagation phase term $e^{i(\omega t - \beta_i z)}$ depends on the angular frequency ω and on a specific mode propagation constant β_i that depends on ω and that is comprised between the wave number k_2 in the cladding and the wave number k_1 in the core:

$$k_2 = \frac{2\pi n_2}{\lambda} < \beta_i < k_1 = \frac{2\pi n_1}{\lambda} \tag{A2.2}$$

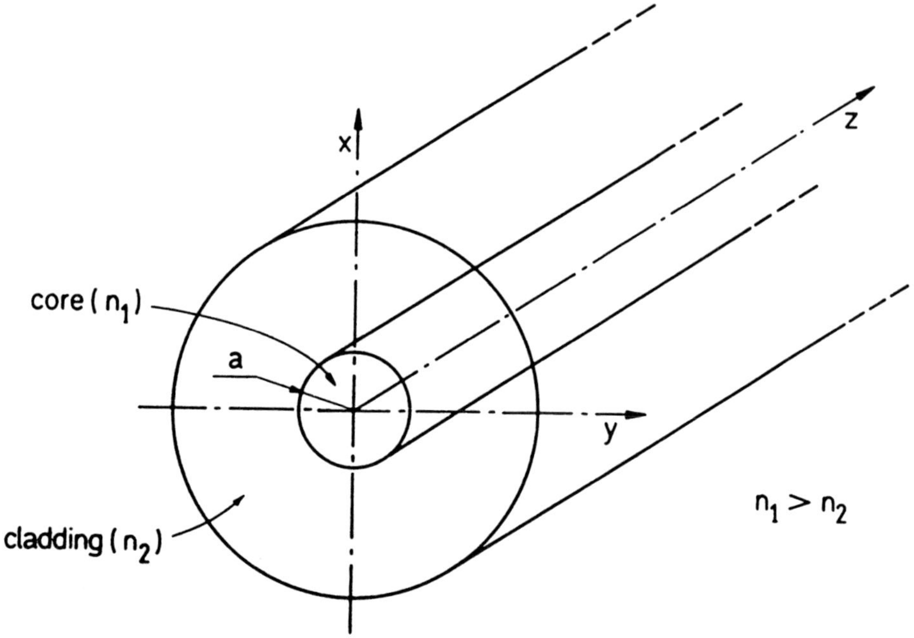

Figure A2.1 Optical fiber.

An equivalent index n_{eq_i} of the mode i is often used with

$$\beta_i = \frac{2\pi n_{\mathrm{eq}i}}{\lambda}$$
$$n_2 < n_{\mathrm{eq}_i} < n_1 \tag{A2.3}$$

The modes are denoted by TE for transverse electric or TM for transverse magnetic, or EH and HE for hybrid electromagnetic, since they have a longitudinal component E_z or B_z in addition to the usual transverse components (E_x, E_y) and (B_x, B_y) of a TEM wave in free space: TE modes have a longitudinal magnetic component, TM modes have a longitudinal electric component, and EH and HE modes have a longitudinal component for both **E** and **B** fields.

In practice, the index difference $\Delta n = n_1 - n_2$ is small: the relative index difference $\Delta = \Delta n/n_2$ is usually on the order of 0.2% to 1%. This yields a negligible longitudinal component, and the modes, then denoted by LP for linearly polarized, can usually be regarded as transverse waves.

The fundamental interest of this modal decomposition is the fact that the set of modes is an orthonormal basis of eigenvectors of the ensemble of all the possible solutions of the propagation equation. This ensemble is, from the mathematical viewpoint, a linear space with a scalar product.

The most familiar linear space is the three-dimensional geometrical space. Any vector $\mathbf{U}$ can be decomposed on the orthonormal basis of eigenvectors ($\mathbf{a}_1$, $\mathbf{a}_2$, $\mathbf{a}_3$):

$$\mathbf{U} = \sum_{i=1}^{3} x_i \mathbf{a}_i \quad \text{(A2.4)}$$

The orthonormal basis is orthogonal; that is, the scalar product of two different eigenvectors is zero:

$$\mathbf{a}_i \cdot \mathbf{a}_j = 0 \text{ if } \mathbf{i} \neq \mathbf{j}$$

and it is also normal; that is, all the scalar square of the eigenvectors are equal:

$$\mathbf{a}_1^2 = \mathbf{a}_2^2 = \mathbf{a}_3^2 = \mathbf{a}_i^2$$

The norm or magnitude or modulus U of a vector $\mathbf{U}$ is defined as the square root of its scalar square, and

$$U = \sqrt{\mathbf{U}^2}$$

There are two important results which may look obvious, or at least very familiar, in a geometrical space, but which are extremely useful in other linear spaces, for which it is not as straightforward:

- The coordinate x_i is

$$x_i = \frac{\mathbf{U} \cdot \mathbf{a}_i}{\mathbf{a}_i^2} \quad \text{(A2.5)}$$

- The square of the modulus is

$$U^2 = \mathbf{U}^2 = (\Sigma\, x_i \mathbf{a}_i)^2 = (\Sigma\, x_i^2)\mathbf{a}_i^2 \quad \text{(A2.6)}$$

In particular, it is very convenient to consider that the ensemble of complex functions $f(x)$, which are said square integrable is a linear space. The infinite integral $\int_{-\infty}^{+\infty} f(x) f^*(x) dx$ is convergent (which is the definition of a square integrable function) and may be regarded as the generalized scalar square $<f|f>$ of the function. A generalized scalar product may be defined with

$$<f|g> = \int_{-\infty}^{+\infty} f(x)g^*(x)dx \tag{A2.7}$$

It is possible to find an orthonormal basis of eigenfunctions (f_i) of this linear space, any function f being accordingly decomposed to:

$$f(x) = \sum_i x_i f_i(x) \tag{A2.8}$$

with $$x_i = \frac{<f|f_i>}{<f_i|f_i>}$$

and $$<f|f> = (\Sigma x_i^2) <f_i|f_i>$$

The fact that the linear space of the functions f has an infinite dimension does not change the generality of the above results, and this definition of the scalar product may be extended to a function of several variables with a multiple integral.

Going back to the fiber, the ensemble of the solutions of the propagation equation is also a linear space, which is the sum of the nonguided solutions and the sum of the guided solutions. The dimension of the ensemble of the nonguided solutions is infinite, but the dimension of the ensemble of the guided solutions is finite: it is equal to the discrete number of modes in the fiber. Any solution **E** of the propagation equation may be decomposed with:

$$\mathbf{E} = \sum_j x_i \mathbf{e}_{mi} + \sum_j x_j \mathbf{e}_{rj} \tag{A2.9}$$

where $\mathbf{e}_{m_i}$ are the normalized guided modes and $\mathbf{e}_{r_i}$ the nonguided modes that are radiated.

The guided modes $\mathbf{e}_{m_i}(x,y,z,t)$ are eigenvectors, and, therefore, their generalized scalar products $<\mathbf{e}_{m_i}|\mathbf{e}_{m_j}>$ are null. Eliminating the z and t dependence, these generalized scalar products yield, for the transverse field distributions $\mathbf{e}_{m0_i}(x,y)$, the so-called overlap integrals which are also null for orthogonal modes:

$$\int_{-\infty}^{+\infty}\int_{-\infty}^{+\infty} \mathbf{e}_{m0_i}(x,y) \cdot \mathbf{e}_{m0_j}{}^*(x,y)\, dx\, dy = 0 \tag{A2.10}$$

The coordinate x_i of **E** on the mode $\mathbf{e}_{m_i}$ is defined with generalized scalar products:

$$x_i = \frac{<\mathbf{E}|\mathbf{e}_{m_i}>}{<\mathbf{e}_{m_i}|\mathbf{e}_{m_i}>} \tag{A2.11}$$

similar to the definition of the coordinate of a vector in geometry with

$$x_i = (\mathbf{U} \cdot \mathbf{a}_i)/\mathbf{a}_i^2 \qquad \text{(A2.12)}$$

With overlap integrals, the (z,t) dependence is eliminated:

$$x_i = \frac{\int_{-\infty}^{+\infty}\int_{-\infty}^{+\infty} \mathbf{E}_0(x,y)\ \mathbf{e}_{m0_i}{}^*(x,y)\ dx\ dy}{\int_{-\infty}^{+\infty}\int_{-\infty}^{+\infty} \mathbf{e}_{m0_i}(x,y) \cdot \mathbf{e}_{m0_i}{}^*(x,y)\ dx\ dy} \qquad \text{(A2.13)}$$

Furthermore, similar to $\mathbf{U}^2 = (\Sigma x_i^2)\mathbf{a}_i^2$ in geometry, there is also:

$$\langle \mathbf{E}|\mathbf{E}\rangle = \sum_i |x_i|^2 \langle \mathbf{e}_{m_i}|\mathbf{e}_{m_i}\rangle + \sum_j |x_j|^2 \langle \mathbf{e}_{rj}|\mathbf{e}_{rj}\rangle \qquad \text{(A2.14)}$$

From the physics standpoint, this last equation shows that the total power of the wave **E** equals the sum of the total powers in each mode, as we might expect. As a matter of fact, the square of the electric field is proportional to the intensity of the wave (i.e., the spatial density of power), and the overlap integral, which is an infinite integral of the power density over the transverse plane *xy*, yields the total power.

This relation applies only to the total powers of the modes. The problem is very different for the local power density in the core. It is the square of the sum of the amplitudes of these modes and is the result of interferences between the various modes. Specifically, there are places without light because of destructive interferences. With a large number of modes, this yields a speckle pattern.

Note that overlap integrals using the "usual" scalar product $\mathbf{e}_{m0_i} \cdot \mathbf{e}_{m0_j}{}^*$ are valid with the assumption of transverse LP modes when the index difference is small. Generally, where there are longitudinal field components, the term $\mathbf{e}_{m0_i} \cdot \mathbf{e}_{m0_j}{}^*$ has to be replaced by the vector product $\mathbf{e}_{m0_i} \cdot \mathbf{b}_{m0_j}{}^*$. This makes calculations more complicated, but the basic principle of orthogonality of the modes is preserved.

A2.2 SINGLE-MODE FIBER

The calculation of the modes is usually carried out with a normalized frequency V instead of the angular frequency ω. It is is defined with:

$$V = a\sqrt{k_1^2 - k_2^2} \qquad \text{(A2.15)}$$

This can be written:

$$V = ak_0 \text{ NA} = \frac{2\pi a}{\lambda} \sqrt{n_1^2 - n_2^2} \tag{A2.16}$$

where $\text{NA} = \sqrt{n_1^2 - n_2^2}$ is the numerical aperture of the fiber as defined in Appendix 1. There is also:

$$V \approx \frac{2\pi n_1 a}{\lambda} \sqrt{2\Delta} \tag{A2.17}$$

The important result is that when $0 < V < 2.405$ a step index fiber is in the single-mode regime, where only the fundamental spatial mode (symbolized HE_{11} or LP_{01}) can be guided. The value 2.405 is the first zero of the J_0 Bessel function. The cutoff wavelength λ_c is defined by:

$$V(\lambda_c) = \frac{2\pi a}{\lambda_c} \sqrt{n_1^2 - n_2^2} = 2.405 \tag{A2.18}$$

that is

$$\lambda_c = 2.6a \text{ NA}$$

and

$$V(\lambda) = 2.405 \frac{\lambda_c}{\lambda}$$

The fiber is single-mode for $\lambda > \lambda_c$. Note that a perfect single-mode fiber can guide any state of polarization with the same propagation constant: the two-dimensional linear space of the polarization modes is said to be degenerated.

The exact definition of the modes require the use of Bessel and modified Bessel functions, but the fundamental mode may be described approximately with a Gaussian distribution, the wave amplitude being (Figure A2.2):

$$\begin{aligned} \mathbf{e}_f(x,y,z,t) &= \mathbf{e}_{f0}(x,y)\, e^{i(\omega t - \beta z)} \\ \mathbf{e}_{f0}(x,y) &= \mathbf{e}_0 e^{-(x^2+y^2)/w_0^2} = \mathbf{e}_0 e^{-r^2/w_0^2} \end{aligned} \tag{A2.19}$$

where $r = \sqrt{x^2 + y^2}$ is the radial coordinate. The mode radius w_0 (at $1/e$ in amplitude and $1/e^2$ in intensity) is for $0.8 < \lambda/\lambda_c < 2$ (Figure A2.3):

$$\frac{w_0}{a} \approx 0.65 + 0.434\left(\frac{\lambda}{\lambda_c}\right)^{3/2} + 0.0149\left(\frac{\lambda}{\lambda_c}\right)^6 \tag{A2.20}$$

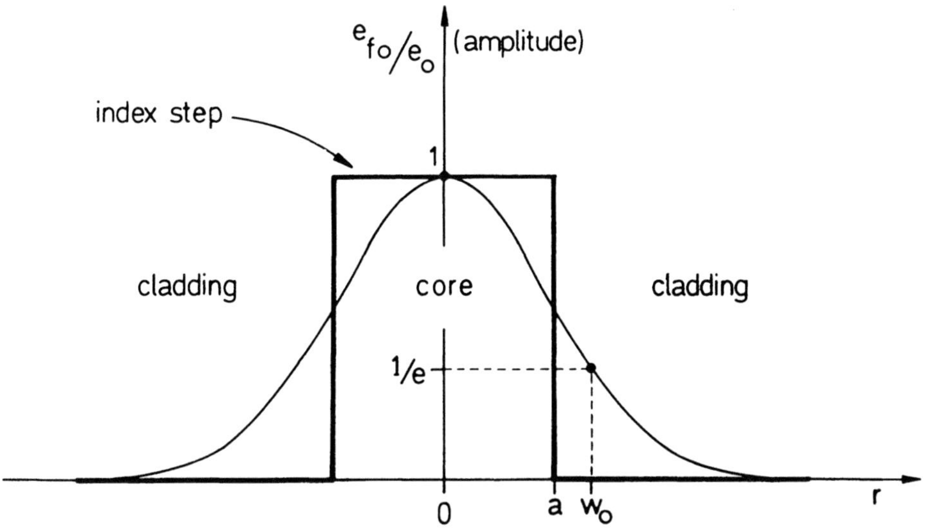

Figure A2.2 Pseudo-Gaussian amplitude of the fundamental LP_{01} mode of a single-mode fiber (case where $\lambda = 1.2\lambda_c$).

An even simpler linear approximation is usually accurate enough in the practical range of use of a single-mode fiber (i.e., $1 < \lambda/\lambda_c < 1.5$):

$$\frac{w_0}{a} \approx 1.1\frac{\lambda}{\lambda_c}$$
$$w_0 \approx 0.42\ \lambda/\mathrm{NA} \tag{A2.21}$$

The propagation constant β varies continuously from k_2 for an infinite wavelength to k_1 for a null wavelength:

- When the wavelength is very large, the mode is very wide and "sees" mainly the cladding and its index n_2,
- When the wavelength is very short, the mode is confined in the core and "sees" mainly the core and its index n_1.

A normalized propagation constant $b(V)$ is often used with:

$$b(V) = \frac{\beta^2(V) - k_2^2}{k_1^2 - k_2^2}$$
$$b(0) = 0 \qquad b(\infty) = 1 \tag{A2.22}$$

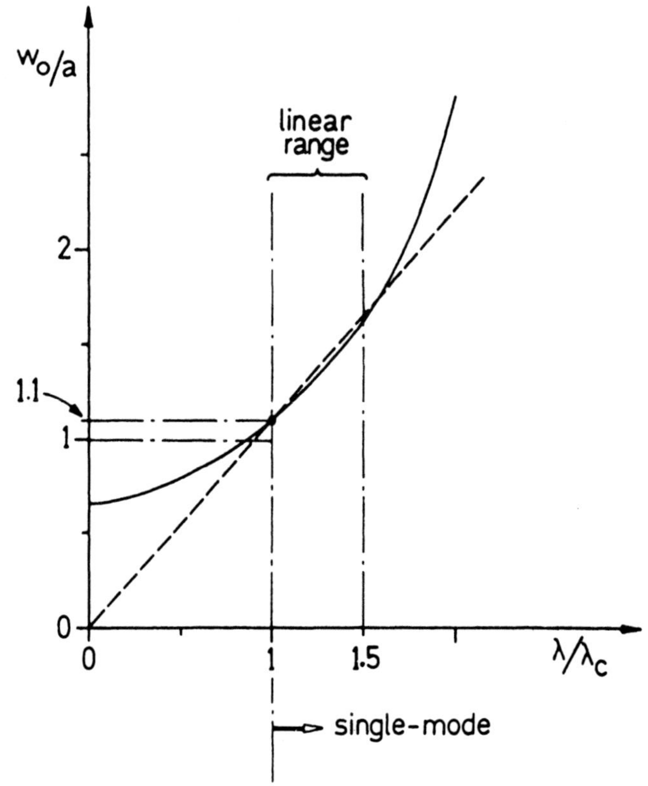

Figure A2.3 Fundamental LP_{01} mode radius w_0 as a function of wavelength λ.

When the relative index difference Δ is small, we can write:

$$\beta \approx k_2\,(1 + b \cdot \Delta) \tag{A2.23}$$

or

$$n_{eq} \approx n_2\,(1 + b \cdot \Delta)$$

For $1 < \lambda/\lambda_c < 1.6$, the value of $b(V)$ can be approximated by (Figure A2.4):

$$b(V) \approx \left(1.1428 - \frac{0.9960}{V}\right)^2$$

$$b(\lambda) \approx \left(1.1428 - 0.4141\,\frac{\lambda}{\lambda_c}\right)^2 \tag{A2.24}$$

$$b(\lambda_c) = 0.53 \approx 1/2$$

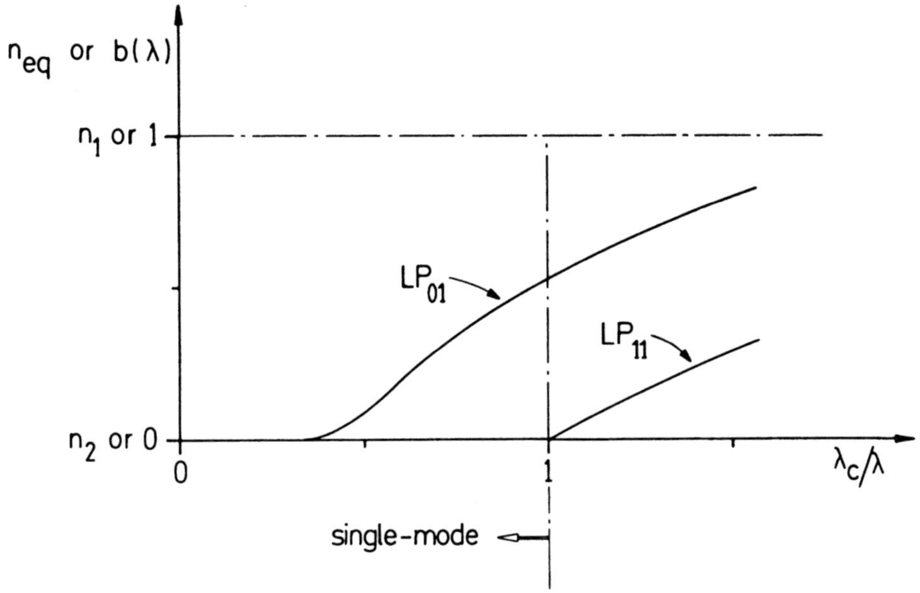

Figure A2.4 Propagation constant $b(\lambda)$ and equivalent index n_{eq} of the LP_{01} and LP_{11} modes.

Then, at the limit of the single-mode regime,

$$\beta(\lambda_c) \approx \frac{k_1 + k_2}{2} \tag{A2.25}$$

$$n_{eq}(\lambda_c) \approx \frac{n_1 + n_2}{2} \tag{A2.26}$$

Note that the wavelength dependence of the propagation constant $b(\lambda)$ yields dispersion effects due to guidance in addition to the proper chromatic dispersion of the material. In particular, modulated signals propagate at the group velocity:

$$v_g = \frac{d\omega}{d\beta} \tag{A2.27}$$

$$v_g = \frac{c}{n_{eq}} \left[1 + \frac{\lambda}{n_{eq}} \frac{dn_{eq}}{d\lambda} \right] \tag{A2.28}$$

with

$$\frac{dn_{eq}}{d\lambda} \approx \frac{dn_2}{d\lambda} + \frac{db}{d\lambda} n_2 \Delta \tag{A2.29}$$

This analysis has been derived for a perfect step-index fiber; however, with some specific manufacturing processes there is some grading of the index profile. Evaluation of the characteristics of such fibers is usually done with the use of an equivalent step-index fiber, which provides simply approximated values of cutoff and mode diameter. In particular, a parabolic index profile of maximum radius a_{max} and maximum relative index difference Δ_{max} may be approximated with an equivalent index step having a radius $a_e = 0.8\, a_{max}$ and a relative index difference $\Delta_e = 0.75\, \Delta_{max}$ (Figure A2.5).

Now, when the fiber is used below its single-mode cutoff, there are higher-order guided modes. In particular, when $0.63\, \lambda_c < \lambda < \lambda_c$ (i.e., $2.4 < V < 3.8$), the antisymmetric LP_{11} mode becomes guided in addition to the fundamental LP_{01} mode. This second-order mode has an odd distribution with respect to one transverse coordinate. The wave amplitude may be approximately described using the normalized derivative of a Gaussian function (Figure A2.6):

$$\mathbf{e}_s(x,y,z,t) = \mathbf{e}_{s0}(x,y)\, e^{i(\omega t - \beta z)}$$
$$\mathbf{e}_{s0}(x,y) = \mathbf{e}_0 \sqrt{2e}\, \frac{x}{w_1\sqrt{2}}\, e^{-(x^2+y^2)/2w_1{}^2} \qquad \text{(A2.30)}$$

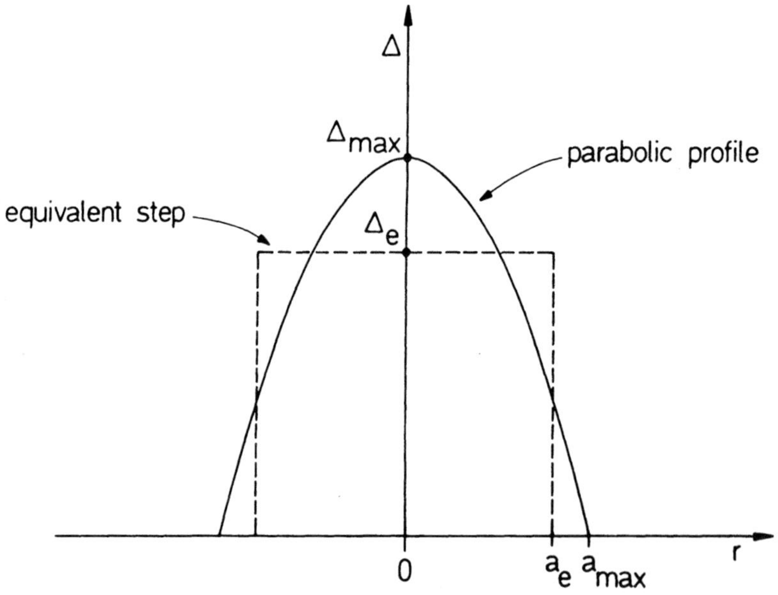

Figure A2.5 Equivalent index step of a parabolic index profile.

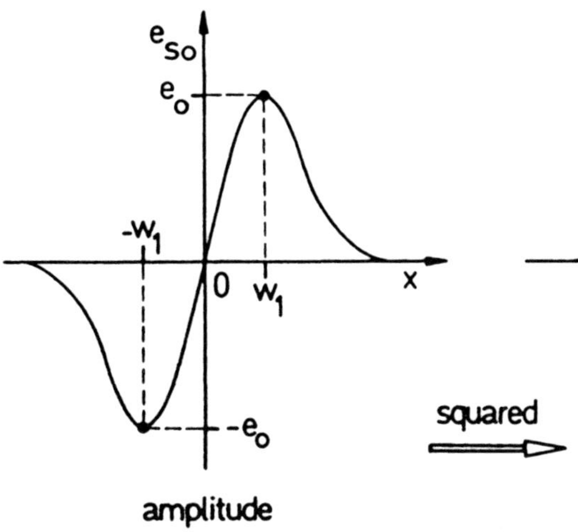

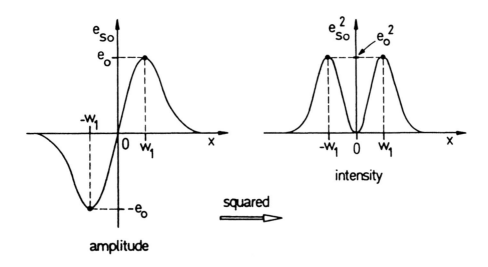

Figure A2.6 Amplitude and intensity distributions of the second-order LP_{11} mode.

This mode is composed of two lobes where the amplitude (or field) has opposite signs (or a π rad phase difference). On the other hand, the intensity (or power), which is the square of the amplitude (or field), is identical and always positive for both lobes. The field $\mathbf{e}_{s0}$ is maximum and equal to $\mathbf{e}_0$ for $x = w_1$ and $y = 0$. The value w_1 is the half-width at the maximum.

The LP_{11} mode is degenerated in terms of polarization as the LP_{01} mode, but it also has a spatial degeneracy, since the lobes may be aligned along any transverse axis (Figure A2.7).

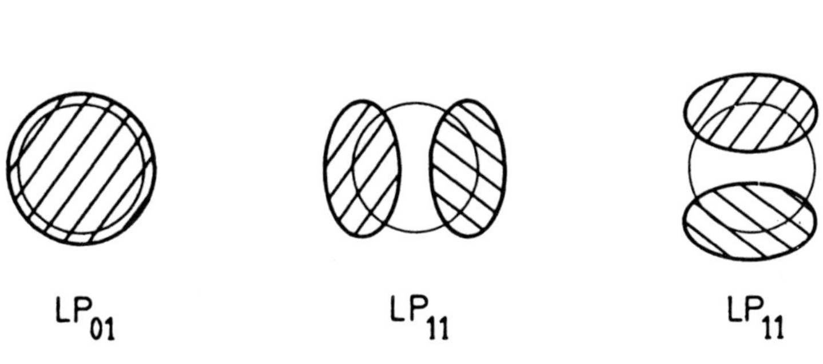

Figure A2.7 Spatial degeneracy of the LP_{11} mode.

A2.3 PRACTICAL SILICA SINGLE-MODE FIBERS

Practical single-mode fibers are made of fused silica (amorphous silicon dioxyde SiO_2), and the most common dopant to increase the core index is germanium. The attenuation is due to three main factors (Figure A2.8):

- Rayleigh scattering loss due to dipolar radiation and inhomogeneities in amorphous glass-like silica. This yields an attenuation per unit length α_R proportional to λ^{-4}.
- Infrared absorption tail of silica that limits the low attenuation range to below 1600 nm.
- Effect of OH impurities that bring residual "water" absorption peaks at 1390, 1240, and 950 nm.

With this attenuation, there are three transmission windows that are commonly used:

- 850-nm window with a typical attenuation of 2 dB/km.
- 1300-nm window with a typical attenuation of 0.4 dB/km. This wavelength also has the important advantage for telecommunications of zero material dispersion, since $d^2n_{SiO_2}/d\lambda^2 = 0$.
- 1550-nm window with the lowest attenuation typically 0.25 dB/km.

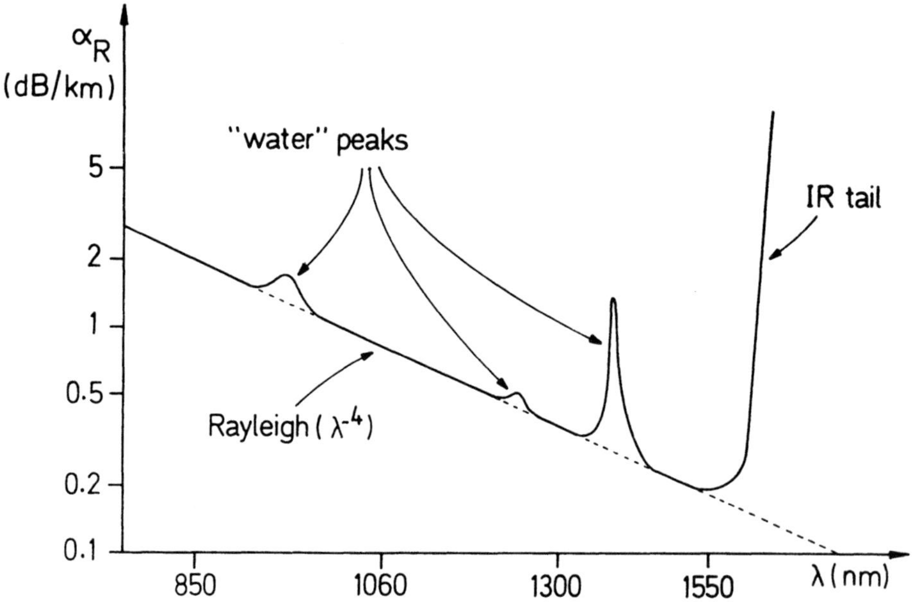

Figure A2.8 Attenuation α_R of silica fiber (log-log scale).

Note that there is also a "Neodymium YAG" window around 1060 nm with an attenuation of 1 dB/km, even if it is not used very often.

The standard single-mode telecommunication fiber has a cladding diameter of 125 μm and a core diameter of 9 μm. The cutoff wavelength is typically λ_c = 1200 nm to be single-mode at 1300 nm and 1550 nm. The index of silica being 1.45 at 1300 nm, this is obtained with a relative index difference Δ of 0.25% (i.e., an index difference $\Delta n = 3.6 \times 10^{-3}$ and a numerical aperture NA = 0.10).

Note that index specification of fibers is sometimes given in terms of numerical aperture $NA = \sqrt{n_1^2 - n_2^2} \approx n_1\sqrt{2 \cdot \Delta}$, even if its simple meaning of condition of total internal reflection in ray optics cannot be directly extended to the case of single-mode fibers, where the mode propagates in the core and the cladding.

The index difference of single-mode telecommunication fibers is relatively low in order to avoid an increase of the attenuation induced by a high level of dopant. However, for fiber gyro application, it is preferable to use a higher index difference to improve the guidance and to tolerate a small coiling diameter. This so-called high-NA single-mode fiber may have a numerical aperture as high as 0.16 (i.e., an index difference of 9×10^{-3}). With such an index difference, a single-mode fiber for 850 nm with λ_c = 750 nm has a diameter $2a$ = 3.6 μm. At 1300 nm with λ_c = 1200 nm, the diameter will be $2a$ = 5.7 μm. Note that gyro fibers usually have a cladding diameter of 80 μm, instead of the standard 125 μm, to reduce the volume of the coil.

To be easily usable, silica fibers need a polymer coating to protect the outside surface of the cladding. The standard coating is a 250-μm diameter of UV acrylate for a standard 125-μm diameter cladding. For gyro fibers, the coating diameter is usually thinner in order to reduce the volume of the coil, and other protecting materials are employed to extend the temperature range of use.

A2.4 COUPLING IN A SINGLE-MODE FIBER

At the output of a single-mode fiber, the pseudo-Gaussian fundamental mode with a radius w_0 at $1/e^2$ is diffracted to form a free-space diverging pseudo-Gaussian beam, called the far-field, with a divergence angle θ_D at $1/e^2$ (Figure A2.9):

$$\theta_D = 2 \cdot \frac{\lambda}{\pi w_0} \tag{A2.31}$$

Since, in the practical range of use ($\lambda_c < \lambda < 1.5\lambda_c$), $w_0/a \approx 1.1\lambda/\lambda_c$, the divergence angle θ_D is about constant:

$$\begin{aligned} \theta_D &\approx 1.8 \frac{\lambda_c}{\pi a} \\ \theta_D &\approx 1.5 \text{ NA} \end{aligned} \tag{A2.32}$$

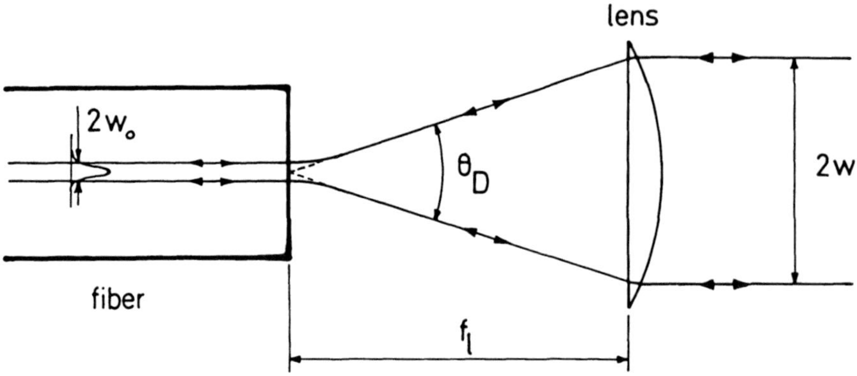

Figure A2.9 Far-field of a single-mode fiber with a collimating lens.

Considering the inverse propagation, a free-space Gaussian beam converging with this same angle θ_D creates a focused Gaussian spot with the same radius w_0, and can be fully coupled in the fiber if the core is centered on the input beam. To get such a converging beam, a parallel Gaussian laser beam with a beam diameter $2w$ at $1/e^2$ has to be focused with a convergent lens that has a focal length f_l such as:

$$2\,w/f_l = \theta_D \tag{A2.33}$$

Misalignments decrease the coupling ratio that can be calculated with the overlap integral between the input wave and the fundamental mode. The input wave $\mathbf{E}_{in}$ may be decomposed on the set of eigenvectors, which comprises a unique guided mode, the normalized fundamental mode $\mathbf{e}_f$, and radiating modes $\mathbf{e}_{rj}$:

$$\mathbf{E}_{in} = x_f \mathbf{e}_f + \sum_j x_j\, \mathbf{e}_{rj} \tag{A2.34}$$

The coordinate x_f is calculated by the generalized scalar products:

$$x_f = \frac{<\mathbf{E}_{in}|\mathbf{e}_f>}{<\mathbf{e}_f|\mathbf{e}_f>} \tag{A2.35}$$

The power or intensity coupling ratio C is the ratio between the generalized scalar square of $(x_f \cdot e_f)$, which is proportional to the power of the coupled wave, and the generalized scalar square of $\mathbf{E}_{in}$, which is proportional to the power of the input wave:

$$C = \frac{<x_f \mathbf{e}_f | x_f \mathbf{e}_f>}{<\mathbf{E}_{\text{in}} | \mathbf{E}_{\text{in}}>} = \frac{|x_f|^2 <\mathbf{e}_f | \mathbf{e}_f>}{<\mathbf{E}_{\text{in}} | \mathbf{E}_{\text{in}}>} \tag{A2.36}$$

$$C = \frac{<\mathbf{E}_{\text{in}} | \mathbf{e}_f>^2}{<\mathbf{e}_f | \mathbf{e}_f> <\mathbf{E}_{\text{in}} | \mathbf{E}_{\text{in}}>} \tag{A2.37}$$

The phase term $e^{i(\omega t - \beta z)}$ may be eliminated in these generalized scalar products, and the power coupling ratio is defined with the overlap integrals of the transverse field distributions:

$$C = \frac{\left[\int_{-\infty}^{+\infty}\int_{-\infty}^{+\infty} \mathbf{E}_{\text{in}0}\,(x,y)\;\mathbf{e}_{f0}{}^*(x,y)\;dx\;dy\right]^2}{\int_{-\infty}^{+\infty}\int_{-\infty}^{+\infty} \mathbf{e}_{f0}\,(x,y)\;\mathbf{e}_{f0}{}^*(x,y)\;dx\;dy \cdot \int_{-\infty}^{+\infty}\int_{-\infty}^{+\infty} \mathbf{E}_{\text{in}0}(x,y)\;\mathbf{E}_{\text{in}0}{}^*(x,y)\;dx\;dy} \tag{A2.38}$$

To go back to the analogy with the three-dimensional geometrical space, the fundamental mode is equivalent to an eigenvector $\mathbf{a}_i$, and the input wave is equivalent to a vector $\mathbf{U}$. Coupling light into the fiber is equivalent to projecting $\mathbf{U}$ on the axis of the eigenvector $\mathbf{a}_i$ to get a projected vector:

$$\mathbf{U}_p = \frac{(\mathbf{U} \cdot \mathbf{a}_i)}{\mathbf{a}_i^2} \cdot \mathbf{a}_i \tag{A2.39}$$

The coupled power is equivalent to the square of the length of $\mathbf{U}_p$, and the power coupling ratio is equivalent to the square of the cosine of the angle θ between $\mathbf{U}$ and $\mathbf{a}_i$:

$$C = \frac{\mathbf{U}_p^2}{\mathbf{U}^2} = \frac{(\mathbf{U}\;\mathbf{a}_i)^2}{\mathbf{a}_i^2\;\mathbf{U}^2} = \cos^2\theta \leqslant 1 \tag{A2.40}$$

When $\mathbf{U}$ is perpendicular or orthogonal to $\mathbf{a}_i$, the coupling ratio C is zero.

Now, with the approximation of Gaussian modes, this power coupling ratio may be calculated using the integral

$$\int_{-\infty}^{+\infty} e^{-(ax^2 + 2bx + c)} dx = \sqrt{\frac{\pi}{a}}\; e^{(b^2/a - c)} \tag{A2.41}$$

Results are often given in decibels with a coupling loss Γ defined by:

$$\Gamma = -10 \log[C] \quad (\text{in dB})$$

Note that the problem of coupling a free-space converging Gaussian beam in a single-mode fiber is identical to the problem of connecting two single-mode fibers with Gaussian fundamental modes. There are several kinds of misalignment that yield coupling loss (Figure A2.10):

- Transverse misalignment $d_\perp$ in the transverse (x,y) plane:

$$\Gamma_\perp = 4.34\left(\frac{d_\perp}{w_0}\right)^2 \quad \text{(in dB)} \tag{A2.42}$$

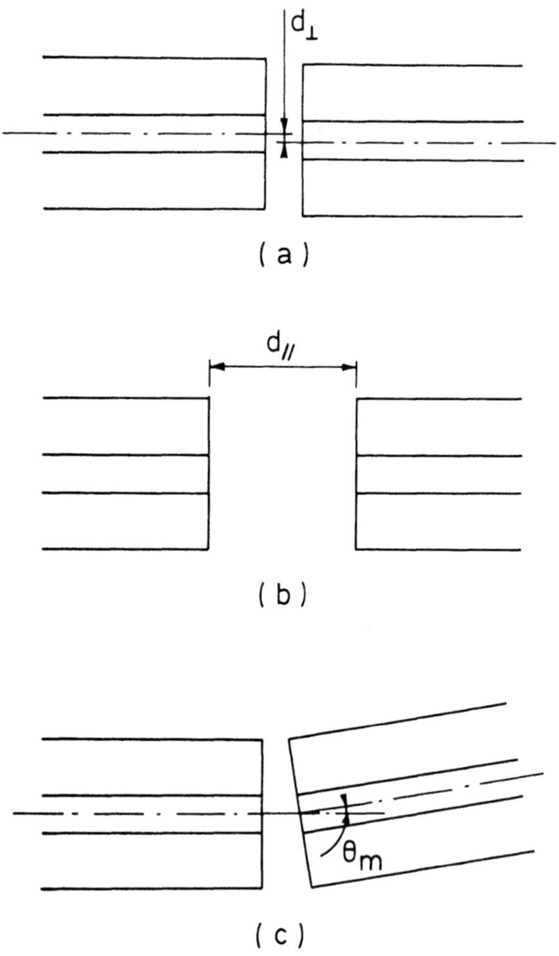

Figure A2.10 Misalignment-induced loss: (a) transverse; (b) longitudinal; (c) angular.

- Longitudinal shift $d_{//}$ in the z propagation direction:

$$\Gamma_{//} = -10 \log\left[\frac{1}{1 + (d_{//}\theta_D/4w_0)^2}\right] \quad \text{(in dB)} \tag{A2.43}$$

- Angular misalignment θ_m:

$$\Gamma_\theta = 4.34 \left[\frac{\theta_m}{\theta_D/2}\right]^2 \quad \text{(in dB)} \tag{A2.44}$$

Note that the transverse and angular misalignments yield the same loss law with the ratio to the mode radius w_0 at $1/e^2$ and the ratio to the half-divergence angle $\theta_D/2$ at $1/e^2$, respectively, because the problem of angular misalignment can be considered a problem of "transverse misalignment" between the "virtual" Gaussian far fields.

Taking again the typical numerical example of the high-NA fiber for 850 nm (λ_c = 750 nm, NA = 0.16, θ_D = 0.24 rad, $2a$ = 3.6 μm, $2w_0$ = 4.5 μm), there is a 0.5-dB (or 10%) loss for:

- $d_\perp/w_0 = 0.34$, i.e., $d_\perp = 0.8$ μm
- $\theta_m/(\theta_D/2) = 0.34$, i.e., $\theta_m = 0.04$ rad $= 2.3°$
- $d_{//}\theta_D/4w_0 = 0.32$, i.e., $d_{//} = 12$ μm

These values are obtained with a "dry" connection (i.e., with a fiber interface in air), and the 4% (0.2-dB) Fresnel reflection loss must be added. With an index-matched connection, there is no Fresnel reflection and the free-space divergence θ_D of the mode has to be replaced in the formulae by the reduced divergence $\theta'_D = \theta_D/n$ in an index-matched medium where the actual wavelength is reduced to $\lambda_m = \lambda/n$. This does not change the transverse effect, but this same 0.5-dB loss is now obtained for an angular misalignment θ_m = 2.3 deg/1.45 = 1.6 deg, or for a longitudinal shift $d_{//}$ = 1.45 × 12 μm = 17.5 μm. As can be seen from this numerical example, the mechanical tolerances of the transverse alignment are very difficult, but the longitudinal and angular alignments are less demanding in comparison.

Another source of coupling loss is a mode diameter mismatch between two different fibers, or a fiber and an integrated optic waveguide, or a fiber and a focused Gaussian beam. The result is also deduced from an overlap integral which can be easily calculated with Gaussian modes. A mismatch between two diameters $2w_0$ and $2w'_0$ at $1/e^2$ yields;

$$C_{dm} = \left[\frac{2w_0 w_0'}{w_0^2 + w_0'^2}\right]^2 \tag{A2.45}$$

and

$$\Gamma_{dm} = -10 \log\left[\left(\frac{2w_0 w_0'}{w_0^2 + w_0'^2}\right)^2\right] \quad \text{(in dB)} \tag{A2.46}$$

A loss of 0.5 dB (or 10%) is induced by a diameter ratio $w_0/w_0' = 1.4$, which shows that the diameter tolerance is not very critical for single-mode fibers.

In the case of elliptical Gaussian modes, the mode amplitude may be written

$$\mathbf{E}(x,y) = \mathbf{E}_0 e^{-\left(\frac{x^2}{w_{0x}^2} + \frac{y^2}{w_{0y}^2}\right)} \tag{A2.47}$$

where w_{0x} and w_{0y} are the half-widths at $1/e^2$ along the minor and major axes instead of the radius w_0, and the loss due to a width mismatch is

$$\Gamma_{am} = -10 \log\left[\frac{2w_{0x}w_{0x}'}{w_{0x}^2 + w'_{0x}{}^2} \cdot \frac{2w_{0y}w'_{0y}}{w_{0x}^2 + w'_{0y}{}^2}\right] \tag{A2.48}$$

Note that when an elliptical mode has to be coupled to a circular mode, the lowest loss is obtained when the radius w_0 of the circular mode is equal to the geometrical mean value of the half-widths of the elliptical mode:

$$w_0 = \sqrt{w'_{0x} \cdot w'_{0y}} \tag{A2.49}$$

For example, with an ellipticity as high as $w'_{0x} = 4w'_{0y}$, then w_0 is optimal when

$$w'_{0x}/2 = w_0 = 2w'_{0y}$$

and the loss is only 2 dB. With a ratio $w'_{0x}/w'_{0y} = 2$, the optimal loss is only 0.5 dB, with $w'_{0x}/\sqrt{2} = w_0 = w'_{0y}\sqrt{2}$. This result for elliptical modes is useful for certain fibers that may have an elliptical core, but also for coupling to semiconductor diodes which usually have a very elliptical emission pattern.

This entire analysis of coupling to a single-mode fiber is assuming an input wave that is spatially coherent; that is, the phases at all the points in a transverse plane are equal or at least correlated. Spatially incoherent sources cannot be efficiently coupled in a single-mode fiber.

Note: This analysis of the coupling of the fundamental pseudo-Gaussian LP_{01} mode can be extended to the case of the second-order "pseudo-Gaussian-derivative" LP_{11} mode. We have seen that an important advantage of the Gaussian function is that it is invariant under Fourier transform ($e^{-\pi x^2}$ and $e^{-\pi\sigma^2}$ form a pair of Fourier transforms), but the Gaussian derivative has similar properties.

As a matter of fact, using the derivative theorem of the Fourier transform, it is found that the Fourier transform of the derivative $\frac{d}{dx}[e^{-\pi x^2}] = [-2\pi x e^{-\pi x^2}]$ is $[i2\pi\sigma e^{-\pi\sigma^2}]$. This means that when the LP_{11} mode is diffracted at the output of a fiber, it forms a free-space diverging beam which keeps the same antisymmetric Gaussian-derivative profile (Figure A2.11). Note, however, that the invariance is not complete, because the additional i term indicates that the Fourier transform of a real odd function is a purely imaginary odd function. Optically, this yields an additional $\pi/2$ phase shift, called the Guoy effect, on the output wave as it is diffracted in free space.

The "pseudo-Gaussian-derivative" beam that diverges in free space follows a law similar to the Gaussian beam with a half-width at the maximum $w'(z)$:

$$w'(z) = \frac{\lambda}{2\pi w_1} z = \frac{\theta_{D1}}{2} \cdot z \tag{A2.50}$$

where w_1 is the half-width at the maximum of the mode, and θ_{D1} is the full divergence angle between both extremes:

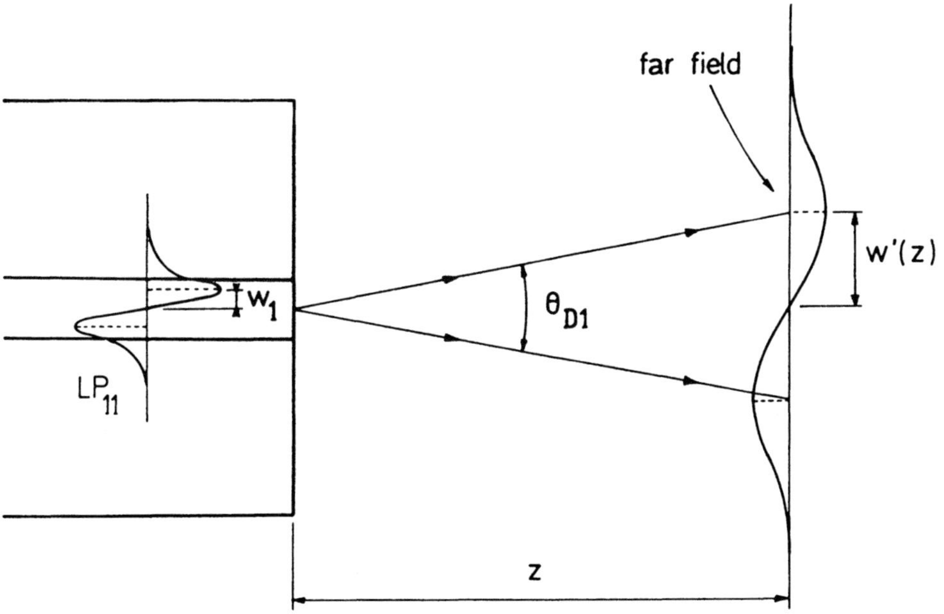

Figure A2.11 Far field of the LP_{11} mode.

$$\theta_{D1} = \frac{\lambda}{\pi w_1} \tag{A2.51}$$

Finally, it is also possible to compute the coupling loss of the LP_{11} mode between two fibers. In particular, a width mismatch yields an intensity coupling ratio:

$$C_{dm1} = \left[\frac{2w_1 w_1'}{w_1^2 + w_1'^2}\right]^4 \tag{A2.52}$$

Note that this formula is similar to those of the fundamental mode, but is at the fourth power instead of the second.

A2.5 BIREFRINGENCE IN A SINGLE-MODE FIBER

So far, the problem of polarization has been obviated in the analysis by considering that the polarization modes are degenerated in a perfect fiber. However, practical fibers have a residual birefringence which modifies the state of polarization as the wave propagates. There are two sources of birefringence of single-mode fibers:

- Shape birefringence induced by a noncircular core;
- Stress birefringence induced by an anisotropic stress through the elasto-optic effect.

The birefringence $\Delta\beta$ is defined by

$$\Delta\beta = \beta_2 - \beta_1 \tag{A2.53}$$

where β_2 and β_1 are the propagation constants of the two eigen orthogonal polarization modes. It may also be expressed with a birefringence index difference Δn_b between the equivalent indices n_{eq2} and n_{eq1}:

$$\Delta n_b = n_{eq2} - n_{eq1} \tag{A2.54}$$

with

$$\Delta\beta = \frac{2\pi\Delta n_b}{\lambda} \tag{A2.55}$$

or a normalized birefringence B:

$$B = \frac{\Delta\beta}{\beta} = \frac{\Delta n_b}{n_{eq}}$$

where β is the mean value between β_2 and β_1 and n_{eq} is the mean equivalent index value between n_{eq2} and n_{eq1}.

The calculation of shape birefringence is difficult, but in the case of a small ellipticity of the core, there is a linear birefringence which can be approximated by

$$B_{sh} = \frac{\beta_y - \beta_x}{\beta} \approx \Delta^2 e_l^2 f(V) \qquad \text{(A2.56)}$$

where $f(V)$ is a term that depends on the normalized frequency V and that is equal to about 0.2 in the practical range of use ($1 < \lambda/\lambda_c < 1.5$), and where $e_l = \sqrt{1 - (a_x/a_y)^2}$ is the ellipticity of a core that has a half-width a_y along the major axis and a_x along the minor axis (Figure A2.12). Note that the shape birefringence is proportional to the square of the normalized index step Δ of the core, and that the fast principal axis is parallel to the minor axis and the slow principal axis is parallel to the major axis.

Birefringence may also be due to an anisotropic normal stress T_n (T_n is positive for tensile stress and negative for compressive stress), which destroys the isotropy

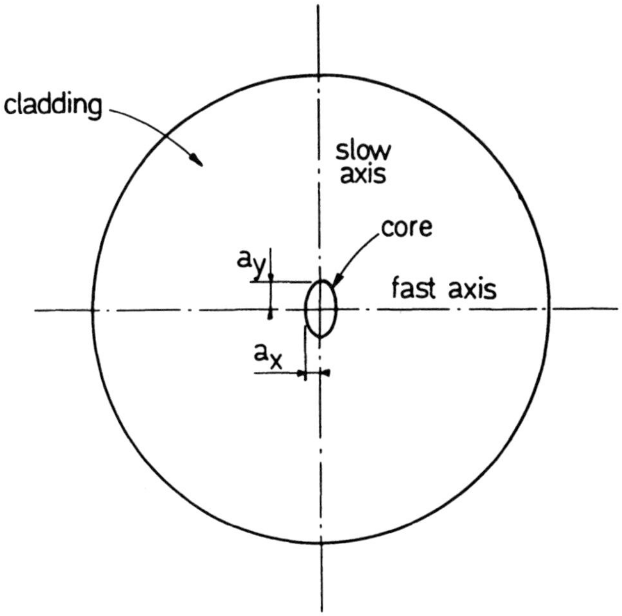

Figure A2.12 Elliptical core fiber.

of amorphous silica. Such stress induces index variations and a uniaxial linear birefringence Δn_b with an extraordinary index variation δn_e for the axis parallel to the stress and an ordinary index variation δn_o for the two other orthogonal axes. Ignoring the effect of the dopant

$$\delta n_e = -\frac{n^3}{2}(p_{11} - 2\nu p_{12})\frac{T_n}{E} \tag{A2.57}$$

$$\delta n_o = -\frac{n^3}{2}(p_{12} - \nu p_{12} - \nu p_{11})\frac{T_n}{E} \tag{A2.58}$$

$$\Delta n_b = \delta n_e - \delta n_o = \frac{n^3}{2}(p_{12} - p_{11})(1 + \nu)\frac{T_n}{E} \tag{A2.59}$$

where $E = 7 \times 10^{10}$ Pa is the Young modulus of silica, $p_{11} = 0.121$ and $p_{12} = 0.270$ are the elasto-optic coefficients of silica, and $\nu = 0.16$ is the Poisson ratio of silica.

In particular, when a fiber is bent, a transverse compressive stress T_{nc} is yielded parallel to the x-axis in the plane of curvature, and is located in the center of the fiber where the light is guided (Figure A2.13):

$$T_{nc} = -\frac{E}{2}\left(\frac{R_f}{R}\right)^2 \tag{A2.60}$$

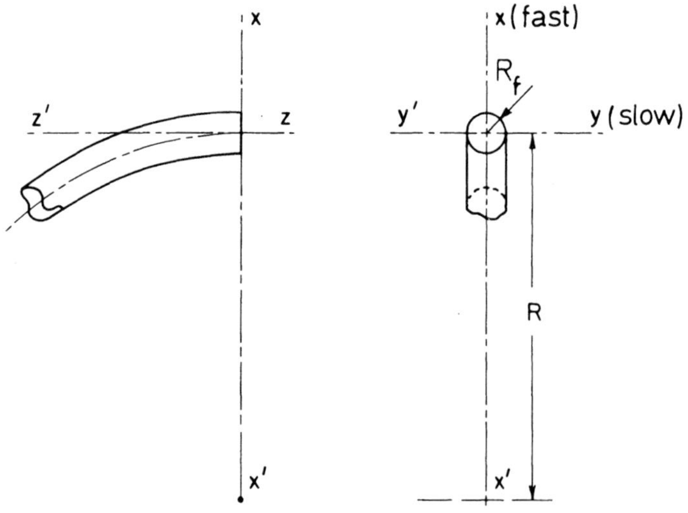

Figure A2.13 Bending-induced birefringence.

where R_f is the radius of the fiber cladding and R the radius of curvature. This compressive stress creates a negative linear birefringence with an index difference:

$$\Delta n_b = \delta n_e - \delta n_o = n_x - n_y = -0.13 \left(\frac{R_f}{R}\right)^2 \quad \text{(A2.61)}$$

Shear stress induced by twisting the fiber also creates birefringent effects. The analysis is more complicated and requires consideration of the longitudinal field component, which is usually ignored in the transverse LP mode approximation. It may be shown that a twist rate t_w (in rad/m) creates a circular birefringence:

$$\Delta\beta_c = 0.14 t_w$$

With circular birefringence (see Appendix 1), the eigen polarization modes are both circular states of polarization, and a linear polarization is "dragged" by the fiber twist. The angle θ_p of rotation of the linear polarization is proportional to the integrated twist angle θ_{tw} of the length L of fiber:

$$\theta_p = \Delta\beta_c \cdot L/2 = 0.07\ \theta_{tw}$$

Note that the twist-induced circular birefringence $\Delta\beta_c$ and the angle of rotation θ_p are wavelength-independent to first order, while for bend-induced linear birefringence, it is the birefringence index difference Δn_b which is wavelength-independent.

When several birefringence effects are combined, the analysis of the resulting effect may be tedious, but the use of a geometrical representation on the Poincaré sphere eases understanding. On a Poincaré sphere the two circular states of polarization are placed respectively at each pole, and the linear states of polarization are placed on the equatorial line (Figure A2.14). The "latitude" gives the degree of ellipticity of the state of polarization, and the "longitude" gives between 0 and 360 deg, the double of the angle of the direction of the major axis of the ellipse with respect to the frame of reference. Two orthogonal states of polarization are opposite on the sphere. The change of state of polarization due to birefringence is given by a rotation around a diameter of the sphere. Circular birefringence is represented by a rotation around the polar diameter, and linear birefringence by a rotation around the equatorial diameter corresponding to the position of the principal axes. A quarter-wave plate yields a rotation of 90 deg and a half-wave plate a rotation of 180 deg.

The Poincaré sphere is used in bulk optics to explain the total effect of several birefringent plates. However, its usefulness is limited because it is well-known that the result of the combination of several successive rotations around nonparallel axes is not straightforward. In single-mode fiber optics, it is a much more powerful tool because it may explain simply the effect of several sources of birefringence

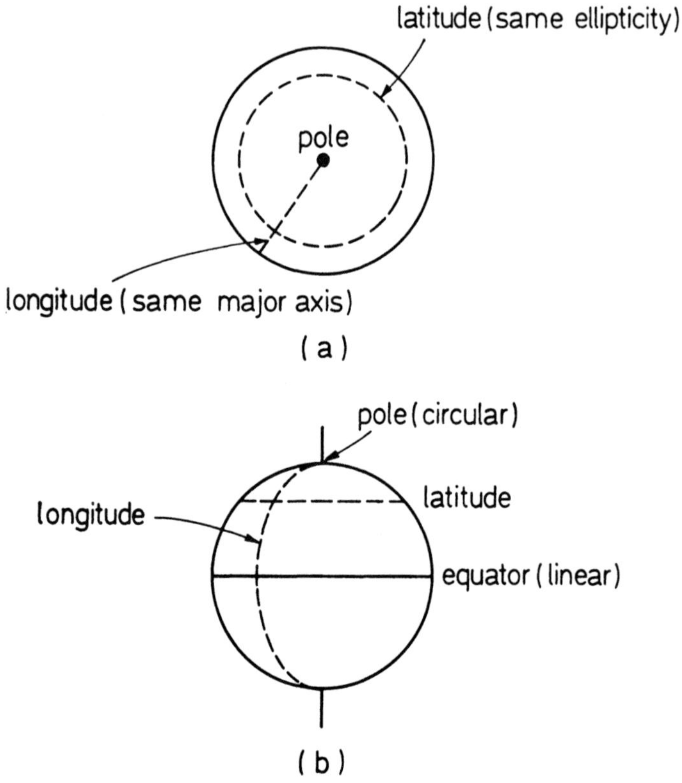

Figure A2.14 Poincaré sphere: (a) top view; (b) side view.

combined locally in the fiber. Geometrically, this does not correspond to the combination of successive rotations anymore, but to the combination of "simultaneous" rotations which can be simply added vectorially. In particular, the combination of linear and circular birefringence is represented by an elliptical birefringence "vector" $\mathbf{\Delta\beta_e}$, which is a vectorial sum (Figure A2.15):

$$\mathbf{\Delta\beta_e} = \mathbf{\Delta\beta_l} + \mathbf{\Delta\beta_c} \quad \text{(A2.62)}$$

and its magnitude is

$$\Delta\beta_e = \sqrt{\Delta\beta_l^2 + \Delta\beta_c^2} \quad \text{(A2.63)}$$

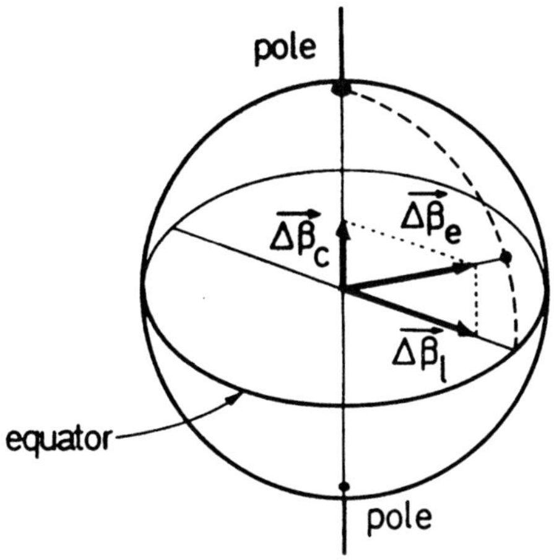

Figure A2.15 Elliptical birefringence resulting from the combination of linear and circular birefringences.

A2.6 POLARIZATION-PRESERVING FIBERS

As we have seen, a standard single-mode fiber has a residual birefringence due to shape effects or spurious stresses. This modifies the state of polarization as the wave propagates, and the output state is not stable over a long period of time. For an interferometric application like the fiber gyro, it is very desirable to use a polarization-preserving fiber, since interferences require two waves in the same state of polarization to get high contrast. Such a conservation of polarization is obtained by creating a strong birefringence in the fiber. When light is coupled in one eigenstate, it will remain in this state. For example, with a high-linear-birefringence fiber, light has to be coupled with a linear polarization that is parallel to one of the two perpendicular principal axes of birefringence.

Polarization conservation may also be obtained with a strong circular birefringence, which will then preserve a circular state of polarization. This technique would have the advantage of eliminating the problem of alignment of principal axes, which is encountered with linear birefringence. However, in practice this is difficult to implement, and most polarization preserving fibers use linear birefringence.

The phenomenon of polarization conservation in high-birefringence fiber is explained by the so-called effect of phase mismatch: when light is coupled in one eigen mode of polarization, a "first" defect will couple some light in the crossed mode. However, the primary wave and the coupled wave travel at different velocities because of birefringence, and the light that will be coupled by the "next" defect will not be in phase with the coupled wave coming from the "first" defect. They do not interfere constructively, and this limits the amount of power transferred in the crossed mode.

Practical polarization-preserving fibers have a strong intrinsic linear birefringence. The first possible solution is the use of a very elliptical core. However, getting a significant birefringence requires a large index step of the core, since B is proportionnal to Δ^2. This has two drawbacks: it requires (1) a high level of dopant, which increases the loss, and (2) a very small core to remain in the single spatial mode regime, which makes the mechanical tolerances of input coupling more severe.

The second possible method, which is now widely generalized, is to use a stress-induced birefringence with additional materials that have a thermal expansion coefficient larger than silica (several 10^{-6}/°C instead of 5×10^{-7}/°C for pure silica). The fiber preform is fabricated with two rods of highly doped silica (usually with boron, phosphorous, or aluminum) located on each side of the core region. After pulling the fiber at high temperature, these highly doped rods will tend to contract on cooling, but their thermal contraction is blocked by the surrounding silica, which has a much lower thermal contraction. This puts the rods under tensile stress and by reaction also induces stress in the core region where light propagates: there is a tensile stress T_n^+ in the axis of the rods and a compressive stress T_n^- in the orthogonal axis. The rods are circular in the case of the so-called "Panda," fiber but the stressing region may have another shape, particularly with "bow-tie" fiber or with an elliptical stressing cladding (Figure A2.16).

The birefringences induced by each stress are added and the total birefringence index difference Δn_b is derived from

$$\delta n_x = -\frac{n^3}{2}(p_{11} - 2\nu p_{12})\frac{T_n^+}{E} - \frac{n^3}{2}(p_{12} - \nu p_{12} - \nu\, p_{11})\frac{T_n^-}{E} \quad \text{(A2.64)}$$

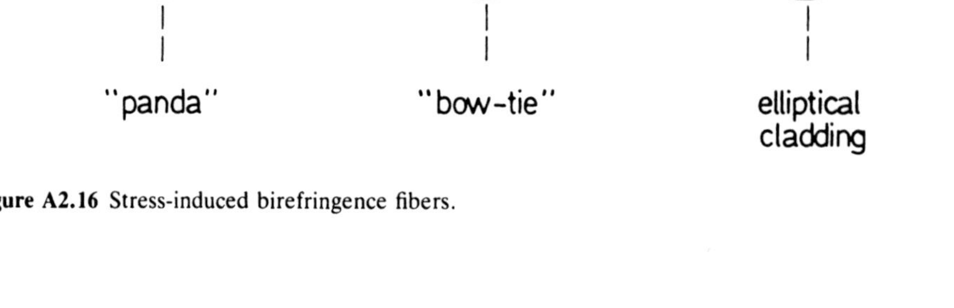

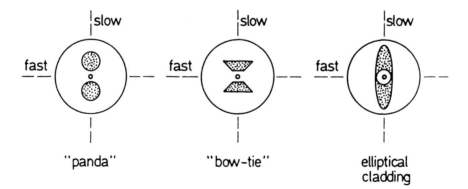

Figure A2.16 Stress-induced birefringence fibers.

$$\delta n_y = -\frac{n^3}{2}(p_{11} - 2\nu p_{12})\frac{T_n^-}{E} - \frac{n^3}{2}(p_{12} - \nu p_{12} - \nu p_{11})\frac{T_n^+}{E} \qquad \text{(A2.65)}$$

and

$$\Delta n_b = \delta n_y - \delta n_x = -\frac{n^3}{2}(p_{12} - p_{11})(1 + \nu)\left(\frac{T_n^+ - T_n^-}{E}\right) \qquad \text{(A2.66)}$$

$$\Delta n_b = -0.26\,\frac{T_n^+ - T_n^-}{E} < 0 \qquad \text{(A2.67)}$$

The fast axis is aligned along yy' and the slow axis is aligned along xx'. Such a fabrication technique yields typical $|\Delta n_b|$ values of 4 to 7×10^{-4}. The birefringence of the fiber is very often expressed in terms of beat length Λ, which has already been defined as $\Lambda = \lambda/\Delta n_b$. At $\lambda = 850$ nm, practical values of Λ are 1.2 to 2 mm. Such a value of the birefringence implies a very high stress in the core region, since the Young modulus $E = 7 \times 10^{10}$ Pa for silica:

$$T_n^+ - T_n^- = |T_n^+| + |T_n^-| \approx 1 \text{ to } 2\ 10^8 \text{ Pa} \qquad \text{(A2.68)}$$

Note that the birefringence index difference does not depend on the wavelength to first order, and therefore the beat length is proportional to the wavelength for a given fiber.

In practice, polarization-preserving fibers are not perfect and yield a residual coupling in the crossed-polarization mode. There are random coupling points along the fiber which can be described with a stochastic process $c(z)$: at a position z, there is a coupling of the amplitude $A_p(z)$ of the monochromatic primary wave in the crossed-polarization state, which yields a crossed amplitude $dA_c(z)$:

$$dA_c(z) = c(z) \cdot A_p(z) \cdot dz \qquad \text{(A2.69)}$$

Along a length L, the total crossed amplitude $A_c(L)$ is the result of the sum integral of all the dA_c terms (Figure A2.17). Taking into account the phase delay due to the propagation constant β_p of the primary mode and the propagation constant β_c of the crossed mode,

$$A_p(z) = A_p(0)\, e^{-i\beta_p z} \qquad \text{(A2.70)}$$

and the crossed-amplitude term $dA_c(z)$ yields at L:

$$dA_{cL}(z) = dA_c(z)\, e^{-i\beta_c(L-z)} \qquad \text{(A2.71)}$$

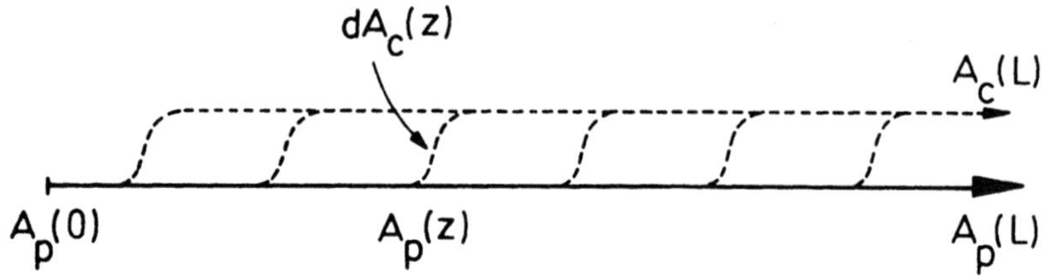

Figure A2.17 Distributed random coupling in the crossed-polarization mode.

Then

$$A_c(L) = \int_0^L dA_{cL}(z) \tag{A2.72}$$

$$A_c(L) = A_p(0)e^{i\beta_c L} \int_0^L c(z)e^{-i\Delta\beta z} dz \tag{A2.73}$$

where $\Delta\beta = \beta_p - \beta_c$ is the birefringence of the fiber.

Since $c(z)$ is a stochastic process, $A_c(L)$ is one also, which is the result of a stochastic integral. Its statistical properties may be calculated simply, when $c(z)$ is a stationary process (i.e., the characteristics of the spurious random couplings are uniform along the fiber). It is possible to define the autocorrelation $R_c(z)$ of the stationary process $c(z)$:

$$R_c(Z) = E\{c(z) \cdot c^*(z - Z)\} \tag{A2.74}$$

where $E\{\}$ denotes the ensemble average. The power spectrum (or spectral density) is the Fourier transform of the autocorrelation:

$$S_c(\sigma) = \int_{-\infty}^{+\infty} R_c(Z)e^{-2i\pi\sigma Z}\, dZ \tag{A2.75}$$

where σ is the spatial frequency. We have

$$A_c \cdot A_c^* = A_p \cdot A_p^* \int_0^L \int_0^L c(z_1)\, c^*(z_2)e^{-i\Delta\beta(z_1 - z_2)}\, dz_1\, dz_2 \tag{A2.76}$$

and

$$E\{A_c \cdot A_c^*\} = A_p \cdot A_p^* \int_0^L \int_0^L E\{c(z_1)\, c^*(z_2)\}e^{-i\Delta\beta(z_1 - z_2)} dz_1\, dz_2 \tag{A2.77}$$

because of the fundamental result of linear transformation, which states that "the ensemble average of the linear transformation $L[x]$ of a stochastic process x is equal to the linear transformation of the ensemble average of this stochastic process x":

$$E\{L[x]\} = L[E\{x\}] \tag{A2.78}$$

Since $A_cA^*_c$ is the intensity I_c of the crossed-coupled wave, and $A_pA^*_p$ is the intensity I_p of the input primary wave, the total intensity coupling ratio C has an ensemble average:

$$E\{C\} = \frac{E\{I_c\}}{I_p} = \frac{E\{A_cA_c^*\}}{A_pA_p^*} = \int_0^L \int_0^L E\{c(z_1)c^*(z_2)\}e^{-i\Delta\beta(z_1-z_2)}dz_1\, dz_2 \tag{A2.79}$$

Applying the change of variable $Z = z_1 - z_2$ gives us

$$E\{C\} = \int_0^L \left[\int_{-z_2}^{L-z_2} R_c(Z)e^{-i\Delta\beta Z}dZ\right]dz_2 \tag{A2.80}$$

Assuming that the length L of the fiber is much longer than the width of the autocorrelation $R_c(Z)$ of the coupling process, we have

$$\int_{-z_2}^{L-z_2} R_c(Z)e^{-i\Delta\beta Z}dZ \approx \int_{-\infty}^{+\infty} R_c(Z)e^{-i\Delta\beta Z}dZ = S_c(\Delta\beta/2\pi) \tag{A2.81}$$

and since $\Delta\beta/2\pi = 1/\Lambda$, where Λ is the birefringence beat length

$$E\{C\} = L \cdot S_c(1/\Lambda) \tag{A2.82}$$

The mean intensity coupling ratio $E\{C\}$ in the crossed state of polarization is proportional to the length L of the fiber and to the value of the power spectrum S_c of the stationary stochastic coupling process $c(z)$ for a spatial frequency equal to the inverse of the birefringence beat length Λ. This term $S_c(1/\Lambda)$ is often called the h parameter of the fiber, with

$$E\{C\} = h \cdot L \tag{A2.83}$$

With stress-induced high-birefringence fibers, the typical values of the h parameter are 10^{-6} to 10^{-4} m^{-1}, which corresponds to a polarization conservation of 30 to 10 dB/km.

This mathematical calculation confirms the explanation of polarization conservation based on phase mismatch. The random coupling $c(z)$ has a power spectrum $S_c(\sigma)$, and its spatial frequency components do not usually yield crossed coupling, except for the frequency $1/\Lambda$, which is the only one to be "phase-matched," because it is spatially synchronized with the propagation delay induced by birefringence.

To carry out this calculation simply, we have assumed that the intensity of the coupled wave remains very small compared to the intensity of the primary wave, and, therefore, that the light which is recoupled in the primary mode may be ignored. When this is not the case, the mean value of C tends to 1/2 because an equilibrium is reached with the same mean intensity in both orthogonal polarizations, and (Figure A2.18)

$$E\{C\} = \frac{1}{2}(1 - e^{-2hL}) \tag{A2.84}$$

The previous calculations were derived using ensemble averages on observations carried out with a monochromatic source. It is well-known with temporal stochastic processes that the statistics of a process may be determined with temporal averaging on a single observation if the process is ergodic:

$$E\{x(t)\} = \langle x(t) \rangle \tag{A2.85}$$

and

$$R(\tau) = E\{x(t)x^*(t - \tau)\} = \Gamma(\tau) = \langle x(t)x^*(t - \tau) \rangle \tag{A2.86}$$

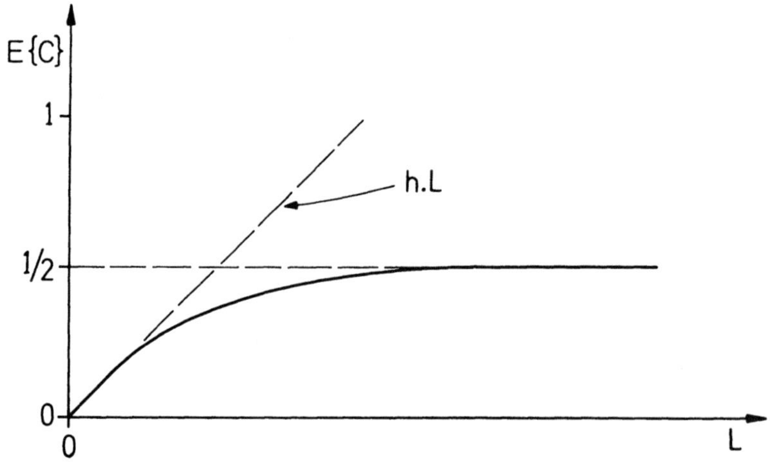

Figure A2.18 Power crossed-coupling $E\{C\}$ along a polarization-preserving fiber.

where the brackets < > denote temporal averaging. With the stochastic coupling process $c(z)$, it is possible to define an equivalent to ergodicity with a broad light spectrum that yields, on a single observation, a stable intensity coupling ratio C_t equal to the ensemble average $E\{C\}$. The broad spectrum source has to be linearly polarized and coupled along one principal axis of the high-birefringence fiber. At the output, most of the input intensity is still in the input eigenstate of polarization, and there is some spurious intensity in the crossed polarization that is not coherent with the primary wave. We have seen (Appendix 1) that the effect of a broad source in an interferometer may be explained by considering the propagation of a wave train, which has a length equal to the decoherence length L_{dc} of the source. When the path difference is larger than L_{dc}, there are two wave trains at the output of the interferometer that do not overlap and cannot interfere.

Similarly, an input wave train coupled on both polarization modes of a high-birefringence fiber propagates at different velocities on each mode, and there are at the output two wave trains which do not overlap if their path difference is larger than L_{dc} (Figure A2.19). The light then becomes statistically depolarized; that is, the state of polarization varies randomly in time, because the phases of both orthogonal mode are not correlated anymore. The length of fiber required to get such a depolarization is called the depolarization length L_d. Since the optical path

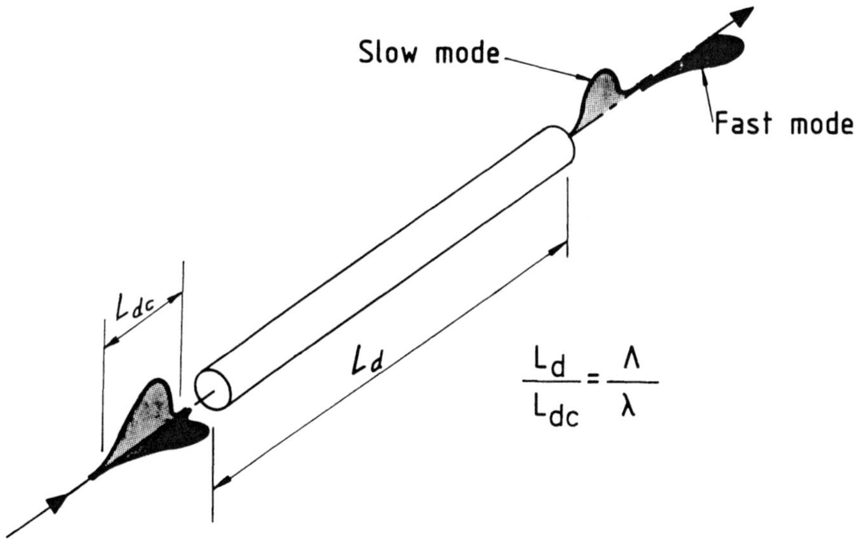

Figure A2.19 Propagation of a wave train along both orthogonal polarization modes of a high-birefringence fiber.

accumulated by one mode is $n \cdot L_d$ and the one accumulated by the other is $(n + \Delta n_b)L_d$, the path difference must be

$$(n + \Delta n_b)L_d - nL_d = L_{dc} \quad \text{(A2.87)}$$

then

$$L_d = L_{dc}/\Delta n_b \quad \text{(A2.88)}$$

or

$$L_d/L_{dc} = \Lambda/\lambda \quad \text{(A2.89)}$$

A fiber may be decomposed in segments with a length equal to L_d. When an input wave train is coupled in one eigen polarization mode, it is possible to consider that a secondary wave train is coupled in the crossed mode for each segment (Figure A2.20). These secondary wave trains are not coherent, and are therefore added in intensity. The total intensity coupling C_t is the sum of all the random intensity couplings C_i along each segment:

$$C_t = \Sigma C_i \quad \text{(A2.90)}$$

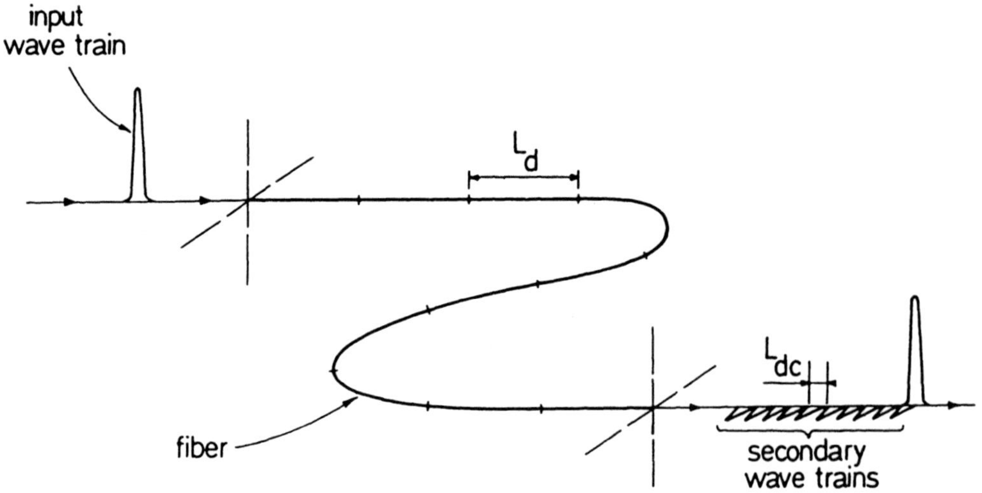

Figure A2.20 Secondary wave trains coupled in the crossed-polarization mode (assuming it is the slow mode).

This sum is actually proportional to the ensemble average $E\{C_i\}$:

$$E\{C_i\} = \frac{\Sigma C_i}{N} \tag{A2.91}$$

where $N = L/L_d$ is the number of depolarization lengths L_d along the fiber length L. Then

$$C_t = \frac{L}{L_d} E\{C_i\} \tag{A2.92}$$

and since $E\{C_i\} = h \cdot L_d$, we find

$$C_t = h \cdot L = E\{C\} \tag{A2.93}$$

It is important to remember that the measurement of the intensity crossed-coupling ratio yields a significant result only if the experiment is carried out with a broad spectrum, because with a single experiment this actually yields an ensemble average of the couplings on all the depolarization lengths along the fiber. With a monochromatic light, the result is a random variable, and a single observation is not sufficient to deduce the statistical properties of the process.

Note: The simple formula $L_d/L_{dc} = 1/(\Delta n_b) = \Lambda/\lambda$ that defined the depolarization length L_d is valid only if the birefringence index difference Δn_b is wavelength-independent, which is the case to first order with stress-induced high-birefringence fibers, which are the most commonly used for getting polarization conservation.

With elliptical core fiber, the birefringence index difference has a significant wavelength dependence, particularly with high core ellipticity, and this simple formula is not valid anymore. The exact definition of the depolarization length L_d is the length of fiber required to have a difference of group, not phase propagation time between both polarization modes equal to the decoherence time τ_{dc}. For modes 1 and 2,

$$\tau_{g1} = \frac{L_d}{v_{g1}} \quad \text{and} \quad \tau_{g2} = \frac{L_d}{v_{g2}}$$
$$\tau_{g1} - \tau_{g2} = L_d\left[\frac{1}{v_{g1}} - \frac{1}{v_{g2}}\right] = \tau_{dc} = \frac{L_{dc}}{c} \tag{A2.94}$$

Since $v_g = d\omega/d\beta$ and $1/v_g = d\beta/d\omega$ (see Appendix 1),

$$\frac{L_d}{L_{dc}} = \frac{1}{c \cdot \dfrac{d(\Delta\beta)}{d\omega}} \tag{A2.95}$$

with the birefringence $\Delta\beta = \omega\Delta n_b/c$. If the birefringence index difference Δn_b is, as usual, wavelength-(or frequency-) independent, then

$$\frac{d(\Delta\beta)}{d\omega} = \frac{\Delta n_b}{c} \tag{A2.96}$$

and the simple formula $L_d/L_{dc} = 1/\Delta n_b$ is retrieved. However, if it is wavelength-dependent, then

$$\frac{L_d}{L_{dc}} = \frac{1}{\Delta n_b + \omega d(\Delta n_b)/d\omega} \tag{A2.97}$$

Note that the case of twist-induced circular birefringence yields a very peculiar result. In fact, we have seen in Appendix 2.5 that polarization is dragged angularly by the twist at a constant rate, independently of the wavelength; that is, the birefringence $\Delta\beta$ is independent of wavelength (or frequency). This implies $d(\Delta\beta)/d\omega = 0$; that is, mathematically, L_d tends to infinity. In practice, because of second-order effects, the difference of group velocity between the two eigen circular polarizations is not zero, but is much smaller than the difference of phase velocity. In this peculiar case, the phenomenon of depolarization induced by the propagation of a broadband source in a high-birefringence fiber is greatly reduced, since the depolarization length is much longer than $L_{dc}/\Delta n_b$.

A2.7 INTERFERENCE WITH SINGLE-MODE FIBERS AND RELATED COMPONENTS

When the phenomenon of interferences in bulk optics has been explained, we have made the assumption of plane waves propagating in bulk interferometers that are mechanically aligned; that is, the orientation of the mirrors and beam splitters is adequate for getting the same phase difference for the whole light wavefront. In practice, these alignments are very delicate and there may be interference fringes or rings. The difficult problem of fringe shape and localization, as well as that of spatial coherence, have been obviated in the explanation, since single-mode fiber optics, with which we are concerned, makes things much simpler: in principle, the fundamental mode of a single-mode fiber is spatially coherent, and with an all-guided interferometer, there is no spatial fringe modulation. Single-mode fiber optics is an ideal technology for making long-path interferometers because problems of mechanical alignments are limited to those of input power coupling. Systemwise, the effect of the transverse dimensions may be forgotten, and only a device with a single curvilinear longitudinal coordinate has to be considered.

As we have already seen, the use of polarization-preserving fibers is suitable for avoiding problems of polarization control, but to take full advantage of this technology, it is desirable to duplicate the various components of a bulk interferometer in a rugged all-guided form. The main function in an interferometer is beam splitting to separate the input wave and recombine the interfering waves. Such a function may be realized in an all-fiber form with evanescent-wave coupling. Light may be coupled between two adjacent cores because the evanescent tail of the bell-shaped fundamental mode extends into the cladding and can excite the mode of the other fiber (Figure A2.21). With two identical cores, the coupled power P_c follows a sine square law with respect to the interaction length L_{int}, and the transmitted power P_t remains complementary:

$$\begin{aligned} P_c &= P_{\text{in}} \sin^2(c_s \cdot L_{\text{int}}) \\ P_t &= P_{\text{in}} \cos^2(c_s \cdot L_{\text{int}}) \end{aligned} \tag{A2.98}$$

where c_s is the coupling strength. The power is completely transferred into the second fiber for the coupling length L_{cp}, with

$$c_s \cdot L_{cp} = \pi/2 \tag{A2.99}$$

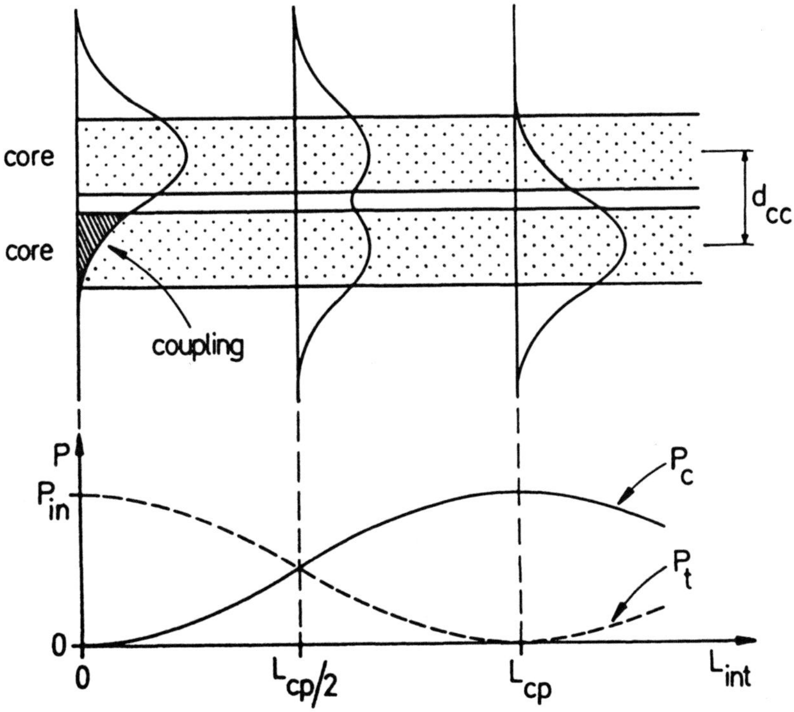

Figure A2.21 Principle of evanescent-wave coupling.

and then starts to come back in the first fiber. To get a 50-50 or 3-dB coupler, the coupling strength c_s has to be adjusted to have the interaction length L_{int} equal to the half-coupling length:

$$L_{int} = \frac{L_{cp}}{2} = \frac{\pi}{4c_s} \quad \text{(A2.100)}$$

The coupling strength is adjusted with the core-to-core distance d_{cc}. A good order of magnitude is $d_{cc} \approx 5a$ (a being the core radius) and $L_{int} \approx 1$ mm (i.e., a few thousands of wavelengths in the fiber). Note that when the wavelength increases, the mode extends further into the cladding, which improves the coupling strength. A good order of magnitude is a relative change $\Delta c_s/c_s$ equal to three or four times the relative wavelength change $\Delta\lambda/\lambda$. This wavelength dependence is used to make all-fiber wavelength multiplexer-demultiplexer.

Such evanescent-field fiber couplers may be fabricated with the so-called side-polishing technique, where the fiber is bonded in a curved groove sawed into a supporting silica block. The ensemble is ground and polished laterally to remove the cladding and get access to the evanescent field. Two identical blocks are then mated with an index-matching bonding to get power splitting (Figure A2.22). Side-polished couplers can be made very small (10 mm long) and have an excellent ruggedness and a good thermal stability. Almost any polarization-preserving fibers may be used, but this requires orienting the stressing-rod axis perpendicularly to the interface to minimize crossed-polarization coupling (Figure A2.23).

An alternative technique of fusion tapering has also been developed. Instead of removing the cladding, two fibers are tapered by fusion and stretching, which reduces the distance between both cores and also increases the diameter of the mode, which becomes "loosely" guided as the core diameter decreases (Figure A2.24). This technique is very advantageous with telecom single-mode fibers, because the fabrication process may be automated and fusion provides an excellent thermal stability. However, with polarization-preserving fibers, this technique requires a specific fiber structure to avoid loss induced by the highly doped stressing rods. Furthermore, the length of a fused coupler is usually larger (20 to 40 mm) than that of a side-polished coupler.

Another important function to fulfill is phase modulation in order to use signal processing techniques that improve signal-to-noise ratio. This can be simply implemented with an optical fiber wound around a piezoelectric tube (Figure A2.25). The driving voltage changes the tube diameter and thus modifies the length L of the fiber. Ignoring mode dispersion effects, the phase change is

$$\Delta\phi = \frac{2\pi}{\lambda}\Delta(nL) = \frac{2\pi}{\lambda}\left[n + L\frac{dn}{dL}\right]\Delta L \quad \text{(A2.101)}$$

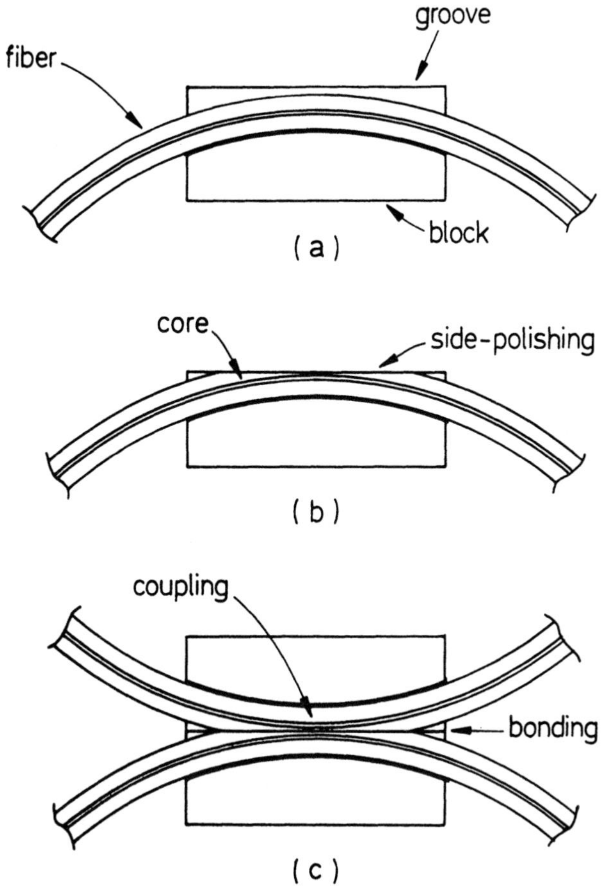

Figure A2.22 Side-polished coupler: (a) half-coupler block; (b) grinding and side-polishing; (c) coupler assembly.

with

$$L \cdot \frac{dn}{dL} = -\frac{n^3}{2}(p_{12} - \nu p_{12} - \nu p_{11}) \tag{A2.102}$$

where, as we have already seen, p_{12} and p_{11} are the elasto-optic coefficients and ν is the Poisson ratio of silica. We have

$$\Delta\phi \approx 1.13 \frac{2\pi \Delta L}{\lambda} \tag{A2.103}$$

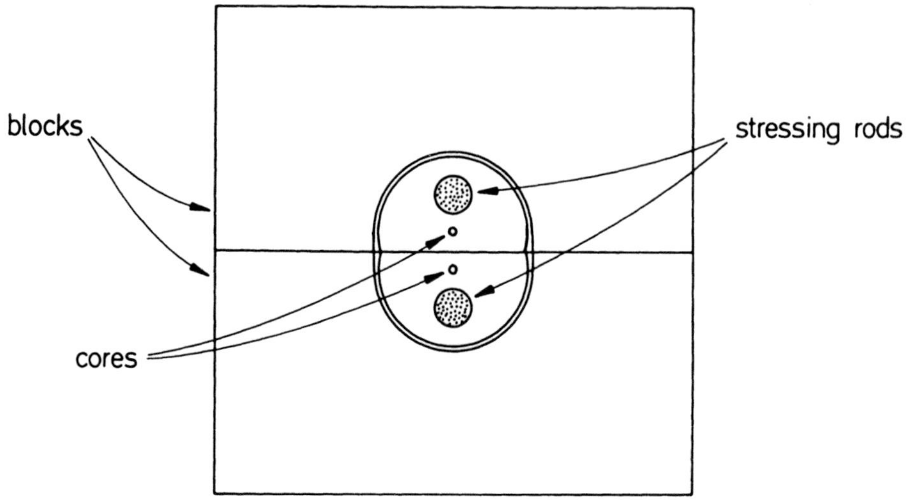

Figure A2.23 Section view of a side-polished coupler with stress-induced birefringence fiber.

Such an all-fiber phase modulator is very simple to fabricate, but its efficiency is limited to the sharp mechanical resonances of the piezoelectric tube. Three kinds of resonance may be used:

- The very efficient loop resonance, which has a typical product frequency × diameter of about 50 kHz · cm (i.e., useful for a few tens of kilohertz);
- The height resonance, which has a medium efficiency and a typical product frequency × height of about 150 kHz · cm (i.e., useful for a few hundred kilohertz);
- The thickness resonance, which has a low efficiency, but may work at a few megahertz, since the typical product frequency × thickness is about 2 MHz · mm.

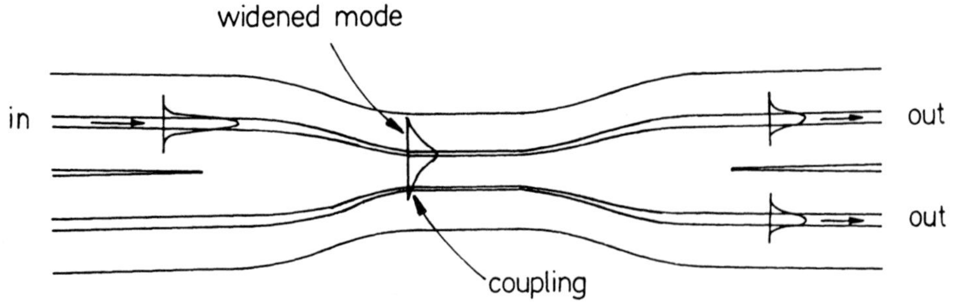

Figure A2.24 Fused-tapered coupler.

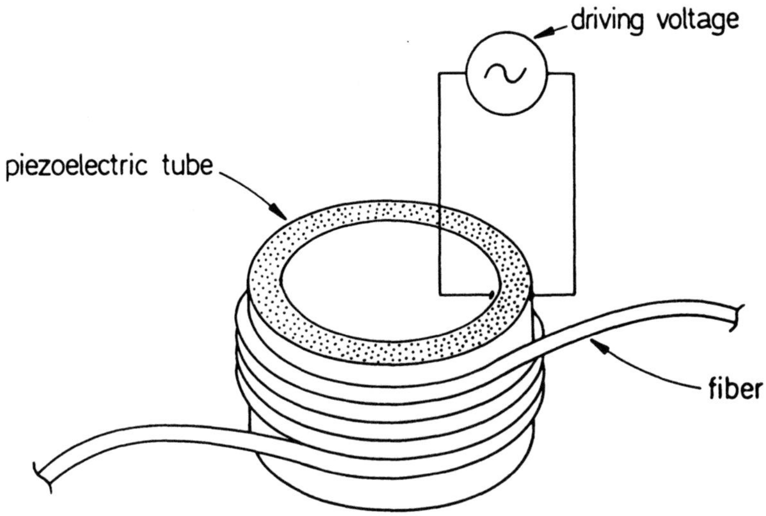

Figure A2.25 In-line phase modulator using fiber wound around a piezoelectric tube.

BIBLIOGRAPHY

About fiber optics:
- Jeunhomme, L. B., Single-Mode Fiber Optics, Marcel Dekker, 1990.
- Okoshi, T., Optical Fibers, Academic Press, 1982.
- Daly, J. C., ed., Fiber Optics, CRC Press, 1984.
- Miller, C. M., Optical Fiber Splices and Connectors, Marcel Dekker, 1986.

About fiber-optic components:
- Dakin, J. C., and B. Culshaw, eds., Optical Fiber Sensor: Principles and Components, Artech House, 1988.

About random functions:
- Papoulis, A., Probability, Random Variables and Stochastic Processes, McGraw-Hill, 1965.

About waveguide theory:
- Marcuse, D., Theory of Dielectric Optical Waveguides, Academic Press, 1974.
- Vassalo, C., Optical Waveguide Concepts, Elsevier, 1991.

Appendix 3
Basics of Integrated Optics

A3.1 INTEGRATED-OPTIC CHANNEL WAVEGUIDE

The concept of integrated optics is based on the use of microlithographic techniques to fabricate optical components with waveguides on a planar substrate. Like integrated electronics, it provides potential for integrating several functions on the same circuit with a mass production process. This also improves compactness and reduces connections.

The basic element of an integrated-optic circuit is the strip or channel waveguide. It is fabricated by increasing the index of refraction with a dopant in a narrow channel defined with microlithographic masking techniques underneath the surface of a substrate. This substrate acts as the equivalent of the surrounding cladding of an optical fiber (Figure A3.1). In particular, single-mode propagation, which is required in an interferometer, is obtained with a waveguide width and depth of a few micrometers and an index variation of a few tenths of a percent. These values are very similar to the characteristics of the core of a single-mode fiber.

However, because of the substrate surface, the cylindrical symmetry of a fiber is lost, which makes the theory more complicated. In particular, the fundamental mode is not degenerated anymore in terms of polarization. Instead of the hybrid HE_{11} fundamental mode of a fiber, the fundamental mode is a transverse electric (TE) mode for the polarization parallel to the substrate surface, and a transverse magnetic (TM) mode for the polarization perpendicular to the substrate surface (Figure A3.2). The longitudinal magnetic component of the TE mode and the longitudinal electric component of the TM mode are usually negligible, as is the case with the longitudinal electric and magnetic components of the HE_{11} mode of a fiber, because the index step of the waveguide is also small.

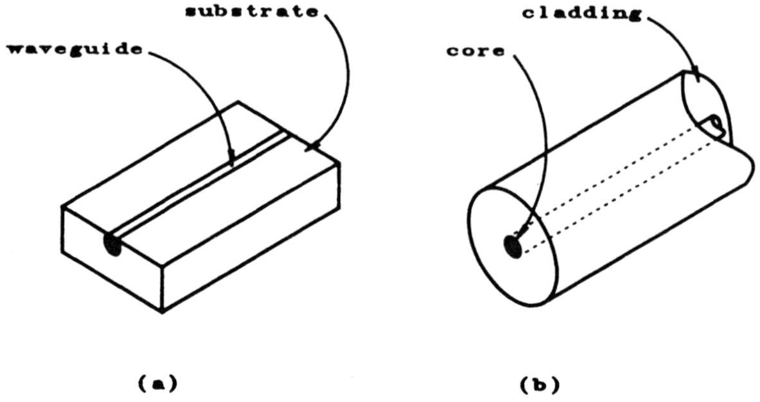

Figure A3.1 Equivalence between (a) an integrated-optic waveguide and (b) an optical fiber.

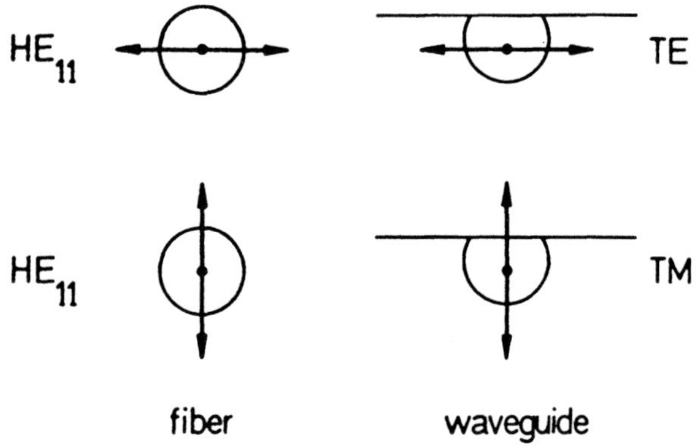

Figure A3.2 HE_{11} modes of a fiber and TE and TM modes of an integrated-optic waveguide.

To couple light into the waveguide, the sides of the substrate are polished with a sharp edge, and fiber pigtails are butted against it with their core facing the waveguide (Figure A3.3). Since the fundamental modes have similar sizes with a fiber core and a waveguide, there is good coupling efficiency (typically 70% to 90%, a connection loss of 1.5 to 0.5 dB). Several methods may be used to ruggedize these connections. In particular, the fiber may be held in a small ferrule that is directly bonded to the substrate side (Figure A3.4), the substrate typically being one millimeter in thickness, while the fiber diameter is on the order of one tenth of a millimeter. Taking into account connection and propagation losses, the ad-

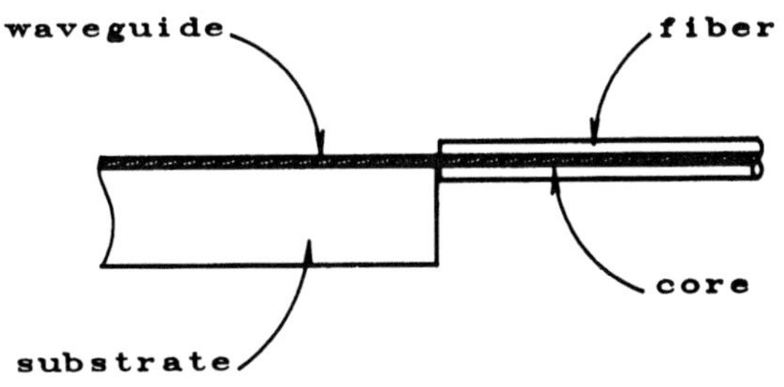

Figure A3.3 End-fire coupling of a fiber with a waveguide.

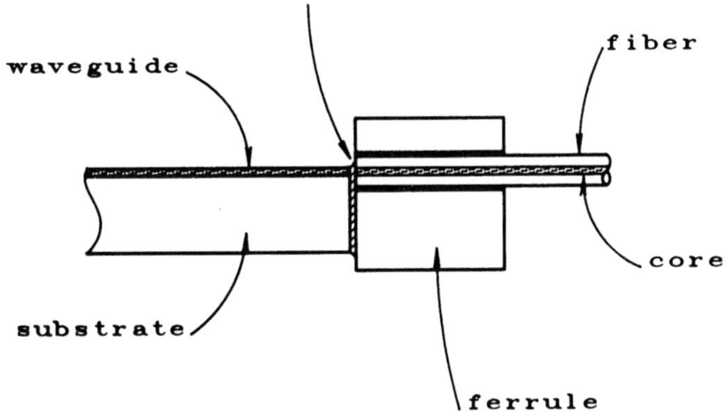

Figure A3.4 Ruggedized coupling of a fiber with a waveguide.

ditional fiber-to-fiber loss of a pigtailed circuit is typically 2 to 15 dB, depending on the complexity of the circuit.

Several materials, such as III-V semiconductors, silica over silicon, or glass waveguides, are potential candidates for integrated-optic circuits; but for fiber gyro applications, the optimal choice is lithium niobate ($LiNbO_3$), whose technology also happens to be the most advanced in terms of development. A fundamental function to fulfill is phase modulation, and $LiNbO_3$ has very good electro-optic properties: by applying an electric **E** field with electrodes, the index of refraction seen by the optical wave may be controlled because of the electro-optic Pockels effect, thus inducing a phase shift. This is actually used to make modulators in bulk form; but integrated optics provides an additional advantage, since electrodes can be placed very close to one another around the waveguide, while, in bulk form, space must be left to avoid diffraction of the light beam. This shortens the length of the electric field line, compared to that of bulk modulators, thus reducing the driving voltage for the required **E** field value in the material (Figure A3.5). The value of V_π (i.e., the voltage required to produce a π rad phase shift) falls into the range of a few volts instead of the hundreds of volts of the bulk form. This makes $LiNbO_3$ integrated optics compatible with low-voltage drive electronics.

A3.2 $LiNbO_3$ INTEGRATED OPTICS

The most common fabrication technique of $LiNbO_3$ integrated-optic waveguides is titanium (Ti) indiffusion (Figure A3.6). With photolithographic masking, very narrow strips of a thin film (several tens of nanometers) of titanium are deposited on the substrate. By heating up the wafer to 900°C to 1100°C for several hours, titanium diffuses into the substrate and locally increases the index of refraction.

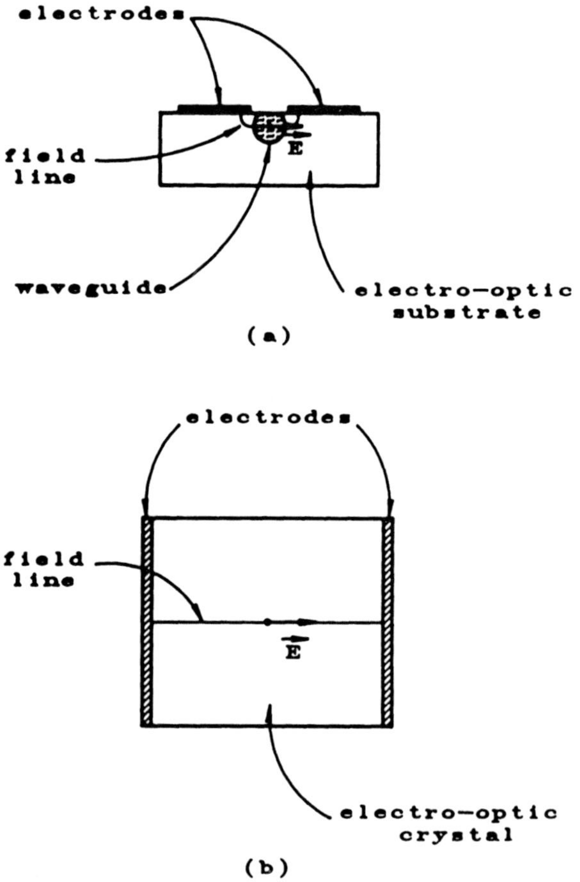

Figure A3.5 Comparison of phase modulators: (a) integrated optics; (b) bulk optics.

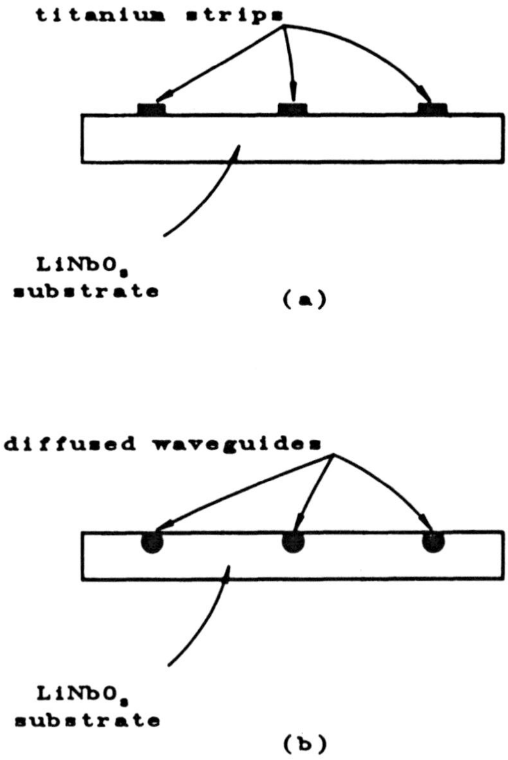

Figure A3.6 Process of Ti-indiffused waveguides in a $LiNbO_3$ substrate: (a) deposition of Ti strips; (b) diffusion.

This provides single-mode guidance in narrow channels without degrading the low attenuation of the bulk material. Because this slow indiffusion process takes place only at high temperatures, Ti-indiffused $LiNbO_3$ waveguides are very stable over time.

The electro-optic Pockels effect used for phase modulation is complex, since $LiNbO_3$ is a uniaxial birefringent crystal, and the electro-optic efficiency depends on the respective orientations of the driving electric field $\mathbf{E}_d$ and the optical electric field $\mathbf{E}_{op}$. The index values of $LiNbO_3$ are summarized in the following table for the most usual "fiber" wavelengths:

$LiNbO_3$	λ = 850 nm	λ = 1300 nm	λ = 1550 nm
Ordinary index $n_0 = n_x = n_y$ (slow axes)	2.25	2.22	2.21
Extraordinary index $n_e = n_z$ (fast axis)	2.17	2.15	2.14
$\Delta n_b = n_e - n_0$	−0,079	−0,075	−0,073

The strongest electro-optic coefficient is the diagonal r_{33} term ($r_{33} = 31 \times 10^{-12}$ m/V); that is, the most efficient phase modulation is obtained when both $\mathbf{E}_d$ and $\mathbf{E}_{op}$ fields are parallel to the extraordinary z-axis (also called C-axis). In this case, the index change δn_z is

$$\delta n_z = -\frac{1}{2} n_z^3 r_{33} E_{dz} \qquad \text{(A3.1)}$$

where E_{dz} is the z-component of $\mathbf{E}_d$.

To get this optimal efficiency, an x-cut substrate (i.e., the x-axis is perpendicular to the substrate surface) with a y-propagating waveguide (i.e., the waveguide is parallel to the y-axis) is needed. Then the TE mode, which has a horizontal E_{op} field parallel to the z-axis, can be efficiently modulated with planar metallic electrodes that are fabricated on both sides of the waveguide in a second step of the photolithographic process. Under the electrodes, the driving $\mathbf{E}_d$ field is vertical because of the electromagnetic boundary conditions on a metal; but the field lines bend under the surface to connect both electrodes, and $\mathbf{E}_d$ is actually parallel to the horizontal z-axis in the waveguiding region (Figure A3.7(a)). The TM mode is also modulated through the crossed r_{13} coefficient, since its optical $\mathbf{E}_{op}$ field is parallel to the x-axis (Figure A3.7(b)), but the value of r_{13} is less than one-third of that of r_{33} ($r_{13} = 9 \times 10^{-12}$ m/V). With such a design, the TE mode has a typical V_π value of 2 volts, and the TM mode has a typical V_π value of 7 volts for a 10-mm modulator length at a wavelength of 850 nm.

Note that working at a longer wavelength increases the V_π value for the same modulator length. As a matter of fact, the width of the waveguide and the electrode spacing have to be scaled up proportionately to the wavelength ratio to keep the same optimal configuration. This increase of electrode spacing, and thus of the field line length, yields an increase of the required driving voltage proportionate to the wavelength ratio. Furthermore, the phase change $\delta\phi$ is inversely proportional to the wavelength for a given index change δn, since $\delta\phi = 2\pi\delta n L/\lambda$: this adds a second wavelength ratio dependence. Then the V_π value is increased proportionately to the square of the wavelength ratio. For the same TE mode and 10-mm length, V_π would be 4.5 volts at 1300 nm instead of 2 volts at 850 nm.

Another important characteristic of integrated-optic phase modulators is their driving bandwidth. They can work with a continuous voltage, even if some problems of long-term drift may arise, and the upper frequency limit is dictated by the residual electrical capacitance of the two electrodes, which are very close to one another and placed on a material that has a very high dielectric permittivity at the usual modulation frequencies ($\epsilon_{LiNbO_3} \approx 30$). A good order of magnitude is a capacitance of 10 pF for a 10-mm length, which yields a bandwidth of 300 MHz with a load resistor of 50Ω in parallel. This value is not limiting for fiber gyro applications. Furthermore, the modulator response is very flat within this bandwidth,

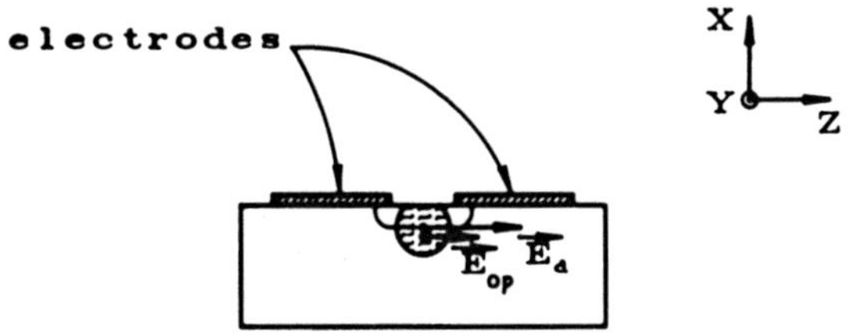

(a)

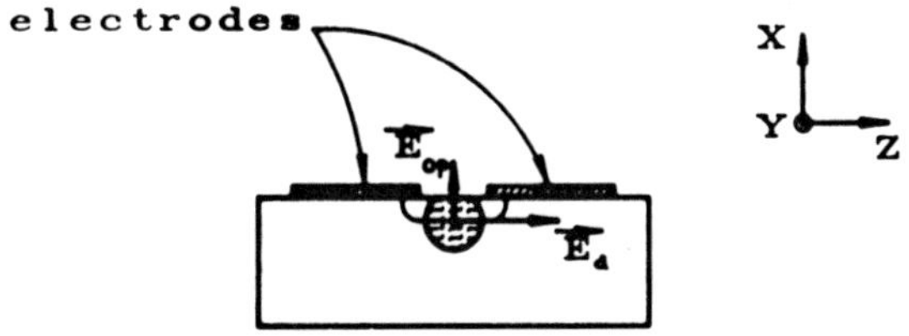

(b)

Figure A3.7 Phase modulation with an x-cut y-propagating waveguide: (a) TE mode; (b) TM mode.

which makes integrated optics the ideal technology to fulfill the important function of phase modulation for fiber gyro signal processing techniques. For telecommunication applications requiring bandwidths of several gigahertz, the modulator design is more sophisticated and needs to use traveling-wave electrodes with matched-line impedance.

This x-cut TE configuration is the most efficient and most typical design, but it may face the problem of outdiffusion of Li_2O at the surface of the substrate during the heating required for the indiffusion of titanium. $LiNbO_3$ is actually a compound of LiO_2 and Nb_2O_5. The stoichiometric composition is $(Li_2O)_{0.5}(Nb_2O_5)_{0.5}$, but the material may withstand a slightly nonstoichiometric composition $(Li_2O)_x(Nb_2O_5)_{1-x}$, with x ranging from 0.48 to 0.5. In particular, the highest material uniformity is obtained in the so-called congruent composition (x = 0.486), for which there is a composition equilibrium between the solid and liquid phases at the melting point (1243°C), where the crystal is grown. Outdiffusion of LiO_2 decreases the value of x, which does not modify the ordinary index n_o, but yields an increase of the extraordinary index n_e. This creates a parasitic planar waveguide for the extraordinary polarization (i.e., the TE mode with x-cut), which may cause optical leakage or cross-talk between the useful indiffused channel waveguides. Outdiffusion may be suppressed with techniques such as wetting the incoming gas flow in the diffusion furnace or saturating the atmosphere with Li_2O powder.

To avoid the effect of outdiffusion, it would be possible to work with the TM mode, which is polarized along the ordinary x-axis with an x-cut substrate, but there is a drawback of lower modulation efficiency. However, this is not desirable for fiber gyro applications, because another important component is the polarizer, and the efficient technique to fabricate such a component is to cover the waveguide with a metallic layer that absorbs the TM mode while it transmits the TE mode (Figure A3.8).

On the other hand, to have this transmitted TE mode polarized along an ordinary axis that is not affected by outdiffusion, it is possible to use a z-cut substrate with a y-propagating waveguide. In this configuration, one electrode must cover the waveguide to get a vertical driving $\mathbf{E}_d$ field parallel to the z-axis in the waveguiding region and to modulate the phase of the TE mode through the r_{13} coefficient (Figure A3.9). Even if it is not optimal, the modulation efficiency remains ac-

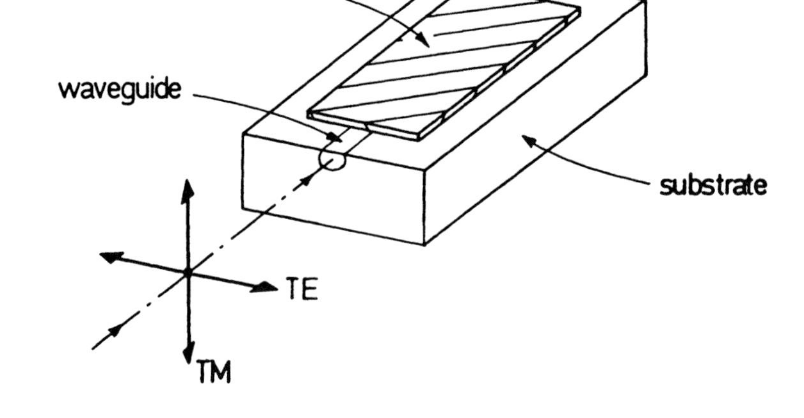

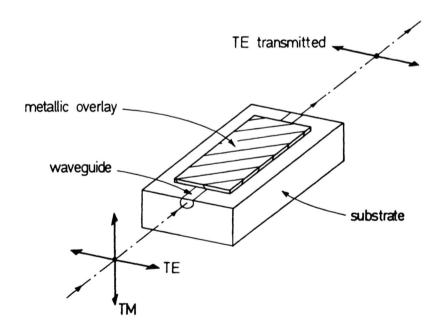

Figure A3.8 Metallic overlay polarizer.

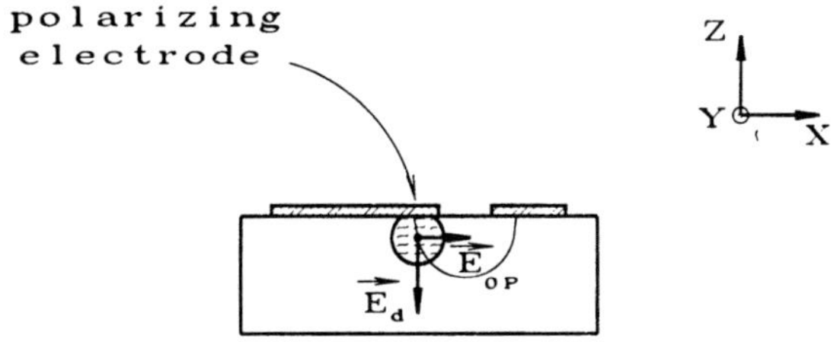

Figure A3.9 Phase modulation of the TE mode with a z-cut y-propagating waveguide, one electrode acting as a polarizer that attenuates the TM mode.

ceptable, and, in addition, the covering electrode acts as a polarizer which may be very useful.

For completeness, we may add that some specific components use alternative orientations and other electro-optic coefficients. In particular, z-propagating waveguides are needed when the effect of the birefringence of $LiNbO_3$ has to be eliminated, but this is not advantageous for fiber gyros, where, on the contrary, birefringence is very beneficial because it induces depolarization.

A3.3 PROTON-EXCHANGED WAVEGUIDES

Ti-indiffused waveguides on a $LiNbO_3$ substrate is a suitable technology for the fiber gyro, but, ideally, a single-polarization waveguide would be preferable to ensure very good polarization filtering. As we have already seen, outdiffusion of Li_2O creates an increase of the extraordinary index, which yields guidance only for a wave polarized along the z-axis. However, this technique is difficult to control for channel waveguide, because the material used for masking must not be diffused, and the induced lack of oxygen ions increases the attenuation.

A similar effect of single-polarization guidance may be obtained with proton exchange, where the $LiNbO_3$ substrate is placed in a melted organic acid and H^+ ions (i.e., protons) replace Li^+ ions in the crystal lattice. This technique does not degrade the attenuation and has the very attractive property of increasing the value of the low extraordinary index while decreasing the value of the high ordinary index. This yields guidance only for the z-polarized mode parallel to the extraordinary axis. Such single-polarization waveguides provide a very high extinction ratio for the nonguided crossed polarization.

Proton exchange, however, is processed at a relatively low temperature (about 200 to 300°C), which avoids the problem of mask indiffusion, but the first experimental demonstrations suffered from poor long-term stability. Annealing techniques have since yielded a significant increase in lifetime, and proton exchange circuits now appear to be a very promising technological choice for fiber gyro circuits, because high polarization rejection is one of the important features required for high performance.

Optimal modulation efficiency is obtained with the same x-cut y-propagating orientation as for Ti indiffusion. Electrodes, placed on both sides of the proton-exchanged waveguide, modulate the transmitted TE mode with the strongest r_{33} electro-optic coefficient (Figure A3.10).

Proton-exchanged waveguides show essentially no sensitivity to optical damage (i.e., index drift under high power illumination), which may be encountered with Ti indiffusion. However, this advantage is not very important for fiber gyro applications, where the optical power is not very high.

Note 1: Lithium tantalate ($LiTaO_3$) could be a possible alternative to $LiNbO_3$ as the substrate material. As a matter of fact, $LiTaO_3$ has electro-optic properties very similar to those of $LiNbO_3$ ($r_{33\ LiTaO_3} \approx r_{33\ LiNbO_3}$), and it is also used very often in bulk-optic modulators. However, both crystals are also ferro-electric, and to get a stable electro-optic efficiency, a ferro-electric crystal has to be electrically poled to orient all the microscopic domains in the same direction (as a ferromagnetic material is magnetically poled to get a permanent magnet); but the poling

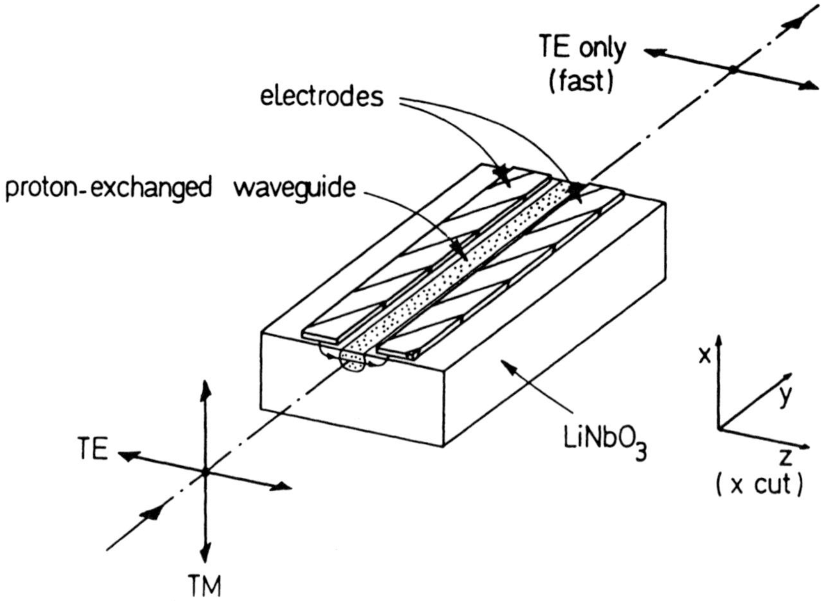

Figure A3.10 Proton-exchanged phase modulator.

is lost when the temperature is raised above Curie temperature T_C. For $LiNbO_3$, $T_C = 1150°C$, and Ti indiffusion remains below T_C; but for $LiTaO_3$, $T_C = 610°C$, which makes metal indiffusion much more complicated because it requires repoling of each processed wafer. However, this drawback of $LiTaO_3$ disappears with proton exchange, since it is realized at only 200 to 300°C.

Note, however, that $LiTaO_3$ has a birefringence that is more than one order of magnitude lower than that of $LiNbO_3$ (Δn is equal to 0.004 instead of 0.08). For fiber gyro applications, this is a disadvantage, since depolarization induced by birefringence is very beneficial.

Note 2: The polarization rejection of a proton-exchanged $LiNbO_3$ circuit is not due to an absorption phenomenon: cross-polarized light is not guided and is diffracted in the substrate. Therefore, some light may be partially coupled back into the output fiber.

Let us take an example of a fiber mode that has a diameter $2w_0 = 5$ μm at $\lambda = 0.85$ μm. Assuming that the crossed polarization is diffracted in a uniform medium, the full divergence angle is $\theta_D = 2\lambda/\pi n w_0$ (see Appendix 1); that is, 0.1 rad in $LiNbO_3$ where $n \approx 2.2$. After a length of 30 mm, the diffraction pattern diameter is $2w_0' = 3$ mm. The recoupling ratio of the crossed polarization in the output fiber would then be (see Appendix 2).

$$\Gamma_{dm} = 20 \log\left(\frac{2w_0}{w_0'}\right) \tag{A3.2}$$

that is, only −50 dB.

Experimental results are much better, but this may be explained simply with an interferometric "Lloyd's mirror" effect on the top interface of the substrate. As a matter of fact, the nonguided wave is in total internal reflection on this interface: this yields a Lloyd's mirror interferometer with interferences between two sources, the input fiber and its virtual image (Figure A3.11). Furthermore, total internal reflection under grazing incidence induces a π rad phase shift; therefore, the central fringe located on the interface is a black fringe, which reduces drastically the power density that gets to the output fiber located just below this interface.

To evaluate this parasitic recoupling ratio more precisely, it is possible to consider that there is actually diffraction of a second-order antisymmetric mode with a real lobe and a virtual image lobe, which is recoupled in the second-order mode of an output waveguide composed of the output fiber and its virtual image. As seen in Appendix 2, this mode can be considered as a "pseudo-Gaussian derivative" mode with a full divergence angle between both extrema:

$$\theta_{D1} = \frac{\lambda}{\pi w_1} \tag{A3.3}$$

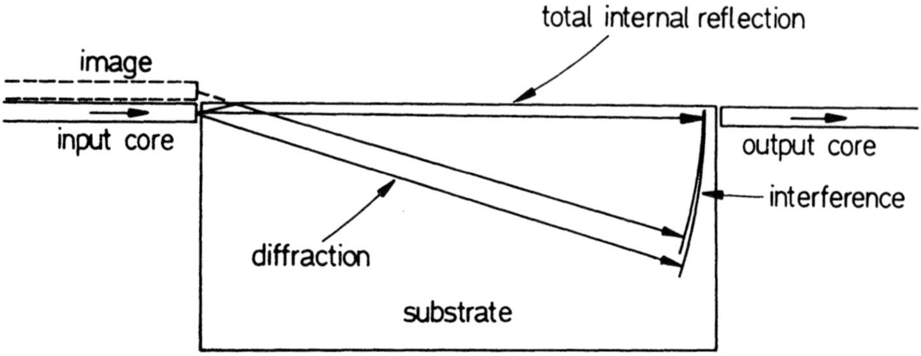

Figure A3.11 Parasitic recoupling of the nonguided crossed polarization.

where w_1 is the half width at the maximum of the input fiber mode. Assuming $w_1 \approx w_0$ the parasitic coupling ratio, because of the antisymmetry of the mode, is now

$$\Gamma_{dm1} = 40 \log\left(\frac{2w_1}{w_1{}'}\right) \tag{A3.4}$$

with $2w_1{}' = \lambda/\pi w_1$. With the same numerical value as above, this yields a theoretical rejection of -85 dB.

However, multiple reflection between both interfaces of the substrate may induce other spurious waves, and the behavior of the radiated crossed polarization has to be analyzed precisely to get the best rejection with proton-exchanged waveguides.

BIBLIOGRAPHY

About integrated optics:

- Hutcheson, L. D., ed., "Integrated Optical Circuits and Components," Marcel Dekker, 1987.

About proton-exchanged $LiNbO_3$ circuits:

- Wong, K. K., ed., "Integrated Optics and Optoelectronics II," SPIE Proceedings, Vol. 1374, 1990.

About waveguide theory:

- Marcuse, D., "Theory of Dielectric Optical Waveguides," Academic Press, 1974.
- Vassalo, C., "Optical Waveguide Concepts," Elsevier, 1991.

Appendix 4

Electromagnetic Theory of the Relativistic Sagnac Effect

A4.1 SPECIAL RELATIVITY AND ELECTROMAGNETISM

Special relativity is based on the principle of equivalence of the so-called inertial frames of reference for all the laws of physics, particularly mechanics and electromagnetism: the main effect being that light velocity c in a vacuum is the same in any inertial frame.

This yields the Lorentz transformation between the spatial coordinates and the time of two reference frames moving with a constant translation velocity v_t (along the x-axis) with respect to each other:

$$
\begin{aligned}
x_L &= \gamma(x - v_t t) \\
y_L &= y \\
z_L &= z \\
t_L &= \gamma\left(t - \frac{v_t x}{c^2}\right)
\end{aligned}
\qquad \text{(A4.1)}
$$

with

$$\gamma = \frac{1}{\sqrt{1 - \frac{v_t^2}{c^2}}}$$

Compared to the Galilean transformation of classical mechanics:

$$\begin{aligned} x_G &= x - v_t t \\ y_G &= y \\ z_G &= z \\ t_G &= t \end{aligned} \qquad \text{(A4.2)}$$

There is a second-order (in v_t/c) difference for the spatial coordinate x and a first-order difference for the time t.

To "simplify" and condense equations, relativity laws are usually expressed with a four-dimensional notation using contravariant and covariant coordinates of four-vectors and tensors. A space-time four-vector ${}^4\mathbf{x}$ has contravariant coordinates defined by

$$x^{\mu} = (x, y, z, ct) \quad \text{with } \mu = 1 \text{ to } 4 \qquad \text{(A4.3)}$$

and covariant coordinates defined by

$$x_{\mu} = (-x, -y, -z, ct) \quad \text{with } \mu = 1 \text{ to } 4 \qquad \text{(A4.4)}$$

Contravariant coordinates are the "usual" coordinates, with

$${}^4\mathbf{x} = \sum_{\mu=1}^{4} x^{\mu}\mathbf{a}_{\mu} \qquad \text{(A4.5)}$$

where $\mathbf{a}_{\mu}$ are the basis eigenvectors. With four-dimensional notation, the Σ term is omitted, and the convention is to sum over any double index appearing once up and once down

$${}^4\mathbf{x} = x^{\mu}\mathbf{a}_{\mu} \qquad \text{(A4.6)}$$

The covariant coordinates are defined with the scalar product:

$$x_{\mu} = {}^4\mathbf{x} \cdot \mathbf{a}_{\mu} \qquad \text{(A4.7)}$$

With the orthonormal basis of a euclidean frame of reference, contravariant and covariant coordinates are equal. Some feeling of the difference between contravariant and covariant coordinates may be obtained with a nonorthogonal basis of a two-dimensional plane (Figure A4.1). Contravariant coordinates are the usual coordinates defined by a parallelogram with side lengths equal to x^1 and x^2. Covariant coordinates defined by the scalar product $\mathbf{x} \cdot \mathbf{a}_1$ and $\mathbf{x} \cdot \mathbf{a}_2$ correspond to the perpendicular projection of $\mathbf{x}$ on each coordinate axis.

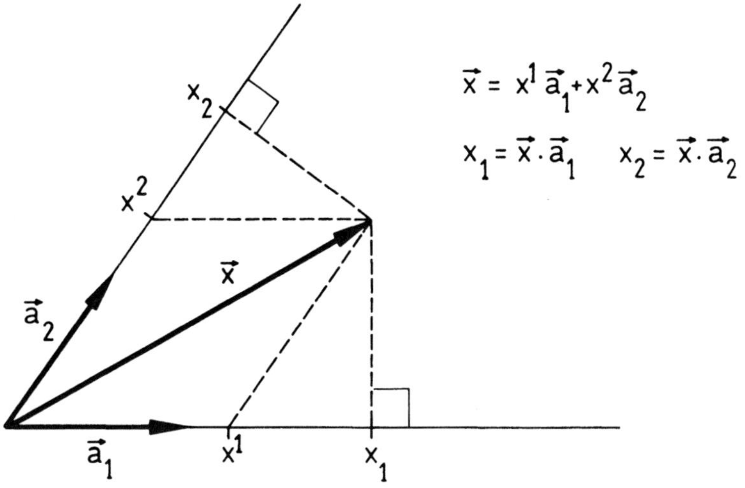

Figure A4.1 Geometrical representation of contravariant coordinates x^μ and covariant coordinates x_μ in a two-dimensional plane.

The covariant coordinates are related to the "usual" contravariant coordinates with a second-rank covariant tensor, the so-called metric tensor $g_{\mu\nu} = \mathbf{a}_\mu \cdot \mathbf{a}_\nu$, which is a characteristic of the frame of reference:

$$x_\mu = g_{\mu\nu}x^\nu \left(= \sum_{\nu=1}^{4} g_{\mu\nu}x^\nu \right) \tag{A4.8}$$

In an inertial frame, there are

$$g_{I\mu\nu} = \begin{bmatrix} -1 & 0 & 0 & 0 \\ 0 & -1 & 0 & 0 \\ 0 & 0 & -1 & 0 \\ 0 & 0 & 0 & 1 \end{bmatrix} \tag{A4.9}$$

Now the Lorentz transformation may be written with

$$x_L{}^\mu = L^\mu_\nu x^\nu$$

where

$$L^\mu_\nu = \begin{bmatrix} \gamma & 0 & 0 & -\gamma\frac{v_t}{c} \\ 0 & 1 & 0 & 0 \\ 0 & 0 & 1 & 0 \\ -\gamma\frac{v_t}{c} & 0 & 0 & \gamma \end{bmatrix} \tag{A4.10}$$

Note that the "length" of a four-vector and its square defined with a scalar product is invariant under Lorentz transform

$$x_L^{\mu} \cdot x_{L\mu} = x^{\mu} \cdot x_{\mu} \tag{A4.11}$$

or

$$c^2 t_L^2 - x_L^2 - y_L^2 - z_L^2 = c^2 t^2 - x^2 - y^2 - z^2 = s^2 \tag{A4.12}$$

where s^2 is the so-called space-time invariant.

Electromagnetism may also be described with this four-dimensional formalism. The vector potential **A** and the scalar potential V form a four-vector ${}^4\mathbf{A}$ with contravariant components:

$$A^{\mu} = \left(A_x, A_y, A_z, \frac{V}{c}\right) \tag{A4.13}$$

and covariant components:

$$A_{\mu} = \left(-A_x, -A_y, -A_z, \frac{V}{c}\right) \tag{A4.14}$$

This four-potential defined the electromagnetic field represented by a second-rank covariant antisymmetric tensor $F_{\mu\nu}$. The usual definitions, $\mathbf{E} = -(\partial \mathbf{A}/\partial t) - \mathbf{grad}\ V$ and $\mathbf{B} = \mathbf{curl}\ \mathbf{A}$, may be written

$$F_{\mu\nu} = \partial_{\mu} \mathrm{A}_{\nu} - \partial_{\nu} \mathrm{A}_{\mu} \tag{A4.15}$$

where ∂_{μ} is covariant and corresponds to the partial derivative with respect to the contravariant space-time coordinate, $\partial/\partial x^{\mu}$. It can be seen from this definition that $F_{\mu\nu}$ is antisymmetric, since $F_{\mu\nu} = -F_{\nu\mu}$. Expressed in terms of the conventional **E** and **B** field components, the covariant components of the field tensor are

$$F_{\mu\nu} = \begin{bmatrix} 0 & B_z & -B_y & E_x/c \\ -B_z & 0 & B_x & E_y/c \\ B_y & -B_x & 0 & E_z/c \\ -E_x/c & -E_y/c & -E_z/c & 0 \end{bmatrix} \tag{A4.16}$$

where μ is the column index and ν the line index. The contravariant components are given by

$$F^{\mu\nu} = g^{\mu\sigma} g^{\nu\rho} F_{\sigma\rho} \tag{A4.17}$$

where $g^{\mu\sigma}$ is the contravariant metric tensor related to the covariant metric tensor $g_{\mu\rho}$, with

$$g^{\mu\sigma} \cdot g_{\mu\rho} = \delta^{\nu}_{\rho} \quad \left[\begin{array}{l}\text{the Kronecker symbol} \\ \text{i.e. } \delta^{\nu}_{\nu} = 1 \text{ and } \delta^{\nu}_{\rho} = 0 \\ \text{if } \nu \neq \rho\end{array}\right] \tag{A4.18}$$

We have

$$F^{\mu\nu} = \begin{bmatrix} 0 & B_z & -B_y & -E_x/c \\ -B_z & 0 & B_x & -E_y/c \\ B_y & -B_x & 0 & -E_z/c \\ E_x/c & E_y/c & E_z/c & 0 \end{bmatrix} \tag{A4.19}$$

The first two Maxwell equations, M_1 and M_2, express that the field is derived from a potential and may be written

$$\partial_\mu F_{\rho\sigma} + \partial_\sigma F_{\mu\rho} + \partial_\rho F_{\sigma\rho} = 0 \quad (\text{with } \mu \neq \rho \neq \sigma \neq \mu) \tag{A4.20}$$

It is also possible to define a four-current vector $^4\mathbf{J}$ with

$$\begin{aligned} J^\mu &= (j_x, j_y, j_c, c\rho) \\ J_\mu &= (-j_x, -j_y, -j_c, c\rho) \end{aligned} \tag{A4.21}$$

The last two Maxwell equations, M_3 and M_4, may be written as

$$\partial_\mu F^{\mu\nu} = \mu_0 J^\nu \tag{A4.22}$$

Note that the first two Maxwell equations use the covariant coordinates $F_{\mu\nu}$ and that the last two use the contravariant coordinates $F^{\mu\nu}$.

Finally, the propagation equation is

$$\partial^\rho \partial_\rho F_{\mu\nu} = 0 \tag{A4.23}$$

where $\partial^\rho = \partial/\partial x_\rho$ and $\partial_\rho = \partial/\partial x^\rho$, that is:

$$\partial^1\partial_1 = -\frac{\partial^2}{\partial x^2}, \partial^2\,\partial_2 = -\frac{\partial^2}{\partial y^2}, \partial^3\partial_3 = -\frac{\partial^2}{\partial z^2} \text{ and } \partial^4\partial_4 = \frac{1}{c^2}\frac{\partial^2}{\partial t^2} \tag{A4.24}$$

The important point is that Maxwell equations and their consequences, like the propagation equation, are invariant under Lorentz transformation. In the new frame, the four-vectors A^{μ} and J^{μ} and the tensor $F^{\mu\nu}$ are transformed accordingly to

$$A_L{}^{\mu} = L^{\mu}_{\nu}A^{\nu},\ J_L{}^{\mu} = L^{\mu}_{\nu}J^{\nu} \text{ and } F^{\mu\nu}_L = L^{\mu}_{\sigma}L^{\nu}_{\rho}F^{\sigma\rho} \qquad \text{(A4.25)}$$

and these quantities $A_L{}^{\mu}$, $J_L{}^{\mu}$, and $F_L{}^{\mu\nu}$ follow the same laws. In particular, the propagation equation remains

$$\partial_{L\rho}\partial_{L\rho}F_{L\mu\nu} = 0 \qquad \text{(A4.26)}$$

with $\partial_{L}\rho = \partial/\partial x_{L\rho}$ and $\partial_{L\rho} = \partial/\partial x^{\rho}_{L}$. This result is consistent with the fact that the velocity of light in a vacuum remains equal to c in any inertial frame.

Note that the equation $F^{\mu\nu}_L = L^{\mu}_{\sigma}L^{\nu}_{\rho}F^{\sigma\rho}$ is the basis of magnetism. With a motionless charge, there is only an electric field **E** and the three magnetic components of $F_0{}^{\sigma\rho}$ in the rest frame are null:

$$F_0{}^{12} = F_0{}^{13} = F_0{}^{23} = 0$$

To take into account the effect of a moving charge, we have to use the Lorentz transformation between the frame where the charge is motionless and the frame where the charge is moving at an opposite velocity $-v_t$ if v_t is the velocity of the moving frame with respect to the rest frame of the particle. Calculating $F^{\mu\nu}_L = \mathrm{L}^{\mu}_{\sigma}\mathrm{L}^{\nu}_{\rho}\mathrm{F}^{\sigma\rho}$, we obtain nonzero terms for $F_L{}^{12}$, $F_L{}^{13}$, or $F_L{}^{23}$; that is, it appears a magnetic component **B**: even if it is sometimes obviated, the magnetic field is a pure relativistic effect. Since the speed of the particles is usually much smaller than c, the pure effect of **B** is usually much smaller than that of **E**. However, if this applies to one particle, in the case of a conductor, then there is cancellation of the **E** effect, because there are as many positive particles as negative ones, while the **B** effect is not cancelled, since the positive and negative particles do not have the same speed. In this case, the global **B** effect becomes predominant despite the fact that the elementary **B** effect for each particle is a lower effect. As will be seen, the problem of the Sagnac effect has some similarities with that of the magnetic field: it is also a first-order effect that is much lower than the zero-order effect; but the zero-order effect may also be nulled out because of reciprocity, and in a ring interferometer the first order effect may become predominant.

The main interest of these notations is that these results remain almost unchanged with any set of coordinates, even if they are not cartesian. In general, we still have

$$\partial_\mu F_{\rho\sigma} + \partial_\sigma F_{\mu\rho} + \partial_\rho F_{\sigma\rho} = 0$$
$$F_{\mu\nu} = \partial_\mu A_\nu - \partial_\nu A_\mu \tag{A4.27}$$

and the last two Maxwell equations are slightly modified:

$$\partial_\mu \left(\sqrt{-g}\, F^{\mu\nu} \right) = \sqrt{-g}\, J^\nu \tag{A4.28}$$

where g is the determinant of the metric tensor. With an inertial frame and cartesian coordinates, we have $g_I = -1$ and $\sqrt{-g_I} = 1$.

The case of the propagation equation is much more complicated, since it uses the "high-index" derivative ∂^ρ in addition to the "low-index" derivative ∂_ρ that is used in the other formulae, and there is no simple generalization of the equation. To avoid this mathematical difficulty, it is possible to define a propagation equation of the four-potential A_μ. As a matter of fact, the equations

$$\partial_\mu(\sqrt{-g}F^{\mu\nu}) = 0$$
$$F^{\mu\nu} = g^{\mu\sigma} g^{\nu\rho} F_{\sigma\rho} \quad \text{(without charge or current)} \tag{A4.29}$$
$$F_{\sigma\rho} = \partial_\sigma A_\rho - \partial_\rho A_\sigma$$

yield a propagation equation:

$$\partial_\mu \left[\sqrt{-g}\, g^{\mu\sigma} g^{\nu\rho} (\partial_\sigma A_\rho - \partial_\rho A_\sigma) \right] = 0 \tag{A4.30}$$

So far, this analysis has been carried out with the assumption that there is a vacuum. If there is a medium, a derived field tensor $G_{\mu\nu}$ composed of the components **D** and **H** has to be used. The first two Maxwell equations that show the relation with a potential are unchanged, but the two last ones become

$$\partial_\mu \sqrt{-g}\, G^{\mu\nu} = J^\nu_{\text{free}} \tag{A4.31}$$

where J^ν_{free} is the free current four-vector.

The derived field tensor $G^{\mu\nu}$ is related to the field tensor $F_{\sigma\rho}$, with

$$G^{\mu\nu} = \chi^{\mu\nu\sigma\rho} F_{\sigma\rho} \tag{A4.32}$$

where $\chi^{\mu\nu\sigma\rho}$ is the constitutive tensor of the material. It depends on the relative permittivity ϵ_r and the relative permeability μ_r, but also on the metric of the frame

of reference, since $G^{\mu\nu}$ is used with its contravariant coordinates and $F_{\sigma\rho}$ is used with its covariant coordinates.
In an inertial frame of reference, there are

$$G^{\mu\nu} = \begin{bmatrix} 0 & H_z & -H_y & -cD_x \\ -H_z & 0 & H_x & -cD_y \\ H_y & -H_x & 0 & -cD_z \\ cD_x & cD_y & cD_z & 0 \end{bmatrix} \tag{A4.33}$$

And the nonzero terms of the constitutive tensor $\chi^{\mu\nu\sigma\rho}$ of a motionless material are

$$\chi^{1212} = \chi^{2121} = \chi^{1313} = \chi^{3131} = \chi^{2323} = \chi^{3232} = \mu^{-1}\mu_0{}^{-1} \tag{A4.34}$$

$$\chi^{1414} = \chi^{4141} = \chi^{2424} = \chi^{4242} = \chi^{3434} = \chi^{4343} = -\epsilon\mu_0{}^{-1} \tag{A4.35}$$

In the most general case, the propagation equation of the four-potential is

$$\partial_\mu\left[\sqrt{-g}\,\chi^{\mu\nu\sigma\rho}(\partial_\sigma A_\rho - \partial_\rho A_\sigma)\right] = 0 \tag{A4.36}$$

As an example of this analysis, we can consider the case of cylindrical coordinates with contravariant (i.e., "usual") values:

$$x_c{}^\mu = (r, \theta, z, ct) \tag{A4.37}$$

The metric tensor $g_{c\mu\nu}$ is:

$$g_{c\mu\nu} = \begin{bmatrix} -1 & 0 & 0 & 0 \\ 0 & -r^2 & 0 & 0 \\ 0 & 0 & -1 & 0 \\ 0 & 0 & 0 & 1 \end{bmatrix} \tag{A4.38}$$

Thus, the determinant $g_c = -r^2$ and $\sqrt{-g_c} = r$ and the covariant coordinates $x_{c\mu} = g_{c\mu\nu}x_c{}^\nu$ are

$$x_{c\mu} = (-r, -r^2\,\theta, -z, ct) \tag{A4.39}$$

The electromagnetic four-potential ${}^4\mathbf{A}_c$ is defined by contravariant coordinates:

$$A_c{}^\mu = \left(A_r, \frac{A_\theta}{r}, A_z, \frac{V}{c}\right) \tag{A4.40}$$

where A_r, A_θ and A_z are the conventional coordinates of **A**, with orthogonal unit vectors $\mathbf{a}_r$, $\mathbf{a}_\theta$, and $\mathbf{a}_z$ parallel to the equicoordinate lines (Figures A4.2). The covariant coordinates are

$$A_{c\mu} = \left(-A_r,\ -r\,A_\theta,\ -A_z,\ \frac{V}{c}\right) \tag{A4.41}$$

For the covariant components of the field tensor, there are, similarly,

$$F_{c\mu\nu} = \begin{bmatrix} 0 & rB_z & -B_\theta & E_r/c \\ -rB_z & 0 & rB_r & rE_\theta/c \\ B_\theta & -rB_r & 0 & E_z/c \\ -E_r/c & -rE_\theta/c & -E_z/c & 0 \end{bmatrix} \tag{A4.42}$$

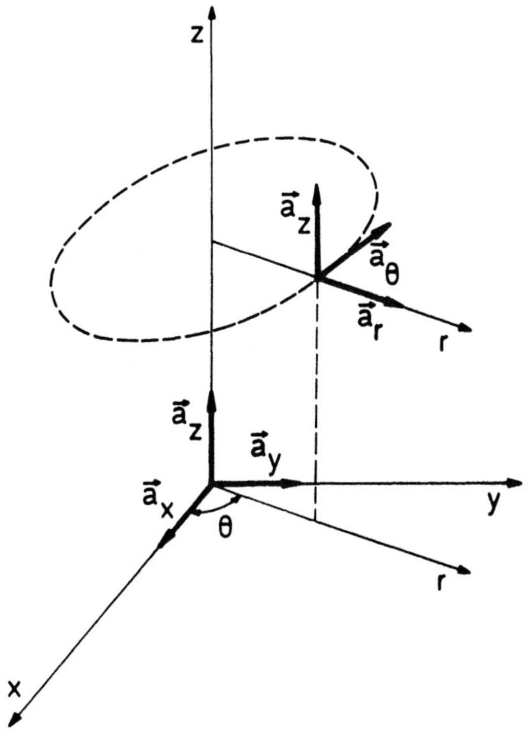

Figure A4.2 Cylindrical coordinates.

and for the contravariant components of the derived field tensor,

$$G_c^{\mu\nu} = \begin{bmatrix} 0 & H_z/r & -H_\theta & -cD_r \\ -H_z/r & 0 & H_r/r & -cD_\theta/r \\ -H_\theta & -H_r/r & 0 & -cD_z \\ cD_r & cD_\theta/r & cD_z & 0 \end{bmatrix} \tag{A4.43}$$

The nonzero terms of the constitutive tensor $\chi_c^{\mu\nu\sigma\rho}$ are

$$\begin{aligned} \chi_c^{2121} &= \chi_c^{1212} = \mu^{-1}\mu_0^{-1}r^{-2} & \chi_c^{4141} &= \chi_c^{1414} = -\epsilon\mu_0^{-1} \\ \chi_c^{3131} &= \chi_c^{1313} = \mu^{-1}\mu_0^{-1} & \chi_c^{4242} &= \chi_c^{2424} = -\epsilon\mu_0^{-1}r^{-2} \\ \chi_c^{3232} &= \chi_c^{2323} = \mu^{-1}\mu_0^{-1}r^{-2} & \chi_c^{4343} &= \chi_c^{3434} = -\epsilon\mu_0^{-1} \end{aligned}$$

Note that, in the case of cylindrical coordinates, the general formulation that has been derived is a way to retrieve the well-known expressions of the various vector operators (gradient, divergence, curl, and laplacian) with these noncartesian spatial coordinates. The constitutive tensor $\chi_c^{\mu\nu\sigma\rho}$ depends on μ and ϵ (i.e., the effect of the material), but also on r (i.e., the effect of the noncartesian spatial coordinates).

A4.2 ELECTROMAGNETISM IN A ROTATING FRAME

From a mathematical point of view, the tensor formalism derived in the previous section may be applied to any kind of coordinates; but we have to define, from the physics point of view, what coordinates can actually be used. Considering a given frame, the coordinates must be measurable with internal experiments relative to this frame. For a translation, it is well known that the new coordinates have to be derived with the Lorentz transformation. A Galilean transformation is valid mathematically, but it yields new coordinates that cannot be measured with internal experiments in the new frame.

In the case of a rotation of rate Ω, the new coordinates could be derived with an equivalent Lorentz transformation, where the tangential speed $r\Omega$ replaces v_t:

$$\begin{aligned} r_{RL} &= r \\ \theta_{RL} &= \gamma(\theta - \Omega t) \\ z_{RL} &= z \\ t_{RL} &= \gamma\left(t - \frac{r^2\Omega}{c^2}\theta\right) \end{aligned} \tag{A4.44}$$

with

$$\gamma = \frac{1}{\sqrt{1 - \frac{r^2\Omega^2}{c^2}}}$$

Locally, this transformation is valid, and the potential A^{μ}_{RL}, the field $F^{\mu\nu}_{RL}$, and the derived field $G^{\mu\nu}_{RL}$ defined with these coordinates correspond to the local values seen particularly by a medium at rest in this rotating frame. With such a motionless medium, the constitutive tensor $\chi_{RL}{}^{\mu\nu\sigma\rho}$ is equal to the constitutive tensor $\chi_c{}^{\mu\nu\sigma\rho}$ defined for an inertial frame using cylindrical spatial coordinates.

However, the angular coordinate is periodic, and, if the whole rotating frame is considered, this periodicity will appear in the time coordinate. To analyze global experiments like the measurement of a phase difference in a ring Sagnac interferometer, the time coordinate must be univocal and thus independent of the periodic angular coordinate, and a Galilean transformation may be used:

$$\begin{aligned} r_{RG} &= r \\ \theta_{RG} &= (\theta - \Omega t) \\ z_{RG} &= z \\ t_{RG} &= t \end{aligned} \tag{A4.45}$$

In contrast to the case of translation, these new coordinates can be measured in the rotating frame. A clock placed on the rotation axis has the same time as the time t in the inertial frame, and it may synchronize the time in the whole rotating frame by sending cylindrical waves that propagate perpendicularly to the displacement and are thus not modified. The angle θ_{RG} may be defined as a constant portion of a full angle (i.e., 360 deg) which remains obviously constant.

Once it is admitted that a Galilean transformation is valid for a rotation, the general tensor formalism may be applied. In particular, the propagation equation of a potential in a vacuum is, with the four-dimensional formalism,

$$\partial_\mu \sqrt{-g_{RG}}\, g^{\mu\sigma}_{RG}\, g^{\nu\rho}_{RG}\, (\partial_\sigma A_{RG\rho} - \partial_\rho A_{RG\sigma}) = 0 \tag{A4.46}$$

where $g^{\mu\nu}_{RG}$ is the metric tensor corresponding to the contravariant coordinates $x^{\mu}_{RG} = (r_{RG}, \theta_{RG}, z_{RG}, ct_{RG})$ and g_{RG} is its determinant. We have

$$g^{\mu\nu}_{RG} = \begin{bmatrix} -1 & 0 & 0 & 0 \\ 0 & -\frac{1}{r^2}\left(1 - \frac{r^2\Omega^2}{c^2}\right) & 0 & -\frac{\Omega}{c} \\ 0 & 0 & -1 & \\ 0 & -\frac{\Omega}{c} & 0 & 1 \end{bmatrix} \tag{A4.47}$$

As can be seen, there are two nondiagonal terms $g_{RG}^{24} = g_{RG}^{42} = -(\Omega/c)$, which yield a difference of propagation velocities between co-rotating and counter-rotating waves. To first order in $r\Omega/c$, the propagation equation in a vacuum becomes

$$\nabla_{RG}^2 A - \frac{1}{c^2}\frac{\partial^2 A}{\partial t_{RG}^2} - \frac{2\Omega}{c^2}\frac{\partial^2 A}{\partial \theta_{RG} \partial t_{RG}} = 0 \tag{A4.48}$$

where the wave amplitude A is any component of the field or of the potential. Compared to the familiar propagation equation in an inertial frame, there is an additional crossed term $-\frac{2\Omega}{c^2}\frac{\partial^2 A}{\partial \theta_{RG} \partial t_{RG}}$ which modifies the wave velocity proportionally to Ω.

With a co-rotating medium, the propagation equation is with four-dimensional formalism:

$$\partial_\mu \left[\sqrt{-g_{RG}}\, \chi_{RG}^{\mu\nu\sigma\rho}(\partial_\sigma A_\rho - \partial_\rho A_\sigma) \right] = 0 \tag{A4.49}$$

where the constitutive tensor $\chi_{RG}^{\mu\nu\sigma\rho}$ replaces $(g_{RG}^{\mu\sigma} \cdot g_{RG}^{\nu\rho})$. This constitutive tensor $\chi_{RG}^{\mu\nu\sigma\rho}$ has to be calculated with the constitutive tensor $\chi_{RL}^{\mu\nu\sigma\rho} = \chi_c^{\mu\nu\sigma\rho}$ in an inertial frame using cylindrical coordinates and the laws of transformation of coordinates between χ_{RL}^{μ} and χ_{RG}^{μ}. To first order in $r\Omega/c$, the propagation equation becomes

$$\nabla_{RG}^2 A - \frac{\epsilon_r \mu_r}{c^2}\frac{\partial^2 A}{\partial t_{RG}^2} - \frac{2\Omega}{c^2}\frac{\partial^2 A}{\partial \theta_{RG} \partial t_{RG}} = 0 \tag{A4.50}$$

It is very important to note that, if ϵ_r and μ_r appear at their usual place with $\partial^2 A/\partial t^2$, none of them appear in the additional term $-\frac{2\Omega}{c^2}\frac{\partial^2 A}{\partial \theta_{RG} \partial t_{RG}}$, which remains identical to the case of a vacuum. The change of wave velocity in a rotating frame, called the Sagnac effect, is independent of the properties of the co-rotating medium, where light propagates.

Note: With this relativistic formalism, it becomes clear that **H** and **D**, involved in Maxwell equations M_3 and M_4, are connected, since they compose the same contravariant tensor $G^{\mu\nu}$ · **E** and **B** represent the basic field, derived from a potential and involved in the first two Maxwell equations, M_1 and M_2, which are independent of matter. They form a covariant tensor $F_{\mu\sigma}$, and the derived field tensor $G^{\mu\nu}$ is connected to $F_{\mu\sigma}$ with the constitutive tensor $\chi^{\mu\nu\rho\sigma}$, which takes into account the properties of the material, but also the properties of the frame of reference.

A4.3 CASE OF A ROTATING TOROIDAL DIELECTRIC WAVEGUIDE

The case of the fiber gyro may be analyzed more precisely by considering a toroidal dielectric waveguide, where R is the radius of curvature and a is the radius of the core (Figure A4.3). As has already been seen (Appendix 2), an approximate solution of the fundamental mode of a straight fiber is the pseudo-Gaussian mode of amplitude:

$$A(x,y,z,t) = A_0 e^{-x^2 + y^2/w_0^2} e^{i(\omega t - \beta z)} \tag{A4.51}$$

When the fiber is bent, the spatial propagation term βz may be replaced by $\beta R\theta$, and there is a centrifugal shift ΔR_c of the mode. The mode amplitude becomes

$$A(r,\theta,z,t) = A_0 e^{-(r - R - \Delta R_c)^2 + z^2/w_0^2} e^{i(\omega t - \beta R\theta)} \tag{A4.52}$$

with

$$\Delta R_c = \frac{n^2 w_0{}^4 \omega^2}{Rc^2} \tag{A4.53}$$

Using a perturbation method, the amplitude in a rotating frame with a co-rotating waveguide is found to be

$$A(r_{RG}, \theta_{RG}, z_{RG}, t_{RG}) = A_0 e^{-(r_{RG} - R - \Delta R_c - \Delta R_R)^2 + z_{RG}{}^2/\omega_0^2} e^{i(\omega t_{RG} - \beta R\theta_{RG} - \Delta K\theta_{RG})} \tag{A4.54}$$

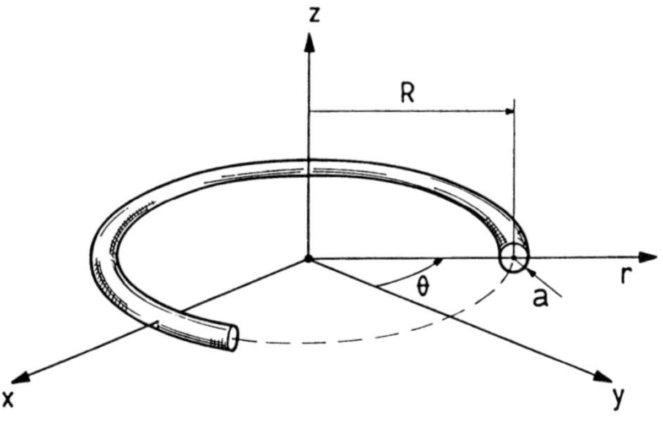

Figure A4.3 Toroidal waveguide.

There is an additional radial shift ΔR_R due to rotation. It depends on the relative direction of propagation and rotation through the sign of the product $\beta \cdot \Omega$:

$$\Delta R_R = -\frac{4\omega w_0{}^4 \beta \Omega}{c^2} \tag{A4.55}$$

but, in practice, it is completely negligible (with typical fiber parameters $\Delta R_R \approx 10^{-11}$ μm for $\Omega = 1$ rad/s). The perturbation ΔK on the propagation constant is

$$\Delta K = \frac{\omega R^2}{c^2} \Omega \tag{A4.56}$$

Its relative value $\Delta K/\beta R$ is also very small, but, when the fiber is placed in a ring interferometer, there is a measurement of the absolute phase difference between the two opposite paths with respect to 2π rad, and it accumulates with the length of the propagation path. After N turns of counterpropagation with the respective propagation constants $\beta + \Delta K/R$ and $-\beta + \Delta K/R$, the phase difference becomes

$$\Delta\phi_R = 4\pi N \Delta K \tag{A4.57}$$

$$\Delta\phi_R = \frac{4N\pi R^2 \omega}{c^2} \Omega \tag{A4.58}$$

This result, known as the Sagnac effect, depends only on the light frequency ω, the vacuum light velocity c, and the total area enclosed by the coil $A = N\pi R^2$. The indexes of the core and the cladding, the phase or group velocities of the mode, and the dispersion of the medium or of the waveguide have no influence.

The Sagnac effect is a pure temporal delay independent of matter, which is measured with a "clock," the frequency of the light source. However, since the wavelength λ in a vacuum is a more familiar quantity than the frequency for optical waves, the formula is usually written

$$\Delta\phi_R = \frac{2\pi LD}{\lambda c} \cdot \Omega \tag{A4.59}$$

where $D = 2R$ is the loop diameter, and $L = N2\pi R$ is the total length of the waveguide coiled over N turns.

BIBLIOGRAPHY

About Relativity and electrodynamics:

- Tonnelat, M. A., Principles of Electromagnetic Theory and Relativity, Gordon and Breach, 1966.
- Landau, L. D., and E. M. Lipshitz, The Classical Theory of Fields, Pergamon Press, 1975.
- Arzelies, H., Cinématique Relativiste, in French, Gauthier-Villars, 1955.

About Electromagnetism theory of the Sagnac effect:

- Post, E. J., "Sagnac Effect," Review of Modern Physics, Vol. 39, 1967, pp. 475–493.
- Post, E. J., "Interferometric Path-Length Changes Due to Motion," Journal of the Optical Society of America, Vol. 62, 1972, pp. 234–239 (SPIE MS8, pp. 79–84).
- Arditty, H. J., and H. C. Lefèvre, "Theoretical Basis of Sagnac Effect in Fiber Gyroscope," Springer-Verlag Series in Optical Sciences, Vol. 32, 1982, pp. 44–51.
- Lefèvre, H. C., and H. J. Arditty, "Electromagnétisme des milieux diélectriques linéaires en rotation et application à la propagation d'ondes guidées," in French, Applied Optics, Vol. 21, 1982, pp. 1400–1409.

Symbols

$A(t)$	amplitude of an optical wave
A^{μ} or A_{μ}	electromagnetic four-potential coordinates
$\mathbf{A}$	electromagnetic vector potential
A	area
$\mathbf{A}$	area vector
a	fiber core radius
a_e	core radius of the equivalent step index fiber
a_h	size of a spatial filtering hole
$a(f)$	frequency component of $A(t)$
B	normalized birefringence
$\mathbf{B}$	magnetic field vector
$b(V)$	normalized propagation constant
C	power coupling ratio
c	velocity of light in a vacuum (2.998×10^8 m $\cdot$ s^{-1})
$c(z)$	random polarization coupling along a fiber
c_s	coupling strength in an evanescent-field coupler
curl	curl operator
D	diameter
$\mathbf{D}$	derived electric field vector
d	distance or thickness
div	divergence operator
E	Young modulus
$E\{\}$	ensemble average
$\mathbf{E}$	electric field vector
e	exp (1) = 2.7183...
e_l	ellipticity
$F^{\mu\nu}$ or $F_{\mu\nu}$	electromagnetic field tensor
$\mathfrak{F}$	finesse of an optical cavity

f	temporal frequency
$\bar{f}$	mean temporal frequency
f_l	focal length
f_m	modulation frequency
f_p	proper or eigenfrequency of the coil
$\mathbf{f}$	force
$G^{\mu\nu}$ or $G_{\mu\nu}$	derived field tensor
$g^{u\nu}$ or $g_{u\nu}$	metric tensor
grad	gradient operator
$\mathbf{H}$	derived magnetic field vector
HE_{11}	fundamental hybrid electromagnetic mode
h	Planck constant (6.63×10^{-34} J · s)
h	h-parameter of a polarization preserving fiber
I	intensity of an optical wave
i	intensity of an electrical current
i	pure imaginary unit value
J_n	Bessel function of order n
J^{μ} or J_{μ}	electrical four-current
$\mathbf{j}$	electrical current density vector
k	Boltzmann constant (1.38×10^{-25} J · K^{-1})
k_0	wavenumber in a vacuum
k_1	wavenumber in the core
k_2	wavenumber in the cladding
k_m	wavenumber in a medium
L	length
L_{bb}	common base-branch length
L_c	coherence length
L_{cp}	coupling length
L_d	depolarization length
L_{dc}	decoherence length
L_{op}	optical path length
L_p	length of a pulse
L_{pc}	polarization correlation length
LP_{01}	fundamental linearly polarized mode
LP_{11}	second order linearly polarized mode
LSB	least significant bit
$\mathbf{M}$	magnetic polarization vector
$M(t)$	modulation signal
N	number of turns
$\dot{N}$	flow of uncorrelated particles
NA	numerical aperture

n	index of refraction
n_1	index of refraction of the core
n_2	index of refraction of the cladding
n_o	ordinary index
n_e	extraordinary index
n_{eq}	equivalent index
P	power
$\mathbf{P}$	electric polarization vector
$\mathfrak{P}$	perimeter
P	degree of polarization
p	pressure
p	parallel (polarization)
p_{ij}	elasto-optic coefficients
ppm	part per million
q	electron charge
R	radius
R	resistance
R_c	autocorrelation (ensemble average)
R_f	radius of the fiber cladding
$\mathfrak{R}$	reflectivity
r	radial coordinate
$\mathbf{r}$	radial vector
r_{ij}	electro-optic coefficients
rms	root-mean-square
S	recapture factor
s	perpendicular (polarization)
T	temperature
T_a	absolute temperature
T_C	Curie temperature
T_m	period of phase modulation
T_n	normal stress
T_{nc}	compressive stress
TE	transverse electric mode
TM	transverse magnetic mode
t	time
$\mathfrak{T}$	transmissivity
$\mathfrak{T}(\lambda)$	transmission of a wavelength filter
V	normalized frequency in a waveguide
V	electromagnetic scalar potential or voltage
V	Verdet coefficient
v	light velocity in a medium

$\mathbf{v}$	speed or velocity
ϑ	visibility of interference fringes
w	radius at $1/e^2$ in intensity of a Gaussian beam
w_0	radius at $1/e^2$ in intensity of the waist of a Gaussian beam or radius at $1/e^2$ of the pseudo-Gaussian fundamental LP_{01} mode
w_1	half width at the maximum of the LP_{11} mode
x,y,z	cartesian spatial coordinates
x_j	coordinate
Z	impedance
Z_0	impedance of a vacuum
α	attenuation per unit length of a fiber
$\alpha(f)$	modulus of $a(f)$
α_F	Fizeau drag coefficient
β	propagation constant
Γ	coupling loss
$\Gamma(\tau)$	autocorrelation function (time average)
γ	$(1 - v_t^2/c^2)^{-1/2}$
$\gamma(\tau)$	normalized autocorrelation function (time average)
Δ	normalized index difference
Δ_e	normalized index difference of the equivalent step-index fiber
Δ	difference of
Δf_{FB}	feedback frequency difference
Δf_{FWHM}	full width at half maximum of the temporal frequency spectrum
Δf_R	rotation-induced frequency difference
Δf_{bw}	counting bandwidth
ΔL	difference of geometrical length
ΔL_{op}	difference of optical path length
Δn_b	birefringence index difference
$\Delta\beta$	birefringence
$\Delta\phi$	difference of phase
$\Delta\phi_e$	error of phase difference
$\Delta\phi_F$	phase difference induced by Faraday effect
$\Delta\phi_{FB}$	feedback phase difference
$\Delta\phi_K$	phase difference induced by Kerr effect
$\Delta\phi_m$	modulation of the phase difference
$\Delta\phi_{PR}$	phase difference induced by phase ramp
$\Delta\phi_R$	rotation-induced phase difference due to Sagnac effect
$\Delta\lambda_{FWHM}$	full width at half maximum of the wavelength spectrum
$\Delta\sigma_{\text{free}}$	free spectral range of an optical cavity
$\Delta\tau_g$	difference of group transit time through the coil
δ	variation or shift of

δn	index variation
$\delta\phi$	phase shift
ϵ	small ratio
ϵ_0	dielectric permittivity of a vacuum (8.854×10^{-12} F m^{-1})
ϵ_r	relative permittivity of a medium
θ	angle
θ_B	Brewster angle
θ_D	full divergence angle at $1/e^2$ of a Gaussian beam in a vacuum
$\theta_D{}'$	full divergence angle at $1/e^2$ of a Gaussian beam in a medium
θ_L	limit angle of total internal reflection
θ_{inc}	angular increment of a gyroscope
Λ	birefringence beat length
λ	wavelength in a vacuum
$\bar{\lambda}$	mean wavelength
λ_c	cutoff wavelength
λ_e	emission wavelength
λ_m	wavelength in a medium
μ_0	magnetic permeability of a vacuum ($4\pi \times 10^{-7}$ H $\cdot$ m^{-1})
μ_r	relative permeability of a medium
ν	Poisson ratio
$\mathbf{\Pi}$	Poynting vector
π	3.14159...
ρ	electrical charge density
ρ_{cr}	cross-coupling polarization intensity
Σ	summation
σ	spatial frequency
$\bar{\sigma}$	mean spatial frequency
1σ	root-mean-square value or standard deviation
τ	temporal delay
τ_c	coherence time
τ_{dc}	decoherence time
τ_g	group delay
τ_ϕ	phase delay
ϕ	phase
$\dot{\phi}$	slope of analog phase ramp
ϕ_b	phase bias
ϕ_m	phase modulation
ϕ_{PR}	phase ramp
ϕ_s	phase step
χ_e	dielectric susceptibility

χ_m	magnetic susceptibility
$\chi^{\mu\nu\sigma\rho}$	constitutive tensor
Ω	rotation rate
$\mathbf{\Omega}$	rotation rate vector
Ω_π	rotation rate inducing a π radian phase difference
Ω_μ	rotation rate inducing a 1-μrad phase difference
ω	angular frequency
$\bar{\omega}$	mean angular frequency
ω_e	emission angular frequency
$< >$	temporal averaging
$< \| >$	generalized scalar product of complex functions
$\| \|$	absolute value
∇	vector differential operator

About the Author

Hervé C. Lefèvre is vice president of research and development for Photonetics, in France. He was born in Paris, France, in 1954. He graduated from Ecole Normale Supérieure de Saint-Cloud, and received the Agrégation de Physique in 1976, the Doctorat de 3ème cycle in 1979, and the Doctorat d'Etat in 1982.

He started working on the fiber-optic gyroscope in 1977 as a graduate research scholar at Thomson-CSF Central Research Laboratory, in France, and from 1980 to 1982 he was at Stanford University as a postdoctoral research associate. He then moved back to Thomson-CSF as a research scientist and became head of the fiber-optic sensor laboratory in 1984. In 1987, he joined Photonetics, which is a company specializing in fiber-optic sensors and photonic instrumentation.

He has authored and coauthored nearly 50 journal and conference publications about the fiber-optic gyroscope and its related technologies, and has been granted 25 patents. He was awarded the Prize Fabry-de Gramont in 1986 by the French Society of Optics and the Prize Esclangon in 1992 by the French Society of Physics.

Index

Breinigsville, PA USA
29 December 2010

252368BV00003B/140/A

9 780890 065372